An Introduction to Thyristors and Their Applications

To my wife
Lakshmi

An Introduction to Thyristors and Their Applications

M. Ramamoorty
Chief of Research and Development
Hindustan Brown Boveri Limited
Baroda, India

First published in India 1977
by Affiliated East-West Press Private Limited

First published in the United Kingdom 1978
Reprinted 1981, 1983

Published by
THE MACMILLAN PRESS LIMITED
London and Basingstoke
Companies and representatives
throughout the world

Printed in Hong Kong

British Library Cataloguing in Publication Data

Ramamoorty, M
An introduction to thyristors and their applications.
1. Electric controllers 2. Thyristors
I. Title
621.31'7 TK2851

ISBN 0-333-25694-8

Contents

Preface

The rapid development of semiconductor technology has revolutionised the art of power modulation. "Thyristor" has become a byword for control in many industrial applications and the subject of several books. Some of these books emphasise its physics and characteristics, and some discuss one or the other of its applications in detail. Few qualify as texts for the undergraduate who needs not the details, which are available in handbooks and journals and/or in the teacher's lectures, but rather the basics.

In an attempt to fill in the lacuna, this text concentrates on the fundamentals of the operation of the thyristor and on its applications. No effort is made to go into the details of system design as a whole. The control circuits described for various applications using thyristors are not meant to be commercially feasible, but provide the basic understanding necessary for synthesising more sophisticated controllers.

The book, as its title indicates, serves as an introductory course on the thyristor and its applications to power control. It is, in fact, based on the lectures of such a course which the author has offered several times over the past few years. The volume is intended primarily for those undergraduates specialising in power control and industrial drives. Such readers have usually had a basic grounding in electronics, electric circuits and machines, and control theory.

The entire content can be covered in a two-semester course. The physics and characteristics of the thyristor, the methods of turning on and off, protection, general applications, phase control, and some fundamental aspects of line-commutated converters and inverters may be allotted to the first semester. The second semester may be devoted to Chapters 7 to 10 which focus attention on industrial controls involving all classes of forced-commutated circuits, and include a brief treatment of some important types of drives and their characteristics and the performance of controllers.

The illustrative problems at the end of the study are intended more to help the student gain an insight into the applications than to test his mathematical skill. A large number of examples are worked out in each chapter to facilitate a grasp of the subject. The list of important references, which identifies the source of material covered, proves useful when additional data is required.

Kanpur
March 1977

M. RAMAMOORTY

Preface

1

Solid-State Power Control

1.1 INTRODUCTION

There are many industrial applications for which control of power input and/or output is required. Examples of such applications are variable speed drives, illumination controllers, and temperature regulators. The initial developments in power-control schemes were based on variable tap-changing transformers and series and shunt regulators (consisting of resistors or reactors) which produce a change in the applied voltage, and thereby vary the power. These schemes either were found to be inefficient or involved costly equipment. Further, the power control was usually in steps. Such drawbacks have since been partially overcome by saturable-core reactors. The advent of magnetic amplifiers paved the way for complete static control of power without moving parts. During the period before World War II, tremendous improvements were made in the performance and design of magnetic power controllers, and the evolution of square-loop materials improved their range of amplification. Magnetic control of power is still adopted for many military and industrial applications where reliability and sturdiness are of prime importance. The major limitation of this method is the bulkiness of the controller which prohibits its use in certain applications. Also, the core-magnetising current and iron losses result in low input power factor and low efficiency.

Parallel with the development of magnetic controllers, work was also done for power control by electronic methods employing thermionic and gas-discharge valves; popular electronic devices for power control include mercury arc converters (used for large direct current (DC) power supplies) and thyratrons (gas-filled triodes used for switching heavy currents). Electronic controllers are more compact and efficient than magnetic devices, but suffer from the disadvantage of being less reliable.

The post-War period has come to be known as the *semiconductor age*. Rapid developments have taken place both in the ratings of power control devices and in their characteristics. Thermionic valves have been replaced by semiconductor devices, resulting in the miniaturisation of electronic circuits. Semiconductor power diodes are used in place of metal and gas rectifiers in many industrial applications. Power transistors are now available with reasonably large ratings for voltage and current. New semiconductor switching devices, called *thyristors*, have been developed with characteristics similar to those of gas-discharge tubes. These new devices have many advantages and are being widely used for power control. Advances in the technology and fabrication of thyristors have resulted in improved reliability, and lower manufacturing costs. Nowadays many industrial power controllers use solid-state devices, and the thyristor has become very popular for such applications.

1.2 HISTORICAL DEVELOPMENT

Pioneering work on the theory and fabrication of the power-switching device, which later came to be known as a *thyristor* (because its characteristics are similar to those of the gas-tube thyratron), was done at the Bell Laboratories in the USA. The first prototype was introduced by the General Electric Company (USA) in 1957. Since then, many improvements have been made, both in the technique of its fabrication and in adapting it to numerous industrial applications. With the development of a number of other devices of similar type and characteristics, the whole family of such power-switching devices has come to be known as 'thyristors'. Since the basic semiconductor material used for the device is silicon, it is also designated as a *silicon-controlled rectifier* (the abreviated form of writing it is SCR, which may also stand for *semiconductor-controlled rectifier*).

The term SCR is often used for the oldest member of the thyristor family which is the most widely-used power-switching device. Its rating has been very much improved since its introduction, and now SCRs of voltage rating 10 kilovolts (kV) and current rating 500 amperes (A) are available, corresponding to a power-handling capacity of about 5 megawatts (MW). This device can be switched by a low-voltage supply of about 1 A and 10 W, which shows the tremendous control capability of the device. Because it is compact and has high reliability and low losses, the SCR has more or less replaced the thyratron and the magnetic amplifier as a switching device in many applications.

1.3 NATURE, CHARACTERISTICS, AND APPLICATIONS

The thyristor has four or more layers and three or more junctions. It has a built-in feature for internal regeneration under specified bias conditions, whereby it goes from the off-state (blocking state) to the on-state (conducting state). This property is known as *switching*. Similarly, the device can also be turned off (conducting state to blocking state). Both states are stable and reversible. This switching property of the device, coupled with its large power-handling capacity, is made use of in power modulation and control. Being a solid-state device, it is compact, has low turn-on and turn-off times, and has negligible losses when fully on or off. Because of these desirable features, the device is used for power control in the following applications:

(a) Speed controllers for alternating current (AC) and direct current (DC) motors.

(b) Temperature and illumination controllers.

(c) AC and DC circuit breakers.

(d) Variable frequency DC-AC inverters.

(e) Variable voltage DC-DC converters.

(f) Variable frequency AC-AC converters.

(g) Variable voltage AC-DC rectifiers.

1.4 COMPARISON OF THYRATRONS WITH POWER TRANSISTORS

The thyratron is a gas tube. Figure 1.1a shows its symbolic diagram and terminal configuration. The device is turned on by applying a positive voltage between the grid and the cathode. Before the development of thyristors, the thyratron was a popular device for many applications in

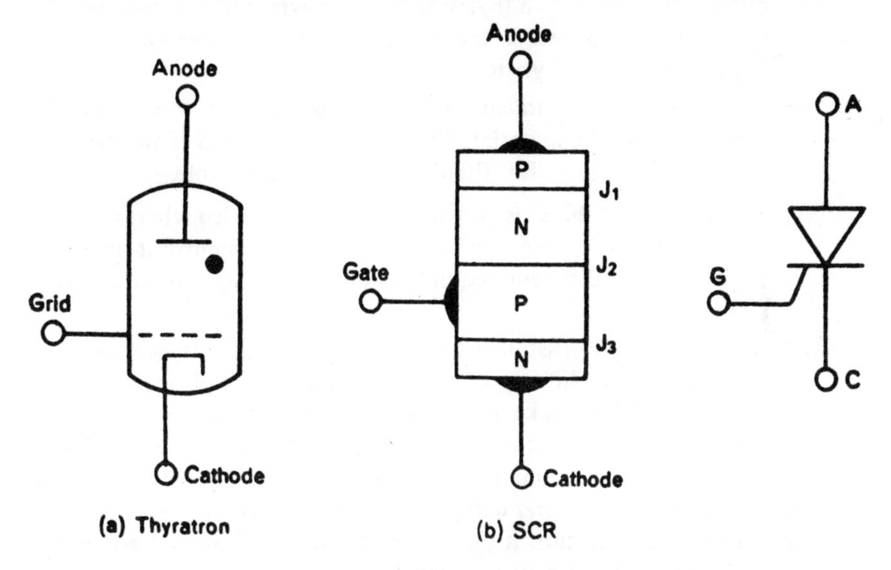

Fig. 1.1 Schematic diagram of a thyratron and an SCR.

industrial controls. Even today, for very large power ratings, arc converters are used for rectification and inversion. Of the two major classes of gas tubes, cold-cathode and hot-cathode, the hot-cathode grid-controlled thyratron has characteristics similar to those of the SCR, and is widely used. Some of the main differences between a thyratron and an SCR are as follows:

(a) A thyratron requires a large anode-to-cathode voltage and a separate filament supply. Some gas tubes require additional power supplies for auxiliary anodes. An SCR requires one main supply and one control signal.

(b) The arc ionising and deionising times for a thyratron are comparatively large and so the device applications are limited to a frequency of 1 kilohertz (kHz). A thyristor can operate over a much greater range of frequency.

(c) The anode-to-cathode arc has a large voltage drop and so the internal losses of a thyratron are higher than those of an SCR. The arc drop is inversely proportional to the molecular weight of the gas used. For large power applications, such as AC-DC conversion and inversion (arc converters), mercury vapour is used. Heavy gases in the interelectrode region produce large turn-on and turn-off times.

(d) Because of the large anode-to-cathode voltage, the anode-to-cathode spacing of thyratrons has to be sufficient to avoid any arc-backs and unwanted flash-overs. This results in an arrangement bulkier than that with thyristors. However, for this same reason, it is possible to design the gas tube for a voltage rating much higher than for a thyristor.

(e) Due to ion bombardment and the consequent sputtering of filament and anode material, the life of a thyratron is shorter than that of a thyristor; also, the thyratron is less reliable.

(f) Basically, a gas tube is a voltage-operated device whereas a thyristor, like any other semiconductor device, is current-operated. This results in different requirements in the design of gate-control circuits.

Even though power transistors with high current and voltage ratings are now available, the basic differences in the fabrication and operation of a thyristor and a transistor make it possible for the former to have much higher voltage and current ratings for a given size than those of the latter. There is no regenerative action in a transistor; it requires a continuous base current to stay in the conducting state. On the other hand, an SCR needs only a gate pulse to turn it on. Further, a thyristor is used only as a switching device, whereas a power transistor is required to operate in the active region in many applications. Therefore, the internal power losses in a power transistor are much higher than those in a thyristor. The rating of a power transistor is in the range of several hundred watts and the rating of an SCR is in the range of kilowatts. However, it must be noted here that these two devices are used in different applications.

1.5 THYRISTORS—SYMBOLIC REPRESENTATION

As already mentioned, the term 'thyristor' pertains generally to the family of semiconductor devices used for power control. The oldest and widely-used member of this family is the SCR. This is a four-layer, three-junction device with three terminals, namely, the anode, the cathode, and the gate. It is a unilateral device and conduction takes place from anode to cathode under proper conditions of bias. Figure 1.1b shows a schematic diagram of this device and also its symbolic representation.

The other members of the thyristor family, in general, are low-power devices, except the *triac* which is a bilateral device with three terminals, and conducts in both directions. So, the terms 'anode' and 'cathode' are not used for the terminals of a triac. The triac is equivalent to two SCRs connected in antiparallel. The *diac* is a two-terminal, four-layer device, which is generally used for triggering triacs. The *silicon-controlled switch* (SCS) is similar to the SCR except for the fact that the device has two gates and can be turned on or off by either gate. The *complementary SCR* (CSCR) or the *silicon unilateral switch* (SUS) has the gate on the anode side. This can be used as a programmable unijunction transistor. A recent addition to the thyristor family is the *light-activated SCR* (LASCR) which is turned on by photon bombardment. This also is a unilateral device with four layers and three terminals. There are many other devices in this family and many more may be added. Table 1.1 lists a few of these devices and shows their respective *V-I* characteristics and symbolic representation. For more details see Chapter 4.

Table 1.1 Device characteristics and symbolic representation

Device	Characteristic	Symbol
SCR (silicon-controlled rectifier)	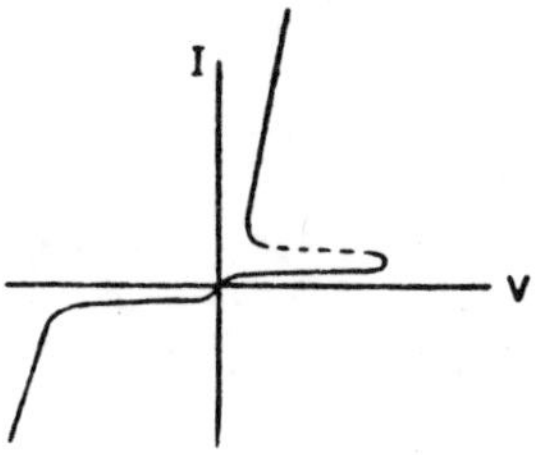 I, V	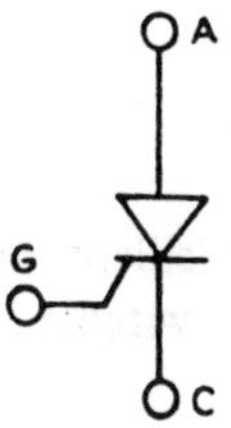 A, G, C
Triac	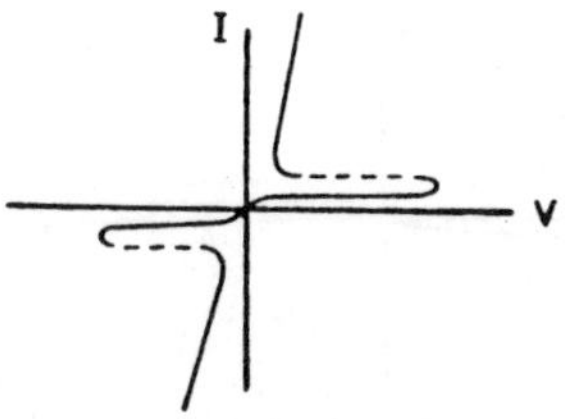 I, V	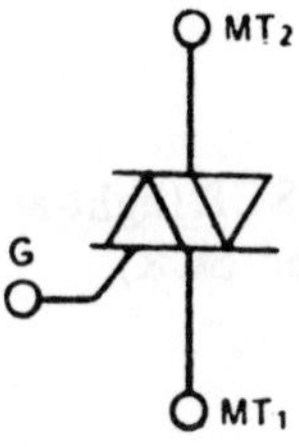 MT_2, G, MT_1

Table 1.1 (*cont.*)

Device	Characteristic	Symbol
Diac	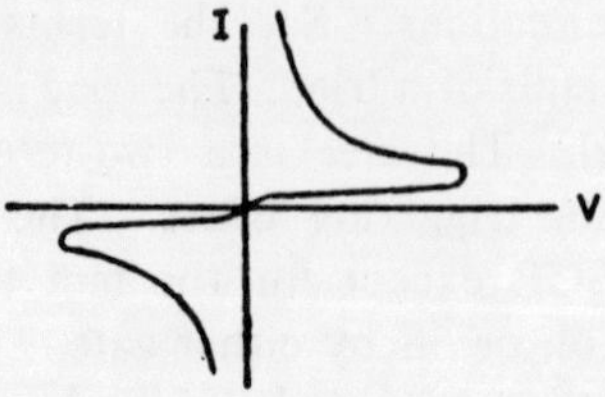	 
SCS (silicon-controlled switch)	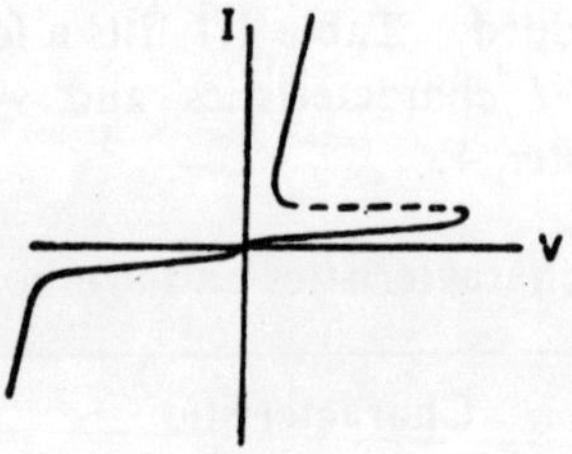	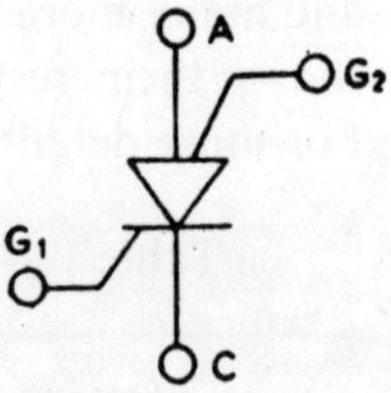
SUS (silicon unilateral switch)	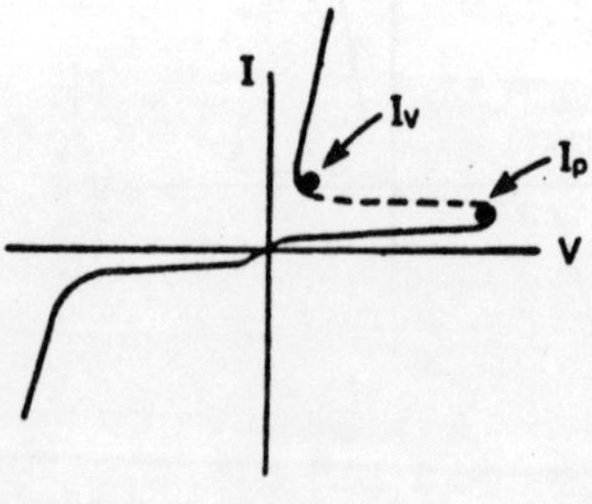	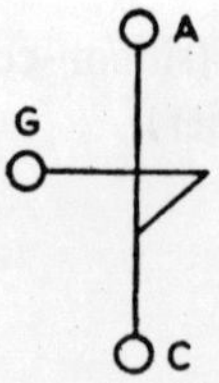
LASCR (light-activated SCR)	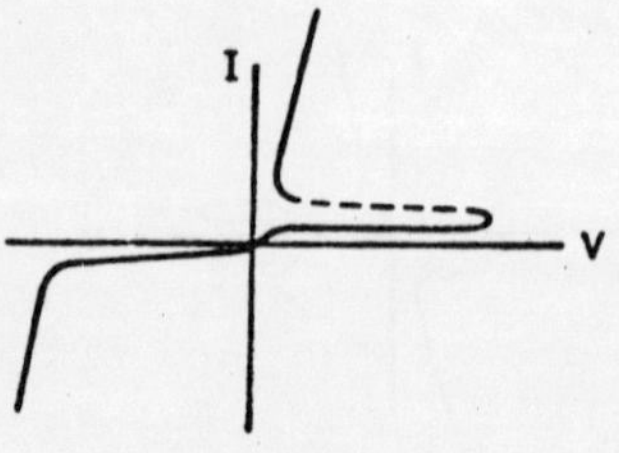	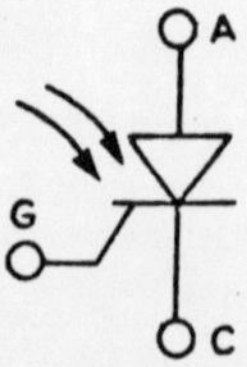

1.6 CONSTRUCTIONAL DETAILS

The cross-sectional view of a typical SCR is shown in Fig. 1.2. Basically, the SCR consists of a four-layer pellet of P- and N-type semiconductor materials. Silicon is used as the intrinsic semiconductor to which the proper impurities are added. The junctions are either diffused or alloyed. The planar construction shown in Fig. 1.2a is used for low-power SCRs. This technique is useful for making a number of units from a single silicon wafer. Here, all the junctions are diffused. The other technique is the mesa construction shown in Fig. 1.2b. This is used for high-power SCRs.

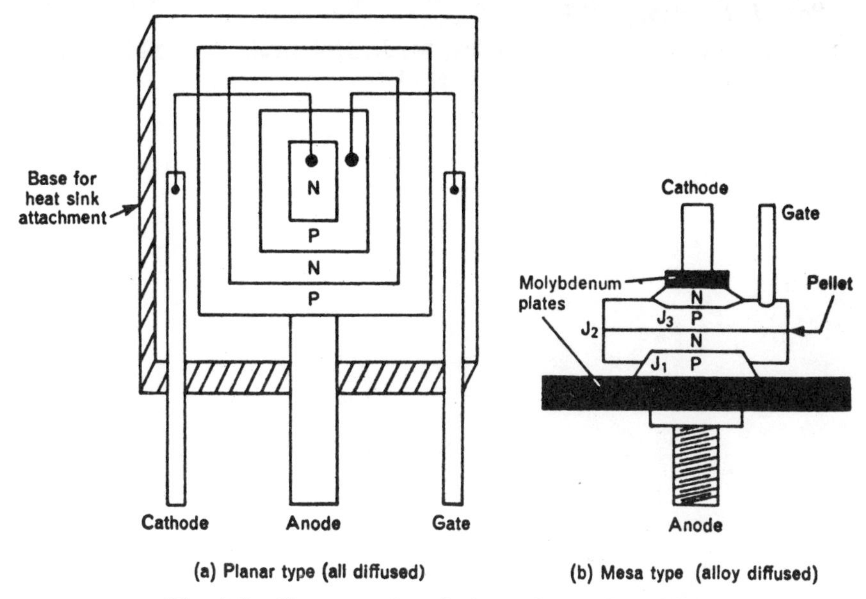

Fig. 1.2 Cross-sectional view of a typical SCR.

Here, the inner junction J_2 is obtained by diffusion, and then the outer two layers are alloyed to it. Because the PNPN pellet is required to handle large currents, it is properly braced with tungsten or molybdenum plates to provide greater mechanical strength. One of these plates is hard-soldered to a copper or an aluminium stud, which is threaded for attachment to a heat sink. This provides an efficient thermal path for conducting the internal losses to the surrounding medium. Details of these fixtures are provided in Section 11.4. The use of hard solder between the pellet and backup plates minimises thermal fatigue when the SCRs are subjected to temperature-induced stresses. For medium- and low-power SCRs, the pellet is mounted directly on the copper stud or casing, using a soft solder which absorbs the thermal stresses set up by differential expansion and provides a good thermal path for heat transfer. When a larger cooling arrangement is required for high-power SCRs, the press-pack or hockey-puck construction is used, which provides for double-sided air or water cooling. For additional information on SCR mountings and cooling arrangements, see

Chapter 11; schematic diagrams and terminal configurations are given in the Appendix.

REFERENCES

Jonscher, A. K., Notes on the theory of four layer semiconductor switches, *Solid State Electronics* (UK), 1961, p. 143.

Mackintosh, I. M., The electrical characteristics of silicon PNPN triodes, *Proc. IRE*, 1958, p. 1229.

Moll, J. L., Tanenbaum, M., Goldey, M. J., Holonyak, N., PNPN transistor switches, *Proc. IRE*, 1956, p. 1174.

2

Thyristors

2.1 SCR CHARACTERISTICS

The basic properties and applications of thyristors have been explained in Chapter 1. Since the SCR is the most widely-used member of the family of thyristors, a more detailed analysis of its operation and characteristics is given here. The other devices will be discussed in Chapter 4.

2.2 PRINCIPLE OF OPERATION

The SCR is a four-layer device with three terminals, namely, the anode, the cathode, and the gate. When the anode is made positive with respect to the cathode (see Fig. 1.1), junction J_2 is reverse-biased and only the leakage current will flow through the device. The SCR is then said to be in the *forward blocking state* or *off-state*. When the cathode is made positive with respect to the anode, junctions J_1 and J_3 are reverse-biased and a small reverse leakage current will flow through the SCR. This is the *reverse blocking state* of the device. When the anode-to-cathode voltage is increased, the reverse-biased junction J_2 will break down due to the large voltage gradient across the depletion layers. This is the *avalanche breakdown*. Since the other junctions J_1 and J_3 are forward-biased, there will be free carrier movement across all three junctions, resulting in a large anode-to-cathode forward current I_T. The voltage drop V_T across the device will be the ohmic drop in the four layers, and the device is then said to be in the *conducting state* or *on-state*. Figure 2.1 shows the characteristics of an SCR. In the on-state, the current is limited by the external impedance. If the anode-to-cathode voltage is now reduced, since the original depletion layer and the reverse-biased junction J_2 no longer exist due to the free movement of carriers, the device will continue to stay on. When the forward current falls below the level of the *holding current* I_h, the depletion region will begin to develop around J_2 due to the reduced number of carriers, and the device will go to the blocking state. Similarly, when the SCR is switched on, the resulting forward current has to be more than the *latching current* I_l. This is necessary for maintaining the required amount of carrier flow across the junctions; otherwise, the device will return to the blocking state as soon as the anode-to-cathode voltage is reduced. The holding current is usually lower than, but very close to, the latching current; its magnitude is of the order of a few milliamperes (mA).

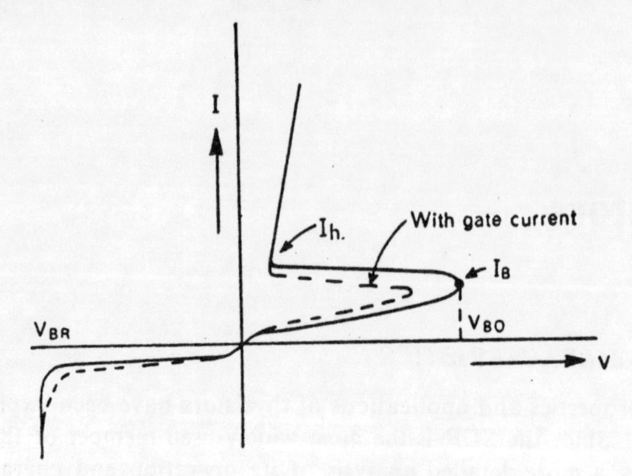

Fig. 2.1 SCR characteristics.

When the SCR is reverse-biased (i.e., the cathode is positive with respect to the anode), the device will behave in the same manner as two diodes connected in series with reverse voltage applied across them. The inner two regions of the SCR will be lightly doped as compared to the outer layers. Hence, the thickness of the J_2 depletion layer during the forward-bias conditions will be greater than the total thickness of the two depletion layers at J_1 and J_3 when the device is reverse-biased. Therefore, the forward breakover voltage V_{BO} will be generally higher than the reverse breakover voltage V_{BR}. The forward current of the device at the breakover point is denoted by I_B (see Fig. 2.1).

From the foregoing discussion, it can be seen that the SCR has two stable and reversible operating states. The changeover from off-state to on-state, called *turn-on*, is achieved by increasing the forward voltage beyond V_{BO}. The reverse transition, termed *turn-off*, is made by reducing the forward current below I_h. A more convenient and useful method of turning on the device employs the gate drive. If forward voltage less than V_{BO} is applied across the device, it can be turned on by applying a positive voltage between the gate and the cathode. This method is known as *gate control*. The *V-I* characteristics of the device with gate drive are also given in Fig. 2.1.

2.3 TRANSISTOR ANALOGY

Figure 2.2 shows the two-transistor model of an SCR. This is obtained by splitting the two middle layers into two separate parts. The corresponding symbolic representation is also given in the figure. The collector current of transistor T_1 becomes the base current for T_2, and vice versa. If α_{b1} and α_{b2} are the common base current gains for transistors I_1 and I_2, respectively, it can be easily derived from Fig. 2.2 that with a forward voltage applied across the anode and the cathode, and the gate terminal left free, the forward blocking current is

$$I_D = \frac{I_{CO}}{1 - (\alpha_{b1} + \alpha_{b2})}, \quad (2.1a)$$

where I_{CO} is the reverse leakage current of the reverse-biased junction J_2 when the two outer layers are not present. In silicon transistors, α_bs are

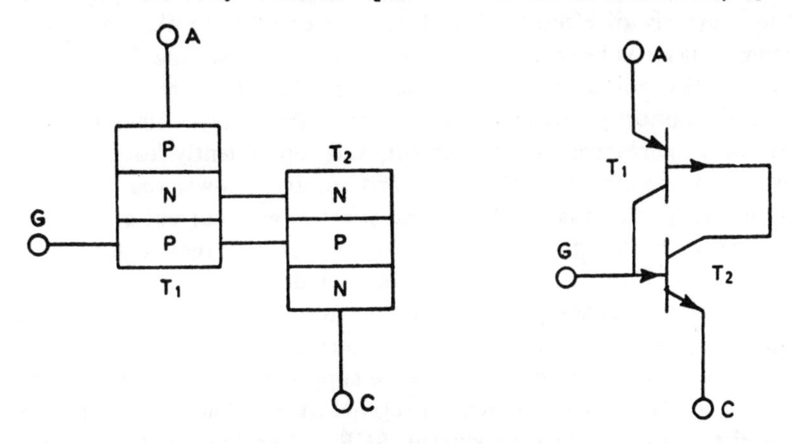

Fig. 2.2 Two-transistor model of an SCR.

dependent on the emitter current. Therefore, initially when the applied forward voltage is small, $(\alpha_{b1} + \alpha_{b2})$ will be less than 1. If the reverse leakage current I_{CO} of junction J_2 is increased, the values of α will also increase, and at some level $(\alpha_{b1} + \alpha_{b2})$ will become equal to 1. Under these conditions, the value of I_D will be equal to I_B. Then, internal regeneration will begin and the device will go to the on-state. This can be seen from Eq. (2.1a). When $(\alpha_{b1} + \alpha_{b2})$ approaches unity, current I_D will tend to infinity. The transistor analogy and Eq. (2.1a) are not valid when the device goes into conduction. The forward current is then limited by the external impedance and not by the base currents of the two-transistor model. The device can go back to the off-state only when the forward current falls below I_h, at which instant the α_bs will be very low, $(\alpha_{b1} + \alpha_{b2})$ will be less than 1, and internal regeneration will stop. In other words, the depletion layer across J_2 will reappear and the forward current will remain low. It can also be seen that this process of internal regeneration is not possible when the SCR is reverse-biased.

The reverse leakage current I_{CO} can be increased by raising the applied forward voltage. As indicated earlier, when this voltage is equal to V_{BO}, the value of $(\alpha_{b1} + \alpha_{b2})$ will become 1 and the device will go to the on-state. A rise in the junction temperature will increase I_{CO} and cause a breakdown. Silicon is used as the intrinsic semiconductor in the fabrication of an SCR to improve its thermal stability and to keep the values of α_b small at normal junction temperatures. With gate current I_g flowing into the base of transistor T_2, Eq. (2.1a) becomes

$$I_D = \frac{(I_g + I_{CO})}{1 - (\alpha_{b1} + \alpha_{b2})}. \tag{2.1b}$$

Thus, the effect of applying a positive voltage between the gate and the cathode, when the device is forward-biased, is that the leakage current through junction J_2 is increased. This is because the resulting gate current consists mainly of electron flow from the cathode to the gate (since the bottom N layer is heavily doped as compared to the gate P layer). Due to the applied voltage gradient, some of these electrons reach region J_2 and add to the minority carrier concentration in the P layer near junction J_2. This raises the reverse leakage current, and consequently leads to a breakdown even though the applied forward voltage is lower than V_{BO} and the junction temperature is normal. The effect of gate current on the forward and reverse blocking currents and breakover voltages is shown by the dashed line in Fig. 2.1. Thus, the gate provides a very convenient method for switching the device from off-state to on-state, with low anode-to-cathode voltages. When the device is turned on, all four layers will be filled with carriers, and even if the gate supply is removed, the device will continue to stay on due to internal regeneration. Therefore, a gate signal is required only for turning on the SCR. For low- and medium-power SCRs, the gate current also is in the milliampere range.

In certain low-power thyristors, it is possible to turn off the device by applying a negative signal to the gate. This outgoing gate current will reduce the base current to the lower transistor T_2 (see Fig. 2.1), which in turn will reduce the base current to T_1. Thus, the loop gain will decrease and the result will be a reduction in the forward current which finally leads to the blocking state of the device. Similarly, there are some thyristors with a gate lead on the anode side. This is known as the *anode gate*, or N gate. Such a device is turned on by applying a negative signal between the gate and the anode. The upper N region is the most resistive layer (due to low doping level) among the four layers of the device. Therefore, a larger voltage is required to turn on a device with an anode gate. When the forward current is low, the device can be turned off by applying a positive voltage across the gate and the anode. This positive voltage will reduce the emitter current of the top transistor, which in turn will reduce the loop gain ($\alpha_{b1} + \alpha_{b2}$) and return the device to the off-state. As far as the SCR is concerned, the gate will not have any control whatsoever once the device is turned on. The process of internal regeneration continues because of the positive feedback, and the device will stay on even if the gate signal is removed. Since there are large current densities in SCRs, the application of a negative voltage to the gate will not reduce the forward current, and thus will not turn off the SCR.

The application of a positive gate-to-cathode voltage, when the device is reverse-biased, raises the reverse leakage current and thereby increases the internal loss. Similarly, when the device is conducting, it is preferable to remove the gate drive since its application will only result in increased losses and higher junction temperatures.

2.4 METHODS OF TURNING ON

As explained in Section 2.3, the SCR can be switched into conduction either by increasing the forward voltage beyond V_{BO} or by applying a positive gate signal when the device is forward-biased. Of these two methods, the latter, called the *gate-control method,* is used as it is more efficient and easy to implement for power control. The following points have to be noted when designing the gate-control circuit:

(a) Appropriate gate-to-cathode voltage must be applied for turn-on when the device is forward-biased.

(b) The gate signal must be removed after the device is turned on.

(c) No gate signal should be applied when the device is reverse-biased.

(d) When the device is in the off-state, a negative voltage applied between the gate and the cathode will improve the characteristics of the device. In such an instance, a large positive voltage will be required to overcome this negative bias for turn-on.

There are three ways of triggering the device by gate control.

Triggering by a DC Gate Signal Here, a DC voltage of proper polarity and magnitude is applied between the gate and the cathode when the device is to be turned on. It must, however, be remembered that the SCR is a current-operated device and it is the gate current (injected carriers) that turns on the device. The drawbacks of this scheme are that the gate signal has to be continuously applied (resulting in an increase in internal power dissipation) and that there is no isolation of the gate-control circuit from the main power circuit.

Triggering by an AC Gate Signal In many power-control circuits that use AC input, the gate-to-cathode voltage is obtained from a phase-shifted AC voltage derived from the main supply. The chief advantage of this scheme is that proper isolation of power and control circuits can be provided. The firing angle control is obtained very conveniently by changing the phase angle of the control signal. However, the gate drive is maintained for one half-cycle after the device is turned on, and a reverse voltage is applied between the gate and the cathode during the negative half-cycle.

Triggering by a Pulsed-Gate Signal Here, the gate drive consists of a single pulse appearing periodically, or a sequence of high-frequency pulses. This is known as *carrier frequency gating.* A pulse transformer is used for isolation. The gate losses are very much reduced since the drive is discontinuous.

For power control in AC circuits, the instant of firing the device is controlled. This is done by varying the phase-shift of the AC voltage applied to the gate (the AC-gate-signal scheme) or by applying the train of high-frequency pulses at the proper time through a logic circuit (the pulsed-gate-signal scheme). Figure 2.3 shows the voltage and current waveforms for the device following the application of a gate signal. The ***turn-on time*** t_{gt}, the *delay time* t_d, and the *rise time* t_r are defined as shown in the figure.

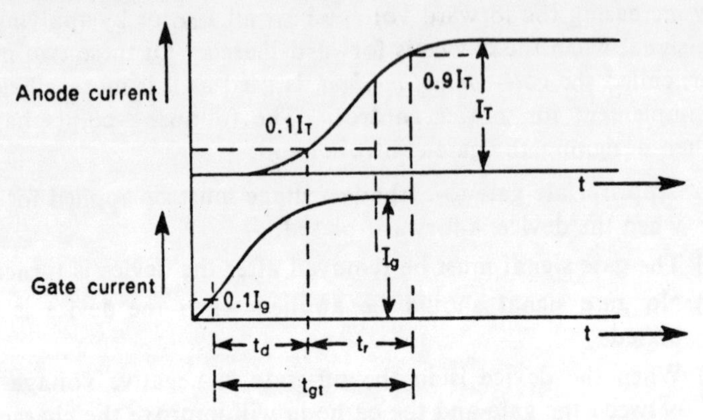

Fig. 2.3 Turn-on characteristics.

It can be observed that the total turn-on time depends on the anode circuit parameters, the gate signal amplitude, and the rise time. The turn-on time is of the order of 2–4 μsec.

2.5 TURN-OFF MECHANISM

When an SCR is turned on by the gate signal, the gate loses control and the device can be brought back to the blocking state only by reducing the forward current to a level below that of the holding current. In AC circuits where the current goes through a natural zero value, the device will be automatically turned off. In such circuits, the SCR has to be triggered synchronously with the zero crossing of the input voltage in every positive half-cycle of the applied AC voltage. In DC circuits where there is no natural zero value of the current, the forward current can be reduced either by shunting the SCR by another device (commutation) or by applying a reverse voltage across the anode and the cathode and forcing the current through the SCR to zero value (forced turn-off). Various methods are available to achieve forced turn-off (see Chapter 8). In all such methods the anode-to-cathode current has to be reduced and kept below the level of the holding current till all the excess carriers in the four layers are swept out or are recombined, and a depletion layer is established around region J_2. In AC circuits, due to the nature of the alternating voltage applied between the anode and the cathode, a reverse voltage will appear across the device immediately after the forward current has gone through the zero value. This reverse voltage will sweep out the excess carriers (electrons from the bottom N layer and holes from the top P layer) from the two outer layers, and thus facilitate turn-off; in this process a reverse recovery current will be set up. It is evident that this negative current can be much greater than the conventional reverse blocking current I_{DR} of the device. The excess carriers in the inner two regions can decay only due to recombination. Thus, the total *turn-off time* t_q required for the device is the sum of the duration for which the reverse recovery current flows after the application of reverse voltage (t_{rr}) and the time required for the recombination of all excess carriers in the inner two layers of the device (t_{gr}). At the end of this turn-off time, a depletion layer develops across junction J_2 and the device can then withstand the forward voltage. Figure 2.4 shows the current waveform following the application of reverse voltage, and indicates the instant when the device can be subjected to forward voltage. It will be observed that the turn-off time is dependent on the anode current I_T,

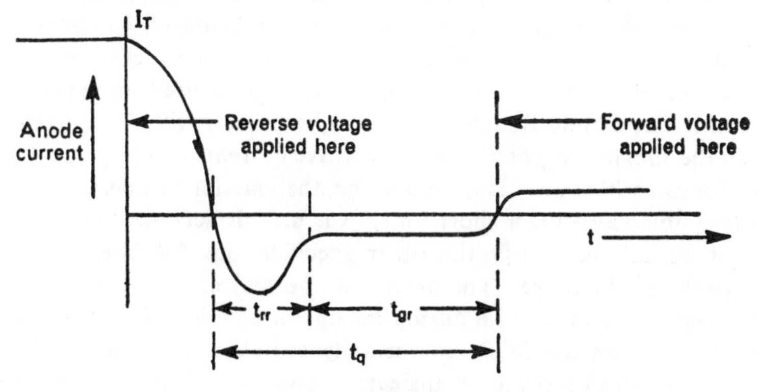

Fig. 2.4 Turn-off characteristics.

the magnitude of the reverse voltage applied, and the magnitude and rate of application of the forward voltage. The turn-off time for normal SCRs is of the order of 50–100 μsec, and for high-frequency SCRs it is about 10–20 μsec.

2.6 DEVICE SPECIFICATIONS, RATINGS, AND NOMENCLATURE

The reliable operation of an SCR, like that of any other semiconductor device, can be ensured only when its ratings are not exceeded during the on- or off-state. The specifications of the device relate to its current-carrying capacity and voltage-withstanding capability. The junction temperature has a direct bearing on the performance of the device since it affects the carrier densities in the four layers. It is for this reason that the forward and reverse breakover voltages will be lower as the temperature increases. Similarly, the turn-off time will increase with temperature for a given forward current, and the minimum gate current required for turn-on will be lower.

The junction temperature will be a function of the internal losses of the device and the efficiency of the heat transfer mechanism. The following factors contribute to internal losses:

(a) The on-state forward voltage drop across the device (V_T).

(b) The off-state forward and reverse currents (I_D, I_{DR}).

(c) The gate current (I_g).

The off-state losses include both forward and reverse blocking conditions. The gate losses can be reduced by pulse-firing. A properly designed heat sink will reduce the temperature rise of the junction. Since the voltage and current ratings of the device depend eventually on the permissible junction temperature rise, there are three types of ratings: continuous, repetitive, and surge or nonrepetitive.

Continuous ratings are expressed in terms of average or RMS values, depending on whether the device is unilateral or bilateral. *Repetitive* and *surge ratings* normally correspond to peak values. The surges are assumed to be sinusoidal in nature and their duration is measured in terms of the period of the operating frequency, i.e., 50 hertz (Hz). Surge ratings correspond to the maximum possible nonrepetitive current or voltage peak which the device can withstand. During a surge, the junction temperature exceeds the permissible value for a short time. It is on this account that the device may not be able to satisfy the other specifications for blocking voltage immediately after a surge. The device can be subjected to surge voltages or currents only once in a while during its operating life. Repetitive ratings are defined because the SCRs are usually switched periodically, and are thereby subjected to repeated transients. Thus, the device is specified by three current ratings in the forward direction and three voltage ratings, in both forward and reverse directions. Similarly, the gate drive is characterised by maximum and minimum gate voltage and current. When the voltages and currents exceed the maximum ratings, the gate-to-cathode junction will be destroyed. On the other nand, it may not be possible to trigger the device if the voltages and currents are less than the minimum ratings. It must, however, be remembered that all these ratings are temperature-dependent. The specifications for the device also include the turn-on time t_{gt} and turn-off time t_q, which have already been defined in Sections 2.4 and 2.5. For SCRs used in high-frequency operation, low turn-on and turn-off times are required. This necessitates a modified design of the gate structure.

Besides the ratings just discussed, the performance of the device is affected by the rate of change of forward current and the rate of application of forward voltage. If the anode-to-cathode current has a short rise time (e.g., for purely resistive loads), the current density in the region of junction J_2 will be high, since only a small area of this region will be conducting immediately after the application of the gate drive. The effective area of conduction across junction J_2 will depend on the diffusion velocity of the carriers injected from the gate circuit. This restriction on the conducting area will cause a hot-spot temperature, which may lead to permanent destruction of junction J_2. Thus, for a given type of construction, the di/dt has to be limited. Similarly, a large dv/dt for the forward off-state voltage V_D will result in high capacitive currents across the reverse-biased junction J_2. This current may turn on the device without a gate signal. To avoid such maloperation, the rate of application of forward voltage must be controlled. This factor is very important in designing proper commutating circuits in which a forward voltage appears across the device after it is subjected to a forced turn-off (see Fig. 2.4). The methods for improving the characteristics of the device and its ratings will be discussed in Section 2.7. Some important device ratings and their notation are listed here for ready reference:

Current at breakover point	I_B
Off-state forward current	I_D
Holding current	I_h
Gate current	I_g
Latching current	I_l
Reverse current	I_R
On-state current	I_T
On-state forward average current	$I_{T\,av}$
On-state forward RMS current	$I_{T\,RMS}$
Repetitive peak forward current	I_{TRM}
Maximum surge forward current	I_{TSM}
Power dissipation in SCR	P_d
Gate power dissipation	P_g
Blocking resistance	R_D
Dynamic forward resistance	R_T
Thermal resistance	R_θ
Delay time	t_d
Junction recovery time	t_{gr}
Turn-on time	t_{gt}
Turn-off time	t_q
Rise time	t_r
Reverse recovery time	t_{rr}
Forward blocking RMS voltage	V_D
Forward breakover voltage	V_{BO}
Reverse breakover voltage	V_{BR}

Repetitive peak off-state forward voltage	V_{DRM}
Nonrepetitive (surge) off-state forward voltage	V_{DSM}
Gate voltage	V_g
Repetitive peak reverse voltage	V_{RRM}
Nonrepetitive peak reverse voltage	V_{RSM}
On-state forward voltage drop	V_T

2.7 IMPROVEMENT OF DEVICE CHARACTERISTICS

In SCRs, the inner two layers are made wide and are lightly doped to increase the voltage rating of the device. This also gives a small value of α_b at low forward current. A reverse-biased junction of two lightly-doped layers can withstand higher voltage because of the wider depletion region. SCRs with voltage ratings upto 10 kV are now available. However, these highly resistive inner layers raise the on-state forward voltage drop due to light doping, and thereby increase the internal power dissipation. This reduces the current rating of the device. One way of increasing the current rating is to lower the internal current densities by providing a greater junction surface area. This will increase the overall size of the SCR. At present, the maximum current rating for an SCR is about 1200 A. A large junction surface reduces the permissible *di/dt*, and increases the turn-on time.

By a slight modification in the gate structure, substantial improvements can be achieved in *di/dt* ratings. The centre-gate construction is superior to side-gate construction because, in the former, a circular area around the gate is activated for conduction immediately after the gate drive is

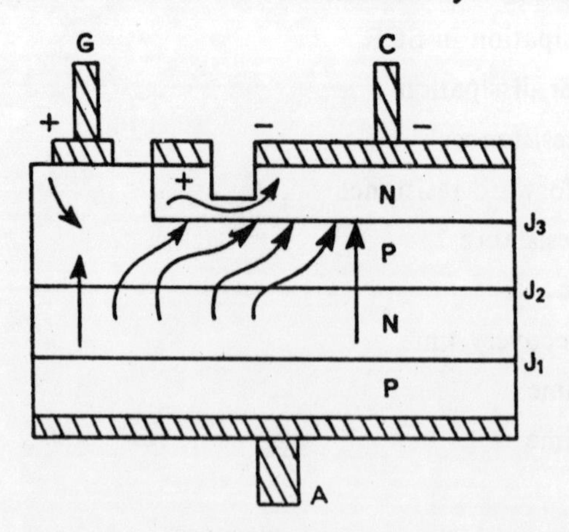

Fig. 2.5 Field-initiated gate structure.

initiated. The N^+ gate, or the field-initiated gate structure, provides for high *di/dt* capability. Here, a definite length of junction J_2 is turned on as soon as the gate signal is applied. The constructional features of a field-initiated gate structure and the movement of carriers from the gate to the cathode are shown in Fig. 2.5. Initially, the left portion of the SCR (minor SCR) is turned on by the gate signal, and its forward current makes the left-hand side of the P region positive. This lateral field drives a current from left to right, which forms the gate drive for the main SCR. Thus, a large junction surface is activated, thereby reducing the current density. The permissible value of *di/dt* with this gate structure is of the order of 100 A/μsec.

The effect of a high value of *dv/dt* for the forward voltage has already been discussed in Section 2.6. In the forward blocking state, all the applied voltage appears across the reverse-biased junction J_2. So, the capacitor-charging current given by $C_j(dv/dt)$, where C_j is the junction capacitance (shown by the dashed lines in Fig. 2.6), appears as a gate drive as far as the top PN layers are concerned. If this current is sufficiently high, the device may be turned on. One way to avoid this problem is to shunt this charging current away from junction J_3. This will reduce the possibility of turn-on by the charging current flowing from the J_2 region. However, a normal gate drive applied between the gate and the cathode will turn on the SCR in the conventional manner. The normal value of *dv/dt* which the device can withstand is of the order of 100 V/μsec. Figure 2.6 shows the structure of the shorted emitter which is used for SCRs with a capability of withstanding high values of *dv/dt*. Here, a part of the thermally-generated

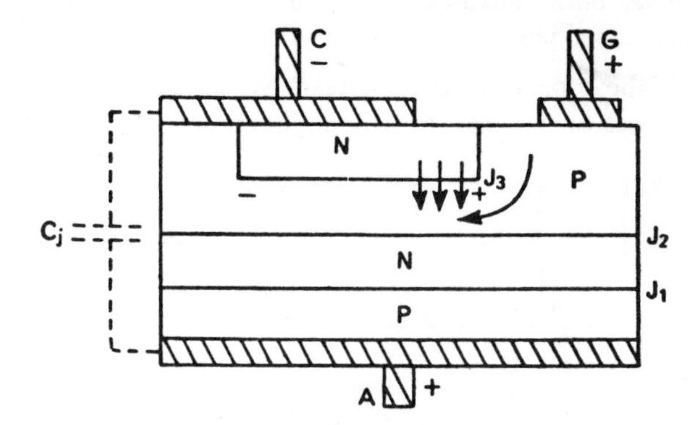

Fig. 2.6 Shorted emitter structure.

leakage current and the charging current across the reverse-biased junction J_2 are diverted around the gate-to-cathode junction by providing an alternative low-impedance path to the cathode as shown in the figure. Because of the side-gate structure and the highly-resistive P layer, the turn-on time of such an SCR is high.

In the SCR, the turn-off time is more than the turn-on time. The turn-off time is limited by the recombination rate of the excess carriers in the inner regions. The total recombination time is reduced by gold doping of the inner N layer, which provides for additional recombination centres.

It has recently been reported that by irradiating silicon with neutrons, the voltage-blocking and current-handling capability of an SCR can be considerably raised.

2.8 GATE CHARACTERISTICS

It has already been explained in Section 2.4 that the most convenient method of switching on a forward-biased SCR is by applying a positive voltage between the gate and the cathode. In some special-purpose SCRs, the gate drive is provided by photon bombardment. This device is called the light-activated SCR. As far as its internal mechanism for turning on is concerned, the process is the same as that of applying a gate drive to a conventional SCR which increases the minority carrier density in the inner P layer, thereby facilitating the reverse breakdown of junction J_2. It may also be mentioned that there are maximum and minimum limits for gate voltage and gate current to prevent permanent destruction of junction J_3 and to provide reliable firing. Similarly, there is a limit on the maximum instantaneous gate power dissipation ($P_{g\,max} = V_g I_g$). The permissible maximum value of $P_{g\,max}$ depends on the type of gate drive. The gate signal can be DC or AC (normally at power frequency), or a sequence of high-frequency pulses. With pulse-firing, a larger amount of instantaneous gate power dissipation can be tolerated if the average value of P_g is within the permissible limits. In other words, the gate can be driven harder (greater V_g and I_g) when pulse-firing is used. This provides for reliable and faster turn-on of the device. It must be noted that all the limits, both maximum and minimum, applied to gate drives are temperature-dependent.

In Fig. 2.7, the aforementioned gate drive limits are shown on the V_g-I_g characteristics of the gate-to-cathode junction. These are the diode

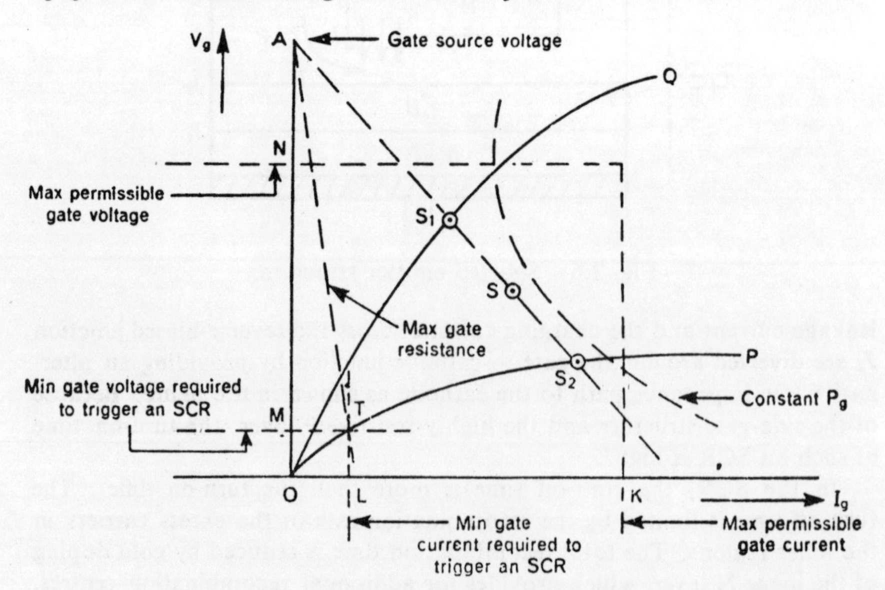

Fig. 2.7 Gate characteristics.

forward characteristics, and curves OP and OQ correspond to the possible spread of the characteristics for SCRs of the same rating. For best results, the operating point S, which may change from S_1 to S_2, must be as close as possible to the permissible P_g curve and must be contained within the maximum and minimum limits of gate voltage and gate current. This will provide the necessary hard drive for the device. If E_g is the source voltage, then the gradient of line AS will give the required gate source resistance R_g. The maximum value of this series resistance is given by line AT, where T is the point of intersection of lines indicating the minimum gate voltage and gate current. For pulse-firing, the value of permissible P_g will be higher, and point S can move further up, thereby providing a harder drive to the gate. The minimum value of gate source series resistance is obtained by drawing line AS tangential to the P_g curve.

On a short-term basis, thyristors may be generally considered to be charge-controlled devices. Therefore, for pulse-firing, the gate trigger current bears an inverse ratio to the pulse width. Also, the gate drive has to be maintained till the forward current has reached the latching-current level to sustain internal regeneration. Normally, for design considerations, the minimum gate pulse width is taken to be equal to the turn-on time of the device. Although the turn-on time is affected to some extent by the peak off-state voltage and the peak on-state current, it is influenced primarily by the magnitude of the gate trigger current. Thus, for a given turn-on time, the minimum gate current required, and the pulse width T can be obtained. If the frequency of firing f is known, the peak instantaneous gate power dissipation $P_{g\,max}$ can be obtained as

$$P_{g\,max} = V_g I_g = \frac{P_{g\,av}}{fT}, \tag{2.2}$$

where $P_{g\,av}$ is the specified maximum permissible average gate power dissipation. Using this value of $P_{g\,max}$ and the specified gate source voltage, the required series resistance is obtained by the construction shown in Fig. 2.7. Having obtained the operating point S, the voltage and current magnitudes V_g and I_g can be computed.

2.8.1 Example

(a) The latching current of an SCR used in a phase-control circuit, comprising an inductive load of $R = 10$ ohms (Ω) and $L = 0.1$ henry (H), is 10 mA. The input voltage is $325 \sin 314t$. Obtain the minimum gate pulse width required for reliable triggering of the SCR if it is gated at an angle $\pi/4$ in every positive half-cycle.

The load current $i(t)$ for the circuit is given by

$$i(t) = \frac{E_m}{\sqrt{(R^2 + \omega^2 L^2)}} [\sin (314t + \pi/4 - \theta) + \sin (\theta - \pi/4) e^{-Rt/L}],$$

where E_m, the peak value of the input voltage, is 325 V, $\tan \theta = 3.14$, and $\omega = 314$ rad/sec.

The gating pulse must be applied until the forward current through the SCR reaches the level of the latching current I_l. Since this current is small, current $i(t)$ can be assumed to have a constant di/dt for a short time after triggering. The value of di/dt at $t = 0$ is given by $(E_m/L) \sin(\pi/4)$. Therefore, the gate pulse width Δt must be larger than

$$I_l L \sqrt{2}/E_m.$$

Hence,

$$\Delta t = \frac{10 \times 0.1 \times \sqrt{2} \times 10^{-3}}{325}$$

$$\approx 4 \ \mu\text{sec.}$$

In practice, the pulse width is made more than the turn-on time of the SCR used in the control circuit.

(b) If the V_g-I_g characteristic (Fig. 2.7) of an SCR is assumed to be a straight line passing through the origin with a gradient of 3×10^3, calculate the required gate source resistance, given that $E_g = 10$ V and allowable $P_g = 0.012$ W.

Referring to Fig. 2.7, the operating point which satisfies $V_g I_g = 0.012$ and $V_g/I_g = 3 \times 10^3$ is given by $V_g = 6$ V and $I_g = 2 \times 10^{-3}$ A. Therefore, the gate source resistance is

$$R_g = 2 \ \text{k}\Omega.$$

2.8.2 Gate Circuit Parameters

From the discussion in Section 2.8, a series resistor R_g is required to be placed in series with the gate source voltage E_g to limit the gate-to-cathode voltage and current, and also to limit power dissipation. A gate-to-cathode shunt resistor R_{gc} is also provided for some SCRs to bypass a part of the junction leakage current and to improve the thermal stability of the device. This shunt resistor, in turn, will increase the required gate trigger current and also raise the levels of the holding and latching currents of the device. A shunt capacitor C_S connected across the gate and the cathode will improve the dv/dt capability of the device. However, with pulse-firing, a larger portion of the gate drive is bypassed by the capacitor; this will increase the delay time and consequently lower the di/dt rating of the device. This shunt capacitor poses one more problem. When the device is turned on, the gate acts as a voltage source and charges the capacitor. This charge can provide enough gate current after the anode current has stopped, and thereby increase the turn-off time of the SCR. If an inductor is connected across the gate and the cathode, the negative gate current will be maintained by the inductance even after the anode current has stopped, and this will facilitate faster turn-off. However, when pulse-firing is used for gating the device, the negative gate current, which continues to flow out of the gate, can possibly turn off the SCR. Thus, gate circuit parameters have to be chosen on the basis of the specific requirement. In general, a resistor R_g in series with the gate source voltage and a small shunt resistor R_{gc} across the gate and the cathode are often used.

The presence of a positive gate current when reverse voltage is applied to the anode will increase the reverse blocking current. As a result, the internal power dissipation will increase. If the type of gate drive used demands this situation, the positive gate current might be bypassed when reverse voltage appears across the device. This can be easily achieved by a diode-clamping circuit as shown in Fig. 2.8. Diode D_1 applies a negative

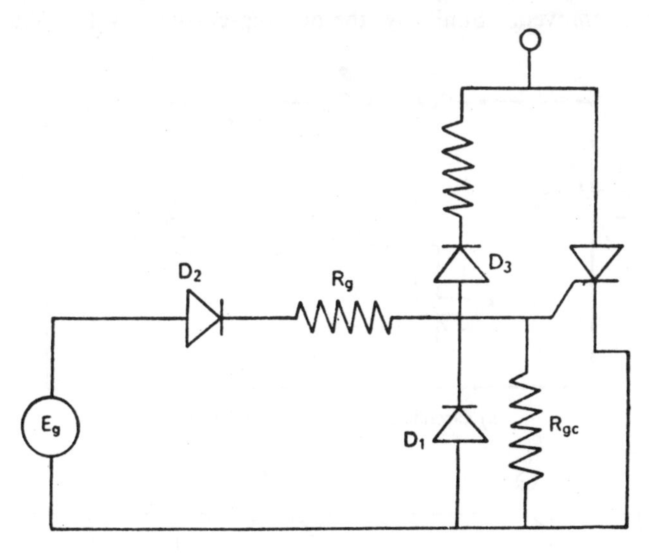

Fig. 2.8 Diode-clamping circuit.

voltage between the gate and the cathode when reverse voltage is applied across the device. This negative gate voltage reduces the reverse blocking current and improves the turn-off mechanism. Diode D_3 must be connected as shown in Fig. 2.8 to block the positive gate current coming from the supply when the device is forward-biased. The clamping diode D_1 also serves to limit the reverse voltage applied between the cathode and the gate if the gate source voltage E_g is alternating. The negative gate current flows through the device while the SCR is on, because diode D_1 will then be reverse-biased. This will increase the dissipation of gate power. A series diode D_2 in the gate circuit will prevent the negative gate source current. However, the shunt resistor R_{gc} can still bypass a portion of the thermally-generated leakage current across junction J_2 when the device is in the blocking state, and thereby improve thermal stability. The use of a negative voltage bias between the gate and the cathode is recommended. This will increase the forward breakover voltage and (dv/dt)-withstanding capability. Similarly, the reverse leakage current will also be reduced by the negative gate bias. The only drawback is that a greater gate source voltage E_g is required to overcome this bias and turn on the SCR.

2.8.3 Measurement of Device Parameters

Figure 2.9a shows a circuit for the measurement of forward and reverse leakage currents, with and without the gate drive. The same circuit can also be used for the measurement of holding and latching currents. The latching-current level is obtained by varying resistance R_s and applying the gate signal repeatedly through S_1. When the forward current is low, the SCR will go to the blocking state as soon as the gate signal is removed. If the anode current exceeds I_l, then the device will stay on even after the gate drive is removed. Similarly, the holding-current level is obtained by

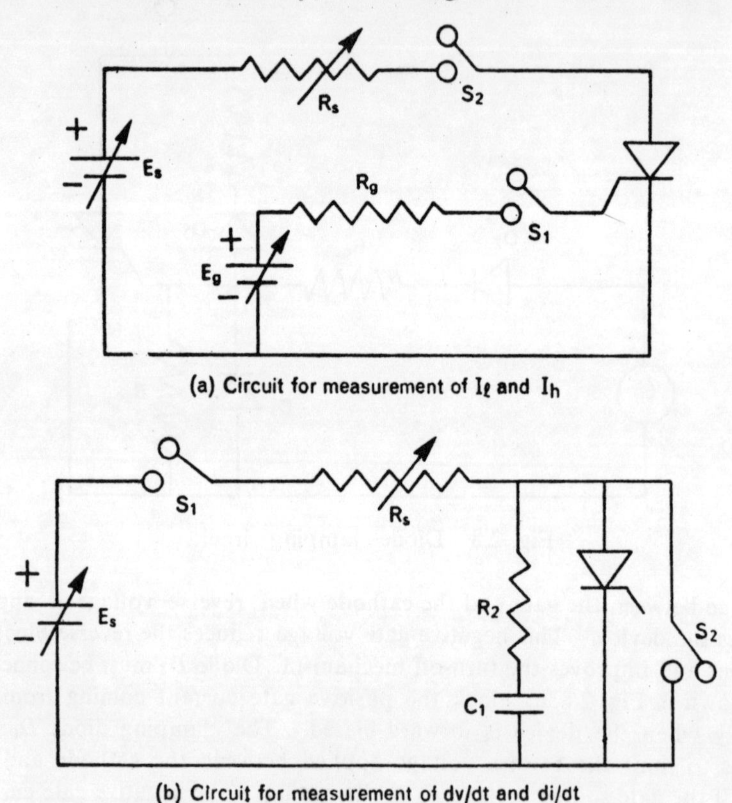

(a) Circuit for measurement of I_l and I_h

(b) Circuit for measurement of dv/dt and di/dt

Fig. 2.9 Measurement of device parameters.

reducing the anode current after the device is fired. This is done by increasing R_s and decreasing the applied voltage E_s. When the anode current falls below I_h, the SCR will go into the blocking state. The effect of temperature on the measurement of the parameters, as described here, can also be investigated. For a given forward blocking voltage, the minimum gate current required for triggering can be obtained from the circuit. The turn-on time of the device as defined in Section 2.4 can be measured by means of this circuit. The anode current is set up for a specific value in the steady state, and the gate drive is properly adjusted to provide reliable firing. The turn-on time can be measured from the oscillograms of the anode and the gate currents.

Figure 2.9b shows the circuit used for measuring the *dv/dt* capability of the device. This is a static *dv/dt* test because the SCR is off before the application of the voltage. A more stringent test is that for determining the applied *dv/dt* capability. Here, the device which was initially on, is subjected to a forced turn-off by an external circuit, and a forward voltage is then applied. Due to the presence of excess carriers in the inner layers, the reapplied *dv/dt* rating is lower than the capability to withstand static *dv/dt*. In Fig. 2.9b, the rate of application of the forward voltage is varied by changing the charging rate of capacitor C_1. The initial value of *dv/dt*, when S_1 is closed and S_2 is open, is given by

$$\frac{dv}{dt} = \frac{E_s R_s}{C_1}. \tag{2.3}$$

If this value is more than the *dv/dt* rating of the SCR, the device will conduct without any external gate signal. It is assumed that resistance R_2 is very small and the initial jump in voltage appearing across the SCR $[E_s R_2/(R_s + R_2)]$ is small enough to maintain the SCR in the off-state. The *dv/dt* rating of the device can be slightly improved by short-circuiting the gate terminal to the cathode terminal. The circuit shown in Fig. 2.9b can also be used for conducting the *di/dt* test on the SCR. For this, a small inductor is connected in series with the anode circuit. The rate of change of anode current, after the device is triggered, is controlled by varying the initial voltage on the capacitor. The specifications for the rise time and the peak current can be obtained from the application notes of the manufacturer. The measurement of the circuit-commutated turn-off time t_q can be made by means of the circuit shown in Fig. 5.3. Here, the SCR is turned off by forced commutation. Details of the circuit are given in Chapter 5. The time t_C for which the main SCR is reverse-biased is varied by changing capacitance C. The voltage waveform across SCR1 is also shown in the figure. When t_C is less than t_q, SCR1 turns on again. This is known as *commutation failure*. More details about forced commutation will be given in Chapter 8.

2.8.4 Example

The reverse-biased junction capacitance of an SCR is 25 picofarads (PF). The device can be turned on if the charging current flowing through the junction capacitor is 5 mA. Calculate the *dv/dt* capability of the device.

The required *dv/dt* to produce a charging current of 5 mA is 2×10^8 V/sec. Since a current of 5 mA in layer P_2 will turn on the device, the *dv/dt* capability is slightly less than 200 V/μsec.

2.8.5 Circuits for Gate Triggering

Figures 2.10a and 2.10b show simple R and RC circuits for triggering SCRs by means of gate control. The gate current magnitude can be changed by varying resistance R. The SCR triggers when there is sufficient gate current. A control on the firing angle can be easily attained when the applied voltage is AC. In Fig. 2.10a, the maximum firing angle can be $\pi/2$ since the gate current is in phase with the applied voltage. A larger variation in the value of the firing angle can be obtained by changing the phase and the amplitude of the gate current. This is achieved in Fig. 2.10b by the RC network.

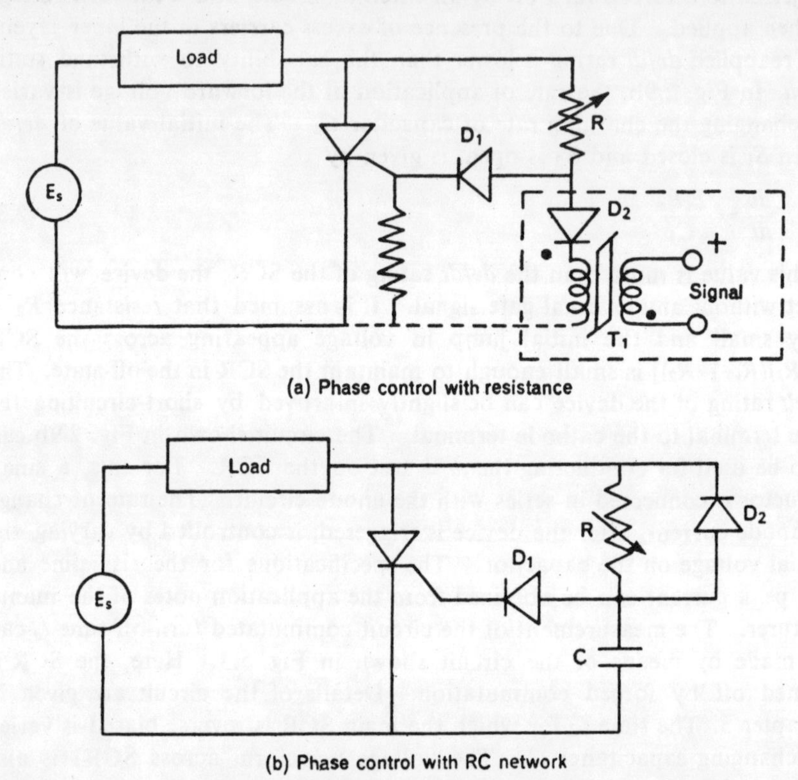

(a) Phase control with resistance

(b) Phase control with RC network

Fig. 2.10 SCR triggering circuit.

Capacitor C gets charged through diode D_2 to the negative peak value of the applied AC voltage during every negative half-cycle. Charging in the positive direction takes place in the following positive half-cycle. The charging rate is controlled by resistance R. When there is sufficient positive voltage across capacitor C, the SCR fires. Diode D_1 is used for preventing reverse breakdown of the gate-to-cathode junction in the negative half-cycle.

The circuit in Fig. 2.10 provides phase control in every positive half-cycle. Figure 2.11 shows the voltage and current waveforms. Power control can also be achieved by using what is known as *on-off control*. Here, the supply is given to the load for some time, called the *on-time*, and is cut off during the *off-time*. By controlling the on-time and off-time durations, power control is obtained. The circuit for this type of control is shown enclosed by dashed lines in Fig. 2.10a. A fixed value for R is used. A saturable-core reactor is shown as T_1 in Fig. 2.10a. The negative current through the primary winding of T_1 is prevented by diode D_2. Thus, the positive half-cycle of the applied voltage raises the flux level in the core, and after a few cycles the core will be saturated. The primary winding will then

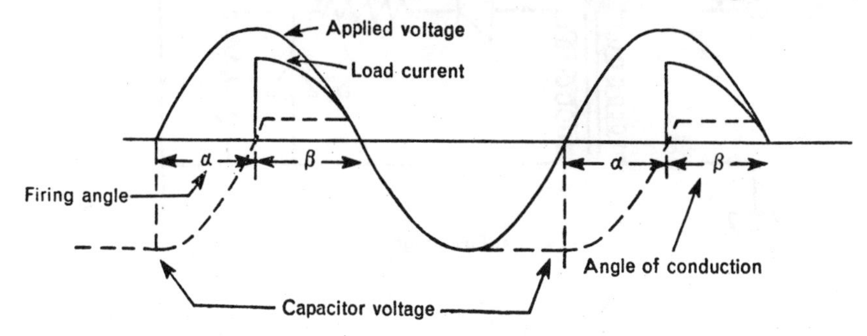

Fig. 2.11 Voltage and current waveforms.

have a very small incremental impedance and bypass the gate-to-cathode current flowing through R. Therefore, the SCR will not fire. The load voltage will be zero, and this is called the off-time. The application of a signal with proper polarity at the secondary winding of T_1 will bring down the flux level. As long as the signal is present, the flux level in the core will be kept low, and so the primary winding will have a high incremental impedance during the positive half-cycles. Therefore, the current through R will now flow through the gate and turn on the SCR. Thus, the SCR will be turned on during every positive half-cycle as long as the signal is applied to the secondary winding of T_1. This is the on-time during which the SCR fires and voltage is applied to the load. When the signal on the secondary side is removed, the core will get saturated after a few cycles and bypass the gate current flowing through resistance R. The SCR will then stop conducting and the load voltage will become zero. More details on the on-off control scheme are given in Chapter 10.

Transformer T can be introduced as shown in Fig. 2.12a to isolate the control circuit from the main power circuit if AC or pulse-firing is used for the gate drive. Diode D_1 prevents negative source current and diode D_2 limits the reverse voltage across the gate and the cathode.

Figure 2.12b shows a unijunction transistor (UJT) relaxation oscillator where the output pulses are coupled to the SCR gate through the isolation transformer T. The pulses here are unidirectional. This requires that the transformer used for isolation and coupling be properly designed. The frequency of the pulses depends on the time constant of the RC circuit. The UJT output can be synchronised with the AC voltage applied to the SCR. More details about this circuit and UJT characteristics are given in Chapter 4. For large SCRs, a low-power SCR can be used to provide reliable triggering, as shown by the dashed lines in Fig. 2.12b. This is known as *slave triggering*. The low-power SCR is triggered by the UJT, using the circuit shown in Fig. 2.12b.

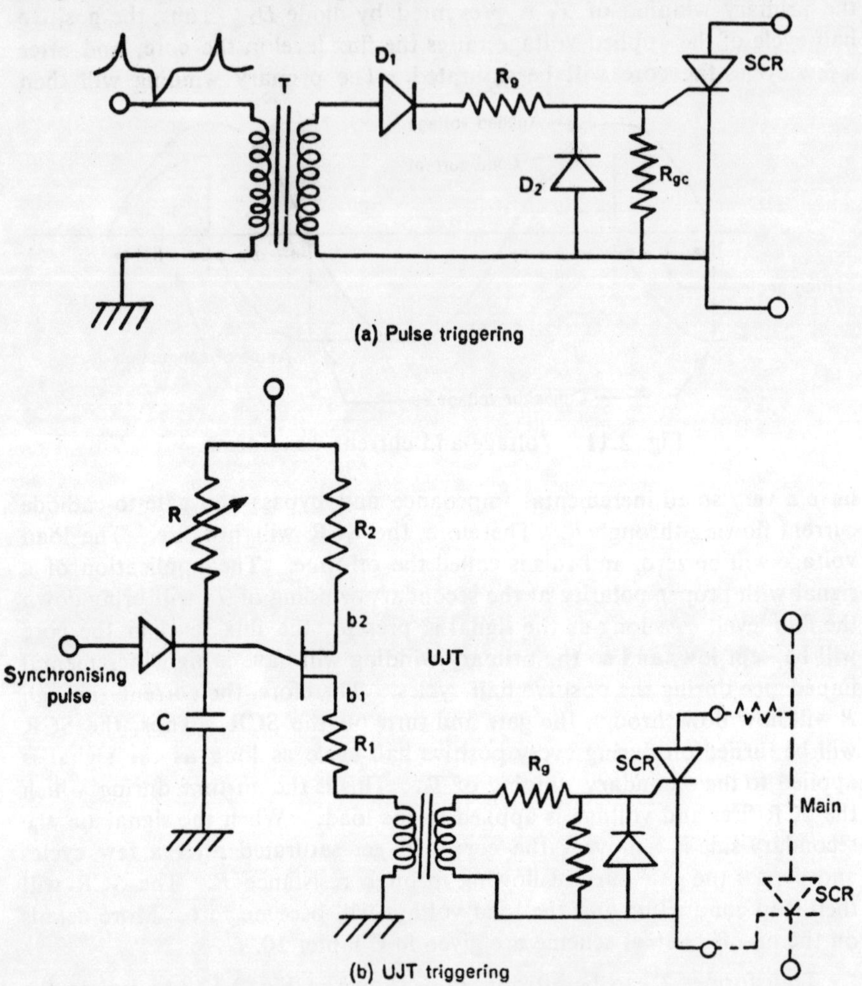

Fig. 2.12 Circuits for gate firing.

The triggering of series- and parallel-connected SCRs is discussed in Chapter 3. The methods of triggering triacs with a diac or a low-power thyristor, such as an SUS, are given in Chapter 4. The General Electric Company has developed an IC chip PA 436 which can be used for triggering both SCRs and triacs. It has a built-in facility for controlling the firing angle and maintaining constant load voltage/current.

2.9 INTERNAL POWER DISSIPATION AND TEMPERATURE RISE

The I^2R loss due to the forward current is the major component of the total power dissipation in the SCR. The switching losses (i.e., the losses during the transition period when the device goes from on-state to off-state, and vice versa) at higher operating frequencies will be greater than those at lower frequencies. However, for the present we may ignore these switching losses and the gate power dissipation. The V-I characteristics for a conducting SCR can be approximated by

$$V_T = V_0 + I_T R_T, \tag{2.4}$$

where R_T is the dynamic resistance of the device and V_0 is a constant. The instantaneous dissipation is then given by

$$P_d = V_T I_T = V_0 I_T + I_T^2 R_T, \tag{2.5}$$

and the average power dissipation is

$$P_{d\,av} = V_0 I_{T\,av} + R_T I_{T\,RMS}^2. \tag{2.6}$$

Thus, the average power dissipation is a function of the average and RMS values of the forward current. The ratio

$$k = \frac{I_{T\,RMS}}{I_{T\,av}} \tag{2.7}$$

depends on the waveform of the current and, therefore, on the applied voltage. If the SCR is operating in a DC circuit, then $k = 1$. If the same device is used for power modulation in an AC circuit by controlling the firing angle α, then the forward current waveform (see Fig. 2.11) will change with the angle of conduction β. The value of k will also change correspondingly. For low conduction angles β, ratio k is high, and when β is 180° (full conduction), $k = 1.57$. Therefore, the average forward current $I_{T\,av}$ at low conduction angles will produce more power loss, and hence higher junction temperatures, than that produced by $I_{T\,av}$ at high conduction angles. In other words, if the maximum junction temperature is limited, then the permissible average forward current of the device has to be lower for low conduction angles. Figure 2.13 shows the allowable on-state

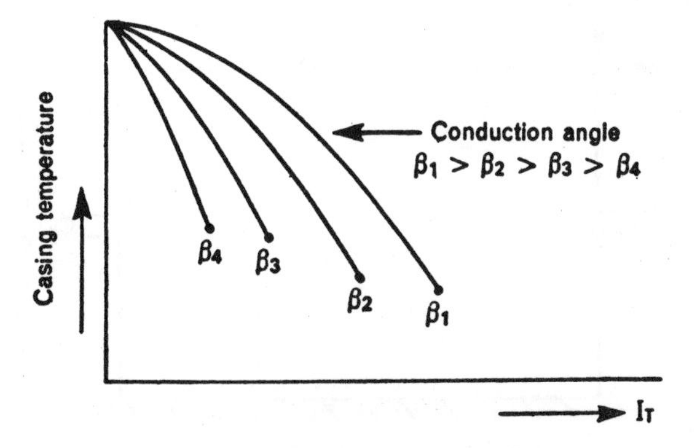

Fig. 2.13 Device current rating.

average current for different casing temperatures and for different conduction angles. As the casing temperature increases, the device rating decreases since the maximum junction temperature is kept constant. For silicon semiconductors, the maximum junction temperature is about 125°C. The end points of all the curves for different conduction angles specify the same $I_{T\,RMS}$. Even though the current rating of the SCR is specified by the average value (since it is a unilateral device), the ratings for the associated leads, connectors, and other passive components are all specified by the RMS current ratings. Thus, whatever may be the conduction angle, the maximum RMS rating should not be exceeded. The curves in Fig. 2.13 are applicable only when the applied voltage is sinusoidal. It is a normal practice to specify the continuous average rating $I_{T\,av}$ of the SCR, and also its $I_{T\,RMS}$ rating, on the assumption that the SCR carries a full half-cycle sinusoidal current at the specified frequency (50 Hz). As the frequency of operation increases, the SCR has to be derated to account for the additional switching losses.

Having computed the internal power dissipation from Eq. (2.6), it is assumed that this dissipation takes place at the internal junctions of the device. The temperature rise of the junctions above that of the casing is obtained from

$$T_j - T_c = P_{d\,av} R_\theta, \tag{2.8}$$

where T_j is the junction temperature in °C, T_c is the casing temperature in °C, and R_θ is the steady-state thermal resistance between the junction and the casing in °C/watt. Equation (2.8) gives the average temperature rise in the steady state. Since the current density across the junction will be more or less uniform in the steady state under normal frequency of operation, this average junction temperature is sufficient for specifying the current rating of the device. Figure 2.14 shows the electrical equivalent of the thermal circuit. Capacitor C_J corresponds to the thermal capacity of the junction. The

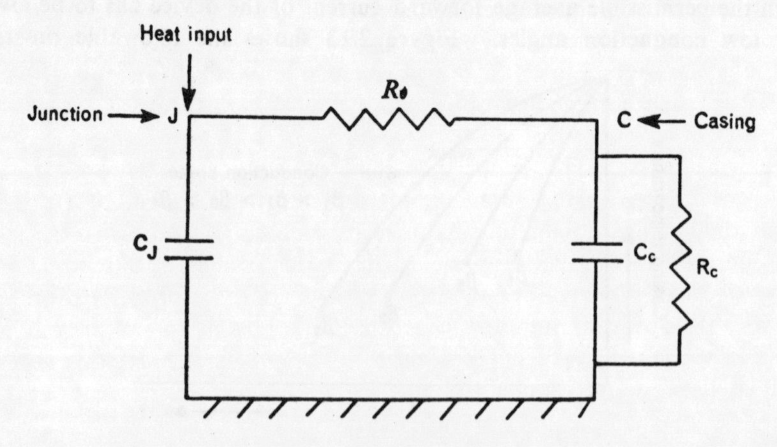

Fig. 2.14 Electrical equivalent of thermal circuit.

thermal resistance between the junction and casing is R_θ. C_c and R_c are the thermal capacity and thermal resistance, respectively, of the surrounding medium. The heat input is represented by a current source. When the potential of node J is such that the current through R_θ is equal to the input current, the steady state is reached. The same is the case in respect of node C. The current flowing through R_c corresponds to the heat flow to the surrounding medium. If the forward current is in the form of pulses, or is discontinuous, the maximum average junction temperature is obtained by using the transient thermal impedance of the device. The transient thermal impedance is the ratio of the instantaneous temperature to the internal power dissipation. To facilitate the computation of temperature rise, the instantaneous power dissipation curve (which is normally a half-rectified sinusoid) is replaced by a rectangular waveform whose magnitude is equal to the actual peak power dissipation, and has the same average value. Because of the intermittent nature of power dissipation, the maximum rise in junction temperature will be less than that for continuous conduction with the same peak power. Therefore, the intermittent repetitive current rating of the device will be higher than its continuous rating.

The recurrent ratings of an SCR for voltage and current ensure that the permissible maximum junction temperature will never be exceeded. However, the device may be subjected to some abnormal operation now and then due to short-circuits and other types of faults. Under such conditions, the device may be required to withstand much higher values of current or voltage for a very short time. These higher values are called the *nonrecurrent* or *surge ratings* of the device. When the device is subjected to a high-current surge, the junction temperature will rise much beyond the normal permissible value. Since the surge will be for a very short duration, the junction will cool down following the decay of the surge current. Therefore, immediately after the surge, the junction temperature will be reasonably higher than the permissible value, and the device will not be able to block any forward voltage. The device can be subjected to such a surge for only a limited number of times during its life. The surge current rating is inversely proportional to the duration of the surge which is measured in terms of the cycles of normal power frequency. For example, a five-cycle surge corresponds to a period of 100 msec consisting of five conducting half-cycles, each followed by an off-period. For durations less than one cycle, the corresponding ratings are called *subcycle surge ratings.* Surge current ratings are to be properly coordinated with the operating time for fuses and circuit breakers which are provided in series with the SCR for its protection. The details of protection schemes will be discussed in Chapter 11.

2.9.1 Example

An SCR has a continuous average current rating $I_{T\,av}$ of 25A and a dynamic resistance R_T of 1 Ω. If the casing temperature is decreased from

40°C to 30°C by efficient cooling, calculate the per cent increase in the device rating. State the necessary approximations.

It is assumed that the total internal power dissipation P_d is due to the forward conduction loss and that the maximum permissible junction temperature is 125°C. Therefore, the P_d of the device is equal to $I^2_{T\,av}R_T$. This is proportional to the rise in junction temperature. Let the new current rating of the device be I. The power dissipation is proportional to I^2 since R_T is the same. Thus, we have

$$25^2R = 125 - 40 = 85,$$
$$I^2R = 125 - 30 = 95.$$

Therefore,

$$I^2 = 95(25^2)/85,$$
$$I = 25\sqrt{\frac{95}{85}} = 26.4 \text{ A}.$$

Hence, the increase in the rating of the device is about 6 per cent.

REFERENCES

Corbyn, D. B., Poller, N. L., The characteristics and protection of semiconductor rectifiers, *Proc. IEE,* 1960, p. 255.

Gutzwiller, F. W., Sylvan, T. P., Power semiconductors under transient and intermittent loads, *AIEE Trans.* (Communications and Electronics), 1960, p. 699.

Mapham, N., The ratings of SCRs when switching into high currents, *IEEE Trans.* (Communications and Electronics), 1963, p. 575.

Newsam, B. U., Transient thermal resistance: Its measurement and use in the rating of thyristor, IEE Conference Publication, No. 53, 1969, p. 85.

Reed, J. S., Dyer, R. F., Power thyristor rating practices, *Proc. IEEE,* 1967, p. 1288.

Rice, J. B., Nickels, L. E., Commutation *dv/dt* effects in thyristor 3-phase bridge converters, *IEEE Trans.* (IGA), 1968, p. 665.

3

Multiple Connections of SCRs

3.1 SERIES-PARALLEL OPERATION OF SCRs

Silicon-controlled rectifiers are now available with a voltage rating upto 10 kV and a current rating upto 1200 A. In many power control applications, the required voltage and current ratings are lower than these maximum limits. Therefore, even though it may be possible to obtain a single SCR of proper voltage and current ratings, on many occasions the designer is forced to use lower-rated SCRs for reasons of economy and availability. In such a situation, lower-rated SCRs have to be connected in series and parallel combinations to suit the voltage and current requirements of the circuit for a particular application. Series and parallel combinations are also often used when it is required to control power in low-voltage high-current circuits or high-voltage low-current circuits because an SCR of suitable voltage and current ratings may not be available.

In this chapter, we will consider the problems associated with series and parallel connections of SCRs, and discuss how these may be overcome so that the SCRs can be utilised to the fullest advantage by improving their string efficiency. This efficiency is defined as the ratio of the total current/voltage rating of the whole string to the individual current/voltage rating of the SCR, multiplied by the number of units in parallel/series in the string. In practice, this ratio is always less than one. For achieving the best efficiency, the characteristics of all SCRs used in the string must be the same. Since it is impossible to get SCRs of identical characteristics, these have to be matched as far as possible. Small deviations in characteristics lead to unequal voltage/current distribution in the units connected in series/parallel. Figure 3.1a shows how voltage is shared by two SCRs of the same voltage rating when connected in series. Since they have to carry the same current, the SCR with the higher blocking resistance R_D $(=dV/dI_D)$ will share a larger portion of the applied voltage. Similarly, Fig. 3.1b shows the dynamic characteristics of two SCRs connected in parallel. Since the voltage drop across the two devices must be the same, the SCR with the lower R_T $(=dV/dI_T)$ will carry more current. The unequal sharing of voltage/current can be corrected by external equalising circuits.

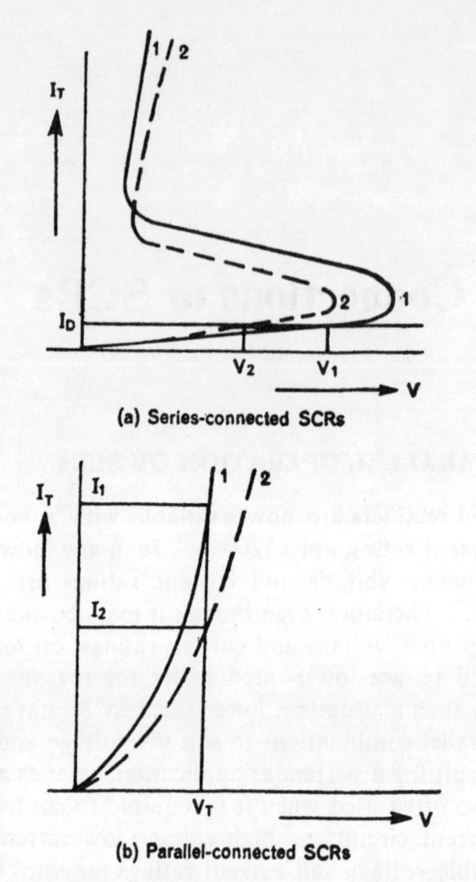

Fig. 3.1 Voltage/current sharing between SCRs.

3.2 SERIES OPERATION

When it is desired to increase the voltage rating of a string, the SCRs can be connected in series as shown in Fig. 3.2a. As already mentioned, these SCRs have to be properly matched as far as possible. Let us suppose that there are some differences in the individual characteristics of the SCRs. For example, assume that SCR1 has a high turn-on time. SCR2 has the lowest forward and reverse blocking currents, and the recovery time t_{rr} of SCR3 is small. Figure 3.2b shows the voltage distribution across the string of SCRs for six different situations: (1) all SCRs in the forward blocking state, (2) immediately after turn-on, (3) all SCRs conducting, (4) reverse voltage applied (all SCRs conducting in the reverse direction), (5) only SCR3 recovered, and (6) all SCRs in the reverse blocking state (all SCRs recovered). For these cases, the forward and reverse voltages applied are taken to be 1200 V.

It can be observed that even though each SCR is identically rated, the voltage distribution is not uniform because of some differences in characteristics, and that the SCR with the lowest voltage will break down, leading to the breakdown of the whole string. Therefore, some external compensating circuit is required to produce a uniform voltage distribution under all

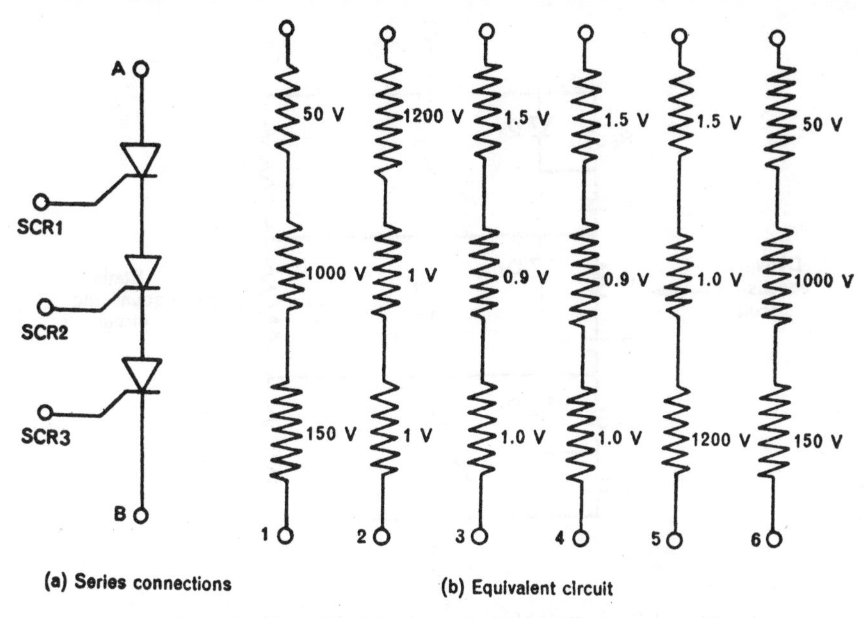

Fig. 3.2 Voltage distribution in series-connected SCRs.

conditions of operation. For conditions (3) and (4), in which all the SCRs are conducting, a small unequal voltage drop will not affect the performance of the series circuit. For conditions (1) and (6), when the string is blocking the forward and reverse voltages, respectively, the voltage distribution can be made more uniform by connecting a shunt resistor across each SCR as shown in Fig. 3.3. This is called the *static equalising circuit.* The value of the shunt resistance R is given by

$$R = \frac{nV_D - V_S}{(n-1)I_B}. \tag{3.1}$$

where n is the number of SCRs in series, V_D is the voltage rating of each SCR, V_S is the total voltage across the string, and I_B is the maximum blocking current of the SCRs at the rated voltage. This shunt resistance reduces the effect of the different blocking resistances of the SCRs, and thus produces a more uniform voltage distribution.

During the turn-on and turn-off periods, corresponding to operating conditions (2) and (5), respectively, because of the transient nature of the voltage and current, a simple resistance divider will not equalise the voltage. Here, the capacitance of the reverse-biased junction controls the voltage distribution, and therefore requires external shunt capacitance, as shown in Fig. 3.3, to provide uniform voltage distribution, and also to improve the dv/dt rating of the device. This is called the *dynamic equalising circuit.* During the period for which any of the devices is in the blocking state, the corresponding capacitor will get charged to the voltage across that unit. When the device is turned on, there will be a heavy

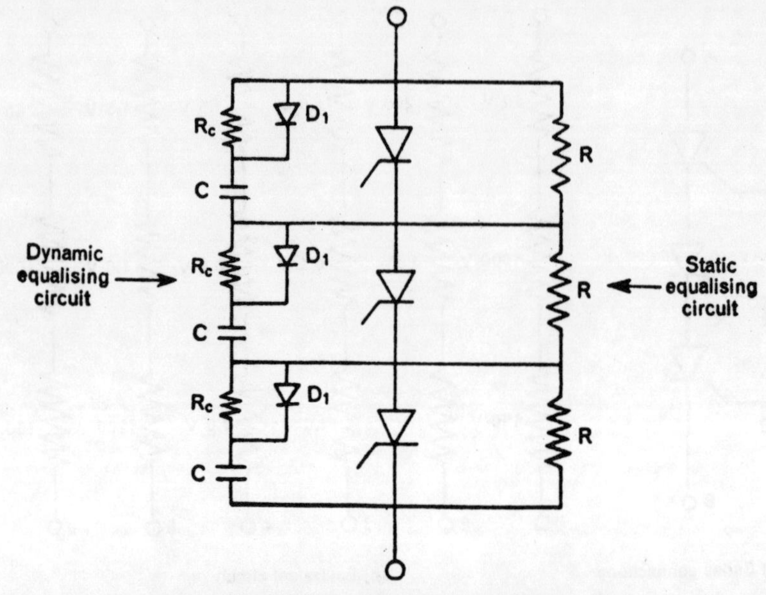

Fig. 3.3 Equalising circuits for series operation.

discharge current from the capacitors. To limit this discharge current, a small resistor R_c is used in series with the capacitor. Diode D_1 will cut off this resistor during the charging time of the capacitor when forward voltage is applied to the string. The value of C is obtained from

$$C = \frac{(n-1)\Delta Q}{nV_D - V_S}, \tag{3.2}$$

where ΔQ is the maximum difference in the recovery charge of the SCRs in the string. The value of ΔQ for various SCRs can be obtained from handbooks or from the application notes of the manufacturers. The choice of R_c depends on the permissible peak repetitive current through the SCR. During the turn-off period, when one SCR in the string has completely recovered, the shunt resistor R will provide an alternate path for the reverse current of the other SCRs to flow through. This facilitates the turn-off of the whole string. Derivations for Eqs. (3.1) and (3.2) are given in Example 3.5.

3.2.1 Triggering of Series-Connected SCRs

In spite of voltage equalising circuits, the sharing of voltage will be non-uniform if all the SCRs are not triggered simultaneously. For proper functioning of the series string, a gate signal of sufficient amplitude should be applied to all the gates of the SCRs at the same time. The small differences in turn-on times are properly corrected by the dynamic equalising circuit. Figure 3.4a shows one method of firing the string. The main triggering pulse is applied to the primary of the transformer. Each of the secondary windings is connected to individual gates of respective SCRs in the string as shown in the figure. To equalise the gate current in each SCR, a resistor R_g is connected in series with the secondary winding for swamping out any difference in the gate-to-cathode impedance of individual units.

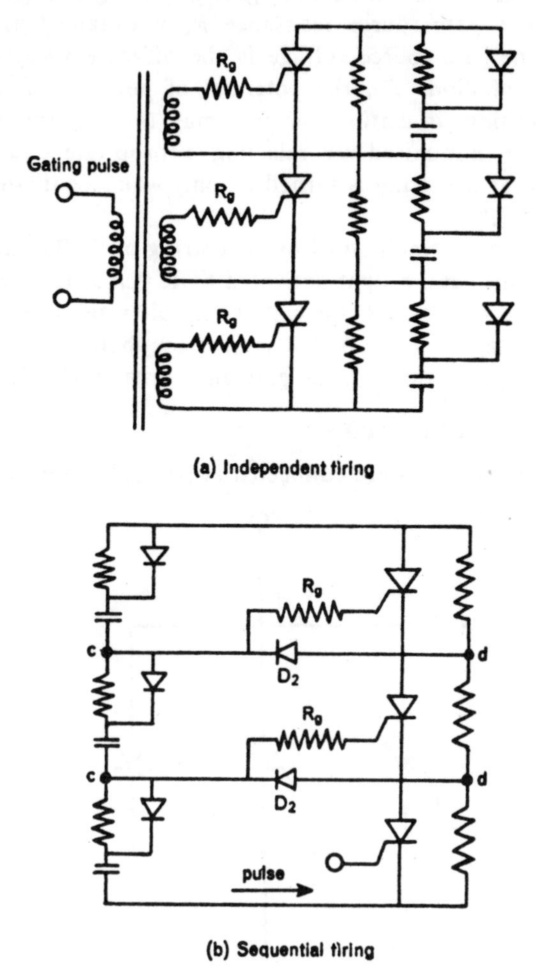

Fig. 3.4 Circuits for triggering series-connected SCRs.

Figure 3.4b shows another method of triggering the series string. Here, only one SCR at the bottom of the string is turned on by the external gate pulse. The discharge current of the shunt capacitor, through the SCR which is turned on, will fire the next SCR in the string. This process takes place rapidly, and all the SCRs are turned on in a very short time. Since the SCRs are turned on in sequence, the topmost SCR in the string will experience an increasing forward voltage. This sequential firing of series-connected SCRs is used for generating impulse voltages. The value of capacitance C, when required to provide uniform voltage distribution and also the necessary trigger current for the SCR, is given (in microfarads, μF) by

$$C = \frac{10}{R_g + V_{GT}/I_{GT}}, \tag{3.3}$$

where V_{GT} and I_{GT} are the maximum gate-triggering voltage and current, respectively. The gate source resistance R_g is obtained as explained in Section 2.8. The gate source voltage is the off-state voltage across each SCR. Because of diode D_2, the potential of points c, d will not be the same. The resulting circulating currents may turn on the SCRs. These currents must be minimised by selecting appropriate equalising circuit parameters so that the string is turned on only when a gate signal is applied to the bottom SCR.

Another convenient method of firing a string of SCRs employs optical triggering. In this case, a light-activated SCR (LASCR) is connected between the gate and capacitor C through a suitable resistance. One LASCR is used for each stage. The LASCR is fired by photon bombardment and capacitor C discharges through the gate and turns on the SCR.

3.3 PARALLEL OPERATION

Silicon-controlled rectifiers are connected in parallel to improve the current

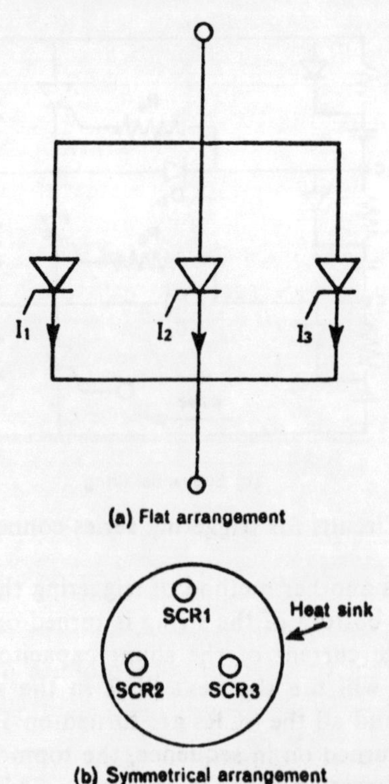

Fig. 3.5 Configurations of SCRs for parallel operation.

rating. Due to the unequal dynamic resistance R_T of the SCRs, the sharing of current will not be equal. This problem leads to what is known as *thermal runaway*. For example, if one of the SCRs in a parallel unit carries more current than the other members, its internal power dissipation will be more, thereby raising the junction temperature and decreasing the dynamic resistance of the SCR. This, in turn, will increase the current shared by this SCR and the process becomes repetitive, the cumulative increase in current results in permanent damage to the SCR, followed by the burning out of other SCRs, one by one. Therefore, one important precaution to be observed when SCRs are operated in parallel is that, as far as possible, all units must operate at the same temperature. This can be done by having a common heat sink. Unequal sharing of current is also the result of the inductive effect on current-carrying conductors. If three SCRs are arranged as shown in Fig. 3.5a, the middle one will have more flux linkages and, therefore, more inductance. In AC circuits, the reactance drop for the central limb will be higher, and so less current will flow through it. The outer two parallel units carry more current. This uneven distribution of current is corrected by having a symmetrical arrangement of SCRs as shown in Fig. 3.5b, where the SCRs are arranged in a circular configuration on the same heat sink.

In DC circuits, the difference in the value of dynamic resistance R_T is compensated by connecting a resistor R_p in series with each SCR, as shown in Fig. 3.6a. If the two SCRs are of identical rating, then the two external series resistors R_{p1} and R_{p2} are chosen such that the total voltage drops are equal, that is, $R_{p1} + R_{T1} = R_{p2} + R_{T2}$, where R_{T1} and R_{T2} are the corresponding dynamic resistances of the two devices at the rated current I_T. If two SCRs of different forward current ratings I_{T1} and I_{T2} are to be operated in parallel, then the same resistance R_p can be used for both units to ensure proper current sharing by the SCRs. Let V_{T1} and V_{T2} be the respective voltage drops across the two SCRs for forward currents I_{T1} and I_{T2}. Since the units are in parallel, their anode-to-cathode voltage drops must be the same. Therefore,

$$V_{T1} + I_{T1}(R_p + R_{T1}) = V_{T2} + I_{T2}(R_p + R_{T2}). \tag{3.4}$$

The value of R_p can be calculated from Eq. (3.4). The disadvantage of this type of compensation is that there is considerable loss of power due to series resistance.

In AC circuits the sharing of current can be made uniform by the magnetic coupling of parallel paths as shown in Fig. 3.6b. If currents I_{T1} and I_{T2} are equal, then the voltage drop in the transformer will be zero because of the mutual cancellation of flux linkages in the coils. If for any reason the currents are unequal, the transformer will produce an induced voltage in the windings which will reduce the difference between I_{T1} and I_{T2}.

3.3.1 Triggering of Parallel-Connected SCRs

The matching of the forward characteristics of SCRs is very important when they are connected in parallel. If one of the SCRs has a large delay time

t_d, then the forward voltage across its anode and cathode will be reduced to a very low value by the other parallel SCRs which would have been turned on. If this voltage is lower than the finger voltage of the blocking SCR (minimum forward voltage required for turning on), this unit will

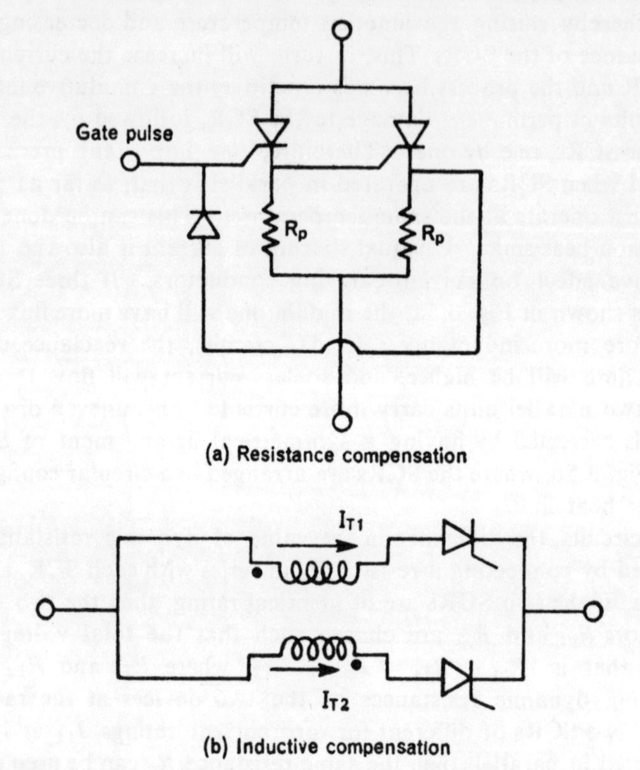

Fig. 3.6 Equalising circuits for parallel-connected SCRs.

never turn on. Not only does this result in unequal sharing of current, but it may also lead to permanent damage of the other SCRs. Such a situation can be avoided by using SCRs which are suitably matched in respect of turn-on time, and by providing sufficient gate drive for each SCR. When resistors R_p are connected in series with the cathodes for ensuring the proper sharing of the total current by individual units, the on-state voltage drop of each device will be more than the finger voltage, and therefore the voltage drop across resistor R_p of the SCR which is turned on can be conveniently used for providing the required gate drive for the other SCRs. This method of triggering is illustrated in Fig. 3.6a.

3.4 STRING EFFICIENCY

In spite of the compensating network used for voltage and current equalisation, the string efficiency will be less than unity, as mentioned in Section 3.1. In order to increase the reliability of the series/parallel string, an extra unit may be added so that the voltage/current applied to each device will be lower than its normal rating. Thus, there is an inherent derating of the units connected in parallel and series configurations. The per cent derating will be

$$\text{per cent parallel derating} = \left(1 - \frac{I_m}{n_p I_T}\right) \times 100, \tag{3.5}$$

where I_m is the total forward current, n_p is the total number of units in parallel, I_T is the forward current rating of each device; and

$$\text{per cent series derating} = \left(1 - \frac{V_S}{n_s V_D}\right) \times 100, \tag{3.6}$$

where V_S is the total voltage applied, n_s is the total number of units in series, V_D is the forward voltage rating of each device.

If the desired per cent derating is known, the number of series and parallel units (n_s and n_p) required can be evaluated; the number of units so determined is then increased by 1 to improve the reliability of the string.

3.5 EXAMPLE

The voltage and current ratings in a particular circuit are 3 kV and 750 A. SCRs with a rating of 800 V and 175 A (General Electric Company designation C 350) are available. The recommended minimum derating factor is 15 per cent. Calculate the number of series and parallel units required. Also obtain the required values of R and C to be used in the static and dynamic equalising circuits if the maximum forward leakage current for the SCRs is 10 mA and $\Delta Q = 20$ microcoulombs (μC). Derive the equations used.

Since a minimum derating factor of 15 per cent is recommended, we have

$$0.15 = 1 - \frac{3000}{n_s \times 800}$$

$$= 1 - \frac{750}{n_p \times 175}.$$

Therefore, the number of series-connected SCRs is

$$n_s = \frac{3000}{800 \times 0.85} = 4.3 \approx 5,$$

and the number of parallel-connected SCRs is

$$n_p = \frac{750}{175 \times 0.85} = 5.05 \approx 5.$$

The values of R and C for the voltage equalising circuits may be derived as follows:

For the series connection of SCRs shown in Fig. 3.1, let there be n_s SCRs and let the SCR at the top of the string have zero blocking current and all the other SCRs have identical blocking currents equal to I_B. Then, voltage V_D across the top SCR will be given by

$$V_D = IR = V_S - (n_s - 1)(I - I_B)R,$$

where R is the shunt resistance to be connected across each SCR for static equalisation of voltages and I is the total current in the string. Voltage V_D must be equal to the forward blocking voltage of the SCR. The value of R obtained from the foregoing equation will be

$$R = \frac{n_s V_D - V_S}{(n_s - 1)I_B} = \frac{(5 \times 800) - 3000}{4 \times 10} \times 10^3 = 25 \text{ k}\Omega.$$

To calculate the value of shunt capacitance C for dynamic equalisation of voltages, the total recovery charge of all the SCRs in the string is required. Considering the worst case when the top SCR has no recovery charge and all the other SCRs have equal recovery charge ΔQ, the V_D appearing across the top SCR is given by

$$V_D = \frac{1}{n_s}\left[V_S + \frac{(n_s - 1)\,\Delta}{C}\right].$$

Therefore,

$$C = \frac{(n_s - 1)\,\Delta Q}{n_s V_D - V_S} = \frac{4 \times 20 \times 10^{-6}}{(5 \times 800) - 3000} = 0.08\ \mu\text{F}.$$

REFERENCES

Hey, J. C., Series operation of SCRs, Application Note 200.40, G. E. Company, Syracuse, USA.

Somos, I., Piccone, D. E., Behaviour of power semiconductor devices under transient condition, IEEE Conference Record (Industrial Static Power Conversion), 1965, p. 243.

Tobisch, G. T., Parallel operation of silicon rectifier diodes, *Mullard Technical Communications* (UK), No. 73, 1964.

4

Power and Switching Devices

4.1 THYRISTORS

As already mentioned in Chapter 1, the term 'thyristor' is used for the family of semiconductor devices that exhibit bistable characteristics and can be switched on and off like the gas-tube thyratron. The SCR, which has been discussed in detail in Chapter 2, is a member of this family, and is widely used because of its large power-handling capacity. The operating characteristics of some of the other important thyristors are described in this chapter. All, except the triac, are low-power devices often used for switching in control circuits. They also find applications as efficient triggering devices for SCRs and triacs, and in digital and pulse circuits.

4.2 LOW-POWER DEVICES

When the forward current in an SCR is low, it can be turned off by applying a negative voltage between the gate and the cathode. The turn-off mechanism can be easily explained by the two-transistor analogy shown in Fig. 2.2. The negative voltage applied to the base of the bottom transistor, when the device is conducting, bypasses a part of the forward current. This reduces the base drive of the bottom transistor; the corresponding decrease in its collector current lowers the base drive of the top transistor. Because of this cumulative action, the α_bs of the two transistors decrease, and when their sum is less than 1, internal regeneration stops and the device goes to the off-state. Thus, the gate can be used for both turning on and turning off the device. This is called the *gate-controlled switch* (GCS) or the *gate turn-off switch* (GTO), and its operation is limited to low anode-to-cathode currents. Its symbolic diagram is the same as that of an SCR.

Using the same two-transistor analogy, an SCR can also be turned on by applying a negative voltage between the emitter (anode) of the top transistor and its base. The base terminal is called the *anode gate*. The negative base current increases the collector current, which, in turn, becomes the base drive for the bottom transistor. This regenerative positive feedback turns the device on. Such a device with an anode gate is called a *complementary SCR* (CSCR). To turn on the device by this method a much harder gate drive is required than that for a device with a conventional cathode gate because the top N_1 layer is highly resistive as its doping level is very low. Since the current flows out of the device through the anode gate, the overall loop gain will also be lowered, and therefore a longer gate signal will be required. The CSCR is often used in the relaxation oscillator circuit as shown in Fig. 4.1. This figure also shows

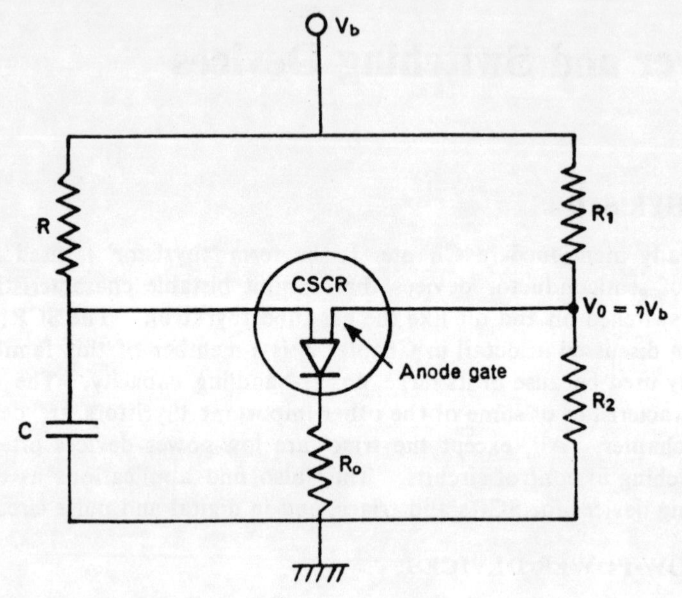

Fig. 4.1 Relaxation oscillator circuit.

the symbolic representation of a CSCR. Resistors R_1 and R_2 provide a fixed potential V_0 ($=\eta V_b$) to the anode gate (Fig. 4.1). The potential ratio η is given by $R_2/(R_1 + R_2)$. The increase in the anode potential will depend on the charging rate of capacitor C. Whenever this potential increases beyond V_0, the device will be turned on and the capacitor will discharge through the output resistance R_o. When the device is conducting, potential V_0 will be very low, and so the capacitor can discharge almost completely. When the discharge current is low ($<I_h$), the device will be turned off, the gate potential will go back to V_0, and the anode potential will again start rising. A similar action will take place when the capacitor voltage reaches V_0. Thus, there will be a periodic pulse output across resistor R_o. The output frequency will depend on the time constant RC of the charging circuit and potential V_0 of the anode gate. The operation of the device is similar to that of a unijunction transistor (UJT), whose characteristics will be discussed in Section 4.3. The only difference between a CSCR and a UJT is the additional degree of freedom in the choice of η for the former. It is for this reason that the CSCR is also referred to as a *programmable unijunction transistor* (PUT).

By providing two gates, a cathode gate as in the conventional SCR, and an anode gate as in the CSCR, an SCR can be turned on by applying a positive signal to the cathode gate, and turned off by applying a positive signal between the anode gate and the anode. Such a device is called a *silicon-controlled switch* (SCS). It is also known as a *unilateral tetrode thyristor*. The symbolic diagram and *V-I* characteristics of an SCS are given in Table 1.1. In this device, all the layers are connected to the outside terminals. It is a low-power device (maximum ratings are 100 V, 200 mA), and so can be turned off by applying a positive signal to the anode gate or a negative signal to the cathode gate. Thus, an SCS can also function as a GCS or a CSCR. This device has many applications in digital and pulse circuits (see Chapter 5).

A *silicon unilateral switch* (SUS) is similar to a CSCR with an anode gate, except for the fact that it has an internally-built low-voltage avalanche diode between the gate and the cathode. Because of this, the device will turn on for a fixed anode gate voltage. The device can also be used as a relaxation oscillator. The output pulses across R_o are used for triggering SCRs. The switching voltage V_0 ($=\eta V_b$) can be varied in the case of a PUT, whereas for an SUS it has a fixed value as determined by the operation of the internal avalanche diode. No separate gate drive is required to turn on the device since it is provided internally when the avalanche diode breaks. The symbolic diagram and the *V-I* characteristics for an SUS are shown in Table 1.1. Because of its very limited negative resistance region [I_p (peak point current) is very close to I_v (valley point current)], the range of output frequency of a relaxation oscillator using an SUS is very small.

The *silicon bilateral switch* (SBS) is a device which essentially comprises two identical SUS structures, arranged in antiparallel, with one gate. As the name indicates, the device conducts in both directions when the applied voltage breaks the internal avalanche diode. The gate is used only for external synchronisation or for proper biasing. Since no external gate drive is utilised for turning on the device, the gate terminal can be removed from the SBS structure. The device is then called a *diac* (or *trigger diode*). The symbolic diagram and characteristics of a diac are shown in Table 1.1. The device exhibits a negative resistance characteristic in both directions when it begins to conduct. This negative resistance region extends over a large range of current. For this reason, the device operates in the relaxation mode over a wide range of frequency. It is often used as a trigger device for triacs which require either positive or negative gate pulses to turn on. It also permits an efficient control of the firing angle α in each half-cycle of the applied AC voltage (phase control) because of the large time interval permissible between consecutive output pulses from the relaxation oscillator. Fig. 4.2 shows a simple phase-controlled circuit using a triac and a diac. During the positive half-cycle (when *A* is positive), the triac requires a positive gate signal for turning it on. This is provided by the capacitor when its voltage is above the breakdown voltage of the diac. The capacitor discharges through the triac gate. When the triac fires, the voltage drop across AB will be zero and the capacitor voltage will be reset to zero. A

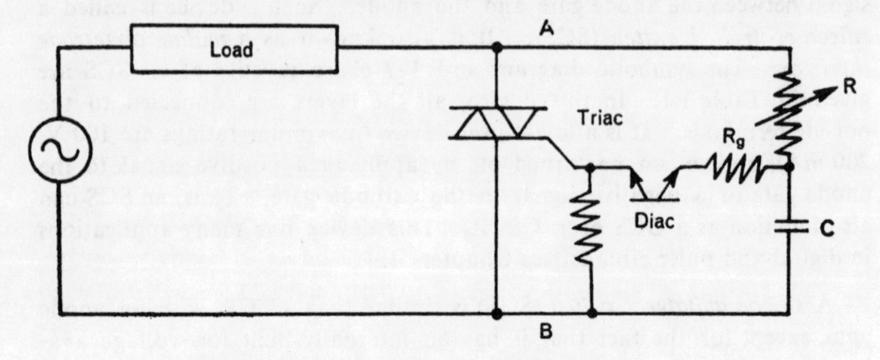

Fig. 4.2 Diac phase-controlled circuit.

similar operation takes place in the negative half-cycle, and a negative gate pulse will be applied when the diac breaks down in the reverse direction. By varying resistance R, the charging rate of capacitor C can be changed and thus the firing angle can be controlled.

4.2.1 Example

A diac of type ST_2 is used for triggering an SC 245 triac in a phase-controlled circuit as shown in Fig. 4.2. Obtain the maximum permissible value of R and the corresponding firing angle.

The ST_2 diac has the characteristics

$$V_{BR} = 30 \text{ V}, \qquad I_{BR} = 100 \ \mu\text{A},$$

where V_{BR} and I_{BR} are its breakdown voltage and breakdown current, respectively. Capacitor C is assumed to be reset to zero voltage at the end of every half-cycle. The SC 245 triac has a voltage rating of 400 V and a current rating of 10 A. The maximum gate trigger voltage is 2.5 V and the gate trigger current is 50 mA. For this data, a capacitor of 0.1 μF and a gate series resistance R_g of 1 kΩ can be used for providing reliable triggering. The input voltage is 230 V single-phase AC supply at 50 Hz.

The maximum firing angle is given by

$$E_m \sin \alpha = V_{BR},$$

where E_m is the peak amplitude of the input AC voltage. Therefore,

$$\alpha_{max} = \pi - \sin^{-1} \left(\frac{30}{230\sqrt{2}}\right) = 174.5°.$$

The value of resistance R for this firing angle is given by

$$R_{max} = \frac{2E_m}{\omega C V_R},$$

where ω is the input supply frequency. Therefore,

$$R_{max} = \frac{2 \times 230\sqrt{2} \times 10^6}{314 \times 0.1 \times 30}$$

$$= 690 \text{ k}\Omega.$$

4.3 UNIJUNCTION TRANSISTOR (UJT)

Although a unijunction transistor does not belong to the family of thyristors as far as its structure is concerned, its characteristics are similar to those of an SUS and a CSCR. It is often used as a relaxation oscillator to obtain sharp pulses with a good rise time in triggering circuits for SCRs. It has a fast switching action. The conventional UJT is made up of an N-type silicon base (with terminals b_1 and b_2) to which is alloyed a P-type emitter with terminal e, as shown in Fig. 4.3a. The symbolic diagram and characteristics of a UJT are shown in Figs. 4.3b and 4.3c, respectively. A complementary UJT is formed by a P-type base and an N-type emitter. Except for the polarity of voltage and current, the characteristics of a complementary UJT are similar to those of a conventional UJT. The characteristics of a conventional UJT will be discussed in this section.

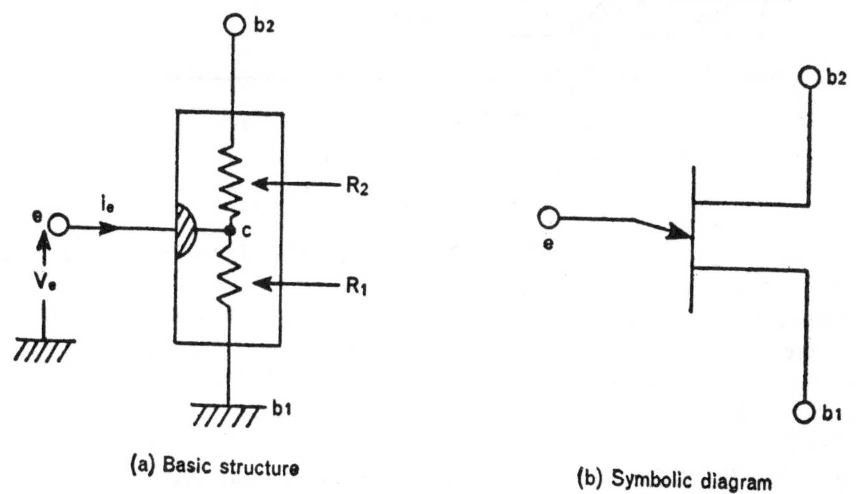

Fig. 4.3 Unijunction transistor (cont.).

With positive voltage V_b applied between b_2 and b_1 (Fig. 4.3a), the potential of point C will be ηV_b, where η, referred to as the *intrinsic standoff ratio*, is determined by the relative magnitudes of the internal resistances R_2 and R_1, respectively, of top (base 2) and bottom (base 1) regions of the device. As long as the emitter voltage V_e is less than the potential of C, the emitter-base 1 diode will be reverse-biased and emitter current i_e will be negative and equal in magnitude to the reverse leakage current as shown by curve PR in Fig. 4.3c. At point A in Fig. 4.3c, when the emitter voltage V_e is equal to $(\eta V_b + V_D)$, where V_D is the forward voltage drop of the diode, current i_e will be positive and the emitter-base 1 junction will begin to conduct. Point A is called the *peak point*, and the corresponding emitter potential and current are denoted by V_p and I_p. The holes injected by the emitter move to the base 1 region due to drift. Because of the increased number of carriers in the bottom region, resistance R_1 will

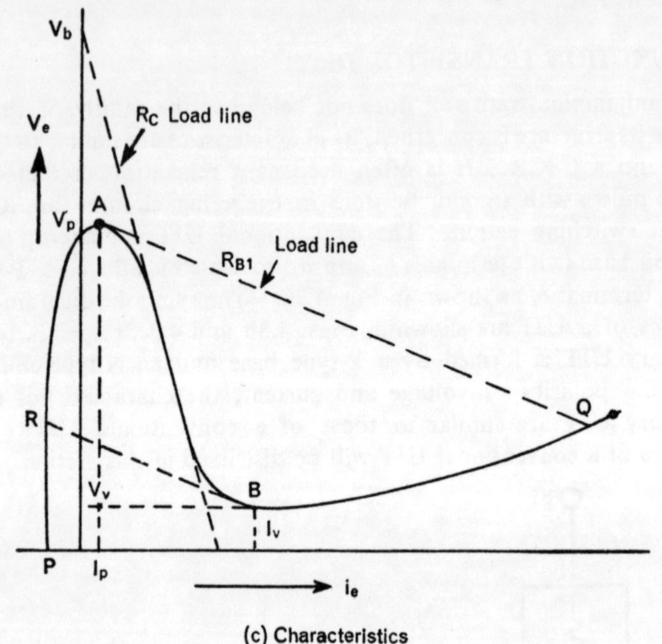

(c) Characteristics

Fig. 4.3 Unijunction transistor.

decrease. So, the potential of point C will fall, the diode will get a greater forward bias, and current i_e will increase. Thus, the device will exhibit a negative resistance region where an increase in current i_e is accompanied by a decrease in the emitter voltage V_e, as shown by curve AB in Fig. 4.3c. At point B, the entire base 1 region will be saturated with carriers and resistance R_1 will not decrease any more. A further increase in i_e will be followed by a rise in voltage. This characteristic is given by curve BQ in Fig. 4.3c. Point B is called the *valley point*; V_v and I_v are the corresponding emitter potential and current.

4.3.1 Relaxation Oscillator using a UJT

Since the UJT exhibits a negative resistance characteristic, it can be used as a relaxation oscillator. Figure 4.4a shows a UJT operating in this mode. The external resistances R_{B1} and R_{B2} are small in comparison with the internal resistances R_1 and R_2 of the UJT base. The emitter potential V_e is varied depending on the charging rate of capacitor C. The charging resistance R_C should be such that the load line intersects the device characteristics only in the negative resistance region AB, as shown in Fig. 4.3c. If the R_C load line intersects the device characteristic either in region PR or in BQ, the resulting operating point will be stable and the circuit will not oscillate. This sets the maximum and minimum limits on the permissible values of R_C, and also of the output frequency.

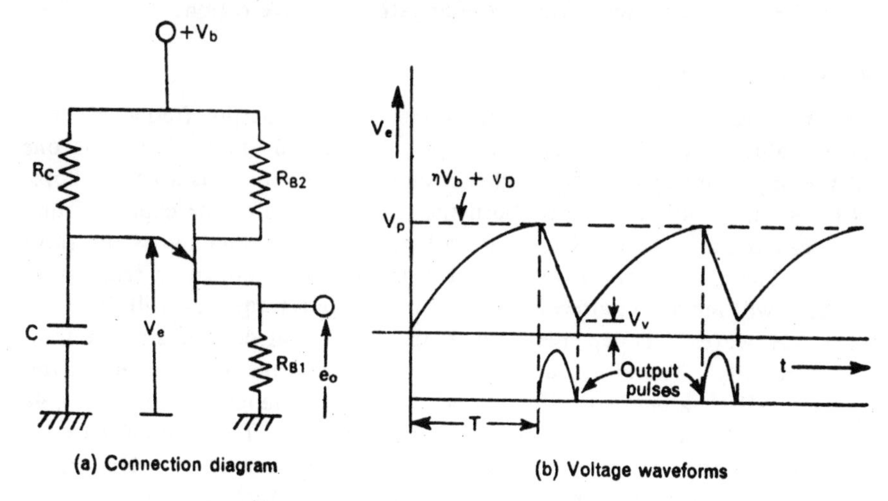

Fig. 4.4 UJT relaxation oscillator.

When the emitter voltage goes to the peak point voltage ($\eta V_b + v_D$), regeneration will start and the capacitor will discharge through resistance R_{B1}. The amplitude of the output pulse is obtained by drawing the dynamic load line for R_{B1} as shown in Fig. 4.3c. Line AQ measured along the voltage axis gives the pulse amplitude. As the capacitor discharges, the operating point Q will move down to B and then to R, when the UJT will again go to the blocking mode. The potential of point c will be ηV_b, and that of the emitter will be V_v, which is close to zero. The rise time of the output pulse will depend on the switching speed of the UJT, and the duration will be proportional to the time constant $R_{B1}C$ of the discharge circuit shown in Fig. 4.4a. The emitter-base 1 diode will again be reverse-biased until the capacitor is charged to voltage ($\eta V_b + v_D$). Then, similar operations will repeat. Figure 4.4b shows the emitter voltage and the output pulses of the relaxation oscillator. The output pulses are of identical magnitude and duration and their period T is given by

$$T = R_C C \ln\left(\frac{1}{1-\eta}\right). \tag{4.1}$$

The value of η is specified for the device. If $\eta = 0.63$, Eq. (4.1) will become $T = R_C C$. If the output pulses are used for triggering the SCR, resistance R_{B1} should be made sufficiently small so that the normal leakage current drop, when the UJT is off, will not trigger the SCR. The total energy dissipated in R_{B1}, in parallel with the gate-to-cathode resistance, is approximately equal to the energy stored in capacitor C (Fig. 4.4a) at the time the UJT is switched. This is equal to $(C\eta^2 V_b^2)/2$. There is a minimum amount of energy ($\int v_g i_g \, dt$) required for reliable turning-on of the SCR. Therefore, for a given V_b, there is a minimum value of capacitance

C. For given values of C, η, and T, the required value of R_C can be obtained from Eq. (4.1). It is important to see whether the R_C load line intersects the device characteristic in the negative resistive region.

4.3.2 Frequency Stability

The UJT relaxation oscillator maintains constant output frequency for a given value of R_C even though the supply voltage fluctuates. This is one of the important advantages of this circuit over the usual astable flip-flops. If the supply voltage changes, both the charging rate of the capacitor and the peak point voltage $(\eta V_b + v_D)$ will change approximately by the same ratio. Therefore, time T required for the emitter voltage to reach peak point A will remain unaltered, and so the output frequency will be constant. However, η is dependent on temperature, and decreases as the temperature falls. Therefore, at constant supply voltage V_b, the output frequency will change with the temperature. Resistance R_{B2} in Fig. 4.4a partly compensates for the effect of temperature. The internal base-to-base resistance of a UJT has a positive temperature coefficient. Hence, for a given value of V_b, the drop across terminals $b1$ and $b2$ of the device (Fig. 4.3a) will increase with temperature as both R_{B1} and R_{B2} are fixed. Since the voltage drop V_{b1b2} will increase, ηV_{b1b2} will remain constant even if η reduces with a rise in temperature. Thus, the output frequency of a UJT relaxation oscillator can be stabilised against variations in the supply voltage and changes in temperature. The approximate value for R_{B2} is $10{,}000/(\eta V_b)$ ohms.

4.3.3 Example

Design a relaxation oscillator using UJT 2N 2646 for triggering SCR 2N 2344 in a phase-controlled circuit. Obtain the limits for the output frequency of the oscillator.

UJT 2N 2646 has the following characteristics:

$\eta = 0.63, \quad I_p = 50\ \mu A, \quad I_v = 4\ mA,$

$V_v = 1\ V, \quad \text{supply voltage} = 15\ V.$

The base-to-base resistance is 6.5 kΩ. The maximum value of capacitance C (Fig. 4.4a) required for the given supply voltage is 0.1 μF (taken from the applications note of the manufacturer) to provide reliable triggering of 2N 2344. Therefore, if we fix the value of C at 0.1 μF, the maximum and minimum values of R_C are

$$R_{C\,max} = \frac{15(1-0.63)}{50} \times 10^6 = 100\ k\Omega,$$

$$R_{C\,min} = \tfrac{14}{4} \times 10^3 = 3.5\ k\Omega.$$

The limits on the output frequency are

$$f_{min} = \frac{1}{R_C C} = \frac{10^3}{100 \times 0.1} = 100\ Hz,$$

$$f_{max} = \frac{10^3}{3.5 \times 0.1} \approx 3 \text{ kHz}.$$

The approximate value of R_{B2} is

$$R_{B2} = \frac{10^4}{15 \times 0.63} \approx 1 \text{ k}\Omega.$$

The leakage current of the UJT is 2 mA. Therefore, if resistance R_{B1} is chosen as 50 Ω, the output voltage when the UJT is off will be 0.1 V. The minimum voltage required to trigger SCR 2N 2344 is 0.2 V. So, the SCR will be gated only when the UJT fires, and the peak voltage across R_{B1} will be about 8 V.

4.4 TRIAC

In the thyristor family, after the SCR, the triac is the most widely-used device for power control. Triacs with reasonably large ratings for voltage and current ($V_D = 500$ V, $I_T = 25$ A) are now available. The triac is a bilateral device with three terminals, and in operation it is equivalent to two SCRs connected in antiparallel. A cross-sectional view of the triac, showing all the semiconductor layers and junctions, is given in Fig. 4.5; for its symbolic

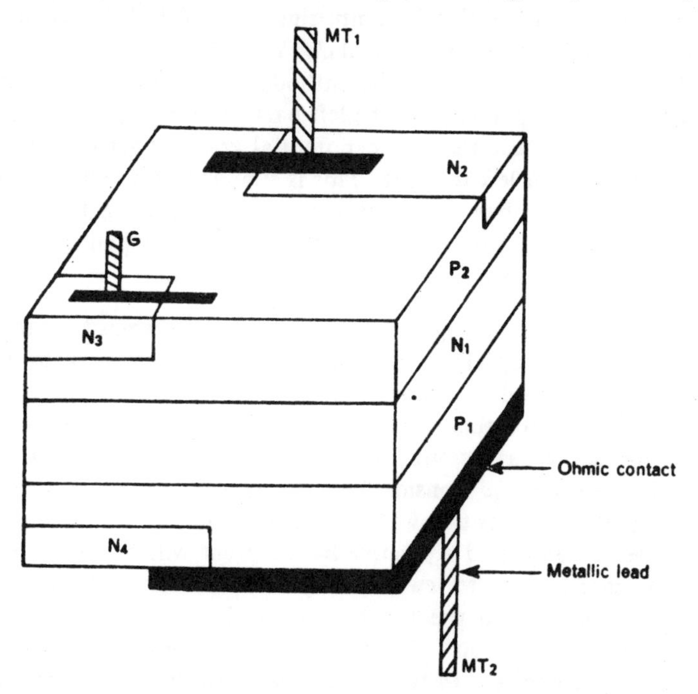

Fig. 4.5 Triac.

representation and *V-I* characteristics, see Table 1.1. The two main terminals are designated MT_2 and MT_1. The gate is near terminal MT_1. When the gate is open, the triac will block both polarities of the voltage applied across MT_1 and MT_2 if the magnitude of the voltage is less than the breakover voltage of the device. The characteristics of a triac are similar to those of an SCR, both in the blocking and conducting states, except for the fact that the SCR conducts only in the forward direction (anode to cathode) whereas the triac conducts in both directions. Because of this disparity the terms 'anode' and 'cathode' are not used for triac terminals. The other difference in operation is in the triggering mechanism. The triac can be triggered into conduction by applying either a positive or negative voltage to the gate with respect to terminal MT_1 whereas the SCR is triggered only by a positive gate signal.

4.4.1 Triggering Mode

When terminal MT_2 is positive and terminal MT_1 is negative, the triac can be turned on by applying a positive voltage between the gate and MT_1. This is a recommended method of triggering the device. In this mode, the triac behaves as a conventional SCR, with four layers $P_1N_1P_2N_2$ and the cathode connected to layer N_2. The device can also be turned on by applying a negative signal to the gate. In this case, the device is switched by an operation called *junction gate operation*. Initially, the left-hand portion of the triac (see Fig. 4.5) comprising layers $P_1N_1P_2N_3$ is turned on by the current flowing from terminal MT_1 to the gate through junction P_2N_3. Terminal MT_1 acts like the cathode gate. When this left-hand portion conducts, the potential of the left part of layer P_2 in contact with N_3 will go up, and because of the potential gradient across layer P_2, the current will flow from left to right. This is similar to the conventional gate current, and the right-hand part of the triac comprising $P_1N_1P_2N_2$ will turn on. The junction gate operation involves high switching losses, and therefore this form of gate drive is not normally used.

When terminal MT_2 is negative and terminal MT_1 is positive, the device can be turned on by applying a positive voltage between the gate and terminal MT_1. In this mode, the device is switched by remote gate operation. The four layers used for this operation are $P_2N_1P_1N_4$. The reverse-biased junction is formed by layers N_1P_1; it will be broken by increasing the carrier concentration in layer N_1. Consider the transistor formed by layers $N_2P_2N_1$. Since the gate is made positive with respect to terminal MT_1, the transistor will be properly biased and a positive base current will flow into layer P_2. This will increase the emitter current and raise the carrier concentration in layer N_1, and thus lead to the breakdown of the reverse-biased junction. The device will then turn on. The same operations will take place even if the gate drive is negative. In that case, layers $N_3P_2N_1$ will form the properly-biased transistor whose base drive is provided by the positive voltage between MT_1 and the gate. The device will turn on due to the increased current in layer N_1. When MT_2 is negative and MT_1 is positive, the recommended mode of triggering is by applying a negative voltage between the gate and terminal MT_1. Both positive and negative triggering pulses to the gate of a triac can be very conveniently obtained using a diac as shown in Fig. 4.2.

4.4.2 Phase Control using a Triac

Figure 4.6a shows the phase control of an inductive load by means of a triac. The power consumed by the load can be changed by varying the firing angle α. The hatched portion in Fig. 4.6b shows the voltage across

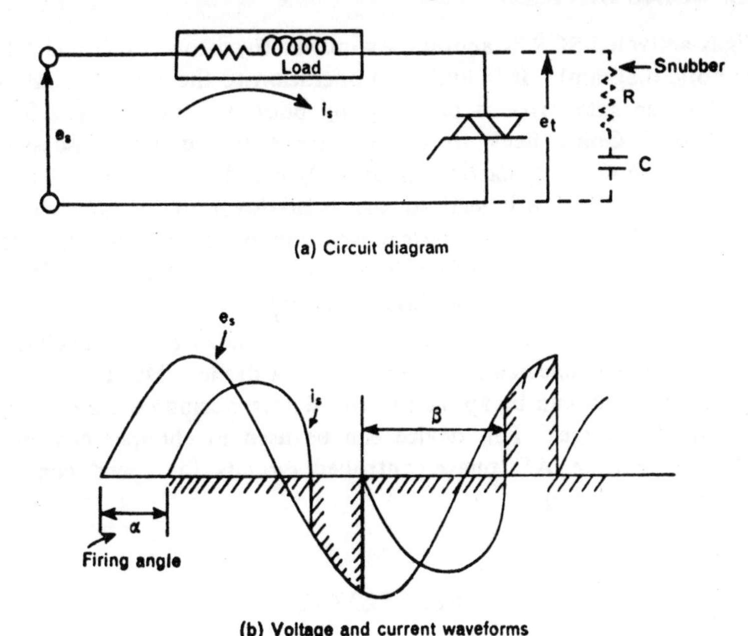

Fig. 4.6 Triac-controlled circuit.

the triac. One of the main problems with triac control is the sudden application of a reverse voltage across the triac immediately after it has stopped conduction. This problem is quite serious with highly-inductive loads where the current flows for a longer duration in each half-cycle for the same firing angle. This high reapplied *dv/dt* can turn on the device, and so the phase control will be lost. To avoid this maloperation, an *RC* snubber circuit (shown by the dashed lines in Fig. 4.6a) is connected in parallel with the triac. This will slow down the rate of change of voltage applied to the triac. The design of a snubber circuit is discussed in Chapter 11.

If the triac is replaced by two SCRs in antiparallel, a greater *dv/dt* can be tolerated. This is because each SCR conducts only in one direction, and as long as one of them is conducting, the other will be in the off-state with a reverse voltage applied. Since the device is initially off before a forward voltage appears across it, the maximum permissible *dv/dt* is related to the static *dv/dt* capability of the device. In the case of a triac, the same device will conduct in both directions, and so the permissible *dv/dt* is related to the reapplied *dv/dt* rating which is lower than the static rating. It is for this reason that SCRs are preferred over triacs for power control of inductive circuits with motor loads. SCRs are also used in circuits, such as controlled rectifiers, inverters, and choppers, where the current flow is only unidirectional.

4.5 SPECIAL DEVICES

The light-activated SCR is another low-power thyristor (see Table 1.1 for its symbolic diagram). It is similar in operation to the conventional SCR, except that the gate drive is provided by photon bombardment from a light source. Going back to the two-transistor analogy discussed in Section 2.3, the minority carriers in layer P_2 can be increased by photon bombardment, which will lead to the breakdown of junction J_2. The main advantage here is the physical isolation of the control and power circuits, whereby remote operation is possible. At present, this device is used in switching circuits or for slave triggering of power SCRs.

The latest addition to the thyristor family is the *reverse-conducting SCR* which is an SCR connected in antiparallel to a diode. Due to the restricted reverse bias that can be applied to this device during commutation, the turn-off time is large. This device can be used in chopper circuits (see Chapter 11) and in AC phase-controlled circuits for speed control of motors (see Chapter 5).

REFERENCES

Aldrich, R. W., Holonyak, Jr., N., Multiterminal PNPN switches, *Proc. IRE*, 1958, p. 1236.

Galloway, J. H., Using triac for control of AC power, Application Note 200.35, G. E. Company, Syracuse, USA.

Gentry, F. E., Bidirectional triode PNPN switches, *Proc. IEEE*, 1965, p. 355.

Howel, E. K., The light activated SCR, Application Note 200.34, G. E. Company, Syracuse, USA.

Spoffard, W. R., Programmable unijunction transistor, Application Note 90.70, G. E. Company, Syracuse, USA.

Stasior, R. A., Silicon controlled switches, Application Note 90.16, G. E. Company, Syracuse, USA.

5

Applications

5.1 POWER CONTROL

Because of the bistable characteristics of semiconductor devices, whereby they can be switched on and off, and the efficiency of gate control to trigger such devices, thyristors have been found to be ideally suited for many industrial applications. As already pointed out in Sections 1.2 and 1.3, their compactness, reliability, low losses, and fast turn-on and turn-off times have given them specific advantages over saturable-core reactors and gas tubes. Out of the several members of the thyristor family described in Chapters 1 and 4, the SCR and the triac have high ratings for voltage and current, and are widely used for power control. The other thyristors are used for low-power applications, and as switching devices in control and digital circuits.

The bistable states of the SCR and the triac and the property that enables fast transition from one state to the other are made use of in the control of power in both AC and DC circuits. Figure 4.2 shows a very simple form of control for AC circuits. This is known as *phase control* because the variation in load power at constant input voltage is obtained by changing the firing angle α. The corresponding load current waveform is shown in Fig. 4.6b. This form of voltage control is used for DC power supplies, temperature regulators, light dimmers, and speed regulation of AC and DC motors. More details about such power controllers are given in Chapter 6. One chief advantage of phase control is that the load current passes through a natural zero point during every half-cycle because of the AC supply. So, the device turns off by itself at the end of every conducting period. No other commutating circuit is required. For DC circuits, since the current does not have a natural zero value, the device has to be turned off by an external commutating circuit (forced commutation). Power control in DC circuits is obtained by varying the duration of on-time and off-time of the device. Such a mode of operation is called *on-off control* or *chopper control*. This method of control can also be used for AC circuits. Load current waveforms for AC and DC circuits with chopper control are given in Fig. 5.1. In certain types of AC choppers, forced commutation is used in every half-cycle. These are discussed in Chapter 10.

For DC circuits using on-off control, the device current is unidirectional, and therefore SCRs are used in such circuits. For AC circuits, the

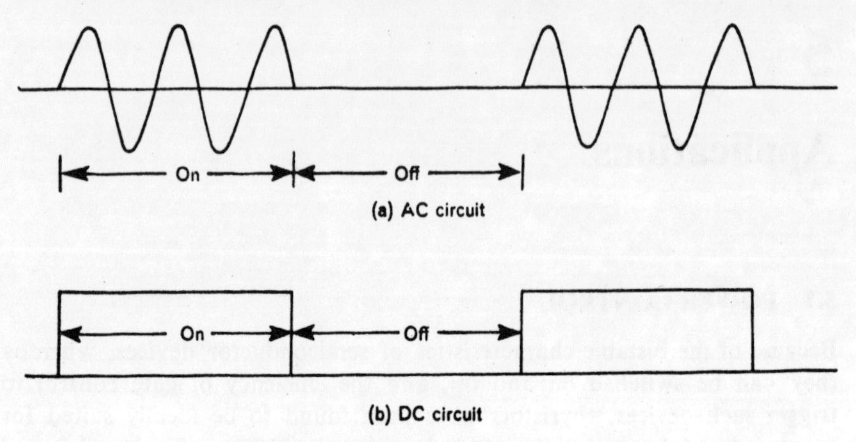

Fig. 5.1 Load current waveforms.

load current has both positive and negative half-cycles during the on-time. Therefore, either a triac or two SCRs in antiparallel are required. More details on on-off control circuits and their applications will be given in Chapter 10. The phase-control circuit shown in Fig. 4.2 is for single-phase supply. For three-phase loads, similar controls can be applied for each phase. To obtain symmetrical output voltages, the SCRs/triacs in each phase are fired in a particular sequence. These schemes will be discussed in Chapter 6. When the load is inductive, it is better to use two SCRs in antiparallel rather than one triac because of *dv/dt* considerations.

Another important application of thyristors is in inverters. *Inversion* is the process of obtaining AC from DC input. The output frequency is related to the triggering frequency of SCRs in the inverters. Thus, variable frequency supply can be easily obtained and used for speed control of AC motors, induction heating, electrolytic cleaning, fluorescent lighting, and many other applications. Because of the large power-handling capacity of the SCRs, the thyristor-controlled inverter has more or less replaced motor-generator sets and magnetic frequency multipliers for generating high frequency at large power ratings. There are many configurations of inverters which will be discussed in Chapters 8 and 9. In all these circuits, the input is DC. So, only the SCRs are used for this operation.

Inverter-fed variable-speed induction motors have characteristics similar to those of a DC shunt motor. The motor speed can be kept constant for all operating conditions, or varied at constant load torque. This form of drive has become very popular in many applications such as blower motors and textile and paper mill drives. The variable speed operation of induction motors is also possible with phase or on-off control. Phase-controlled drives are inexpensive, but as they are inefficient, they are applied only to motors of low horsepower. Because of the problems in maintaining commutators and brush gear, and cost conside rations, many DC motor drives are being replaced by thyristor-controlled variable-speed induction motors. This is particularly so for applications in mines where the use of DC motors is dangerous because of sparking at brush contacts. Since the speed of an induction motor can be kept constant for all values of load torque by controlling the input frequency, speed drives need not depend only on synchronous motors for speed control.

A cycloconverter is another unit for converting frequency from a fixed-frequency AC supply to a variable-frequency AC output. Such an output can be used for driving an induction motor. One important limitation of this method is that the maximum output frequency must be lower than the input frequency; for the conventional inverter discussed in this section, a wide range of output frequency is possible. Cycloconverters find applications in low-speed drives and are specifically used for controlling linear motors in high-speed transportation systems. Line-commutated converters and inverters with thyristors are being increasingly used for reversible DC drives and DC transmission. More details about these line-commutated circuits are given in Chapter 7.

Some of the other applications of thyristors are described in the following sections.

5.1.1 Example

(a) In an on-off control circuit using single-phase 230 V, 50 Hz supply, the on-time is 10 cycles and off-time is 4 cycles. Calculate the RMS value of the output voltage.

The output voltage is equal to the supply voltage during on-time and zero during off-time. Therefore, the total RMS value of the output voltage is given by

$$\begin{aligned} V_{\text{RMS}} &= \{[(230)^2 \times 10 + 0 \times 4]/14\}^{1/2} \\ &= 230 \times 0.84 = 193 \text{ V}. \end{aligned}$$

(b) In a phase-controlled circuit using single-phase 230 V, 50 Hz supply, the firing angle is adjusted to be $\pi/4$ in both half-cycles. Obtain the RMS value of the output voltage. Assume the load to be resistive.

The output voltage is equal to the supply voltage from $\pi/4$ to π in each half-cycle, and for the remaining period, when the SCRs are blocked, it is zero. Therefore, V_{RMS} is given by

$$\begin{aligned} V_{\text{RMS}} &= \left\{\frac{1}{\pi}\int_{\pi/4}^{\pi} (230\sqrt{2}\sin\theta)^2\, d\theta\right\}^{1/2} \\ &= \frac{230\sqrt{2}}{\sqrt{(2\pi)}}\left\{\left(\theta - \frac{\sin 2\theta}{2}\right)_{\pi/4}^{\pi}\right\}^{1/2} = 219 \text{ V}. \end{aligned}$$

(c) A two-pole single-phase induction motor is driven by a variable frequency inverter. The output frequency range is 15–100 Hz. Calculate the range of no-load speed for the motor.

The no-load speed $N_S \approx 120f/P$, where f is the frequency and P is the number of poles. Therefore, the speed range is 900–6000 rpm.

5.2 STATIC CIRCUIT BREAKER

Figure 5.2 shows a circuit in which the SCRs are used for making and breaking a circuit. The input voltage is alternating. The trigger pulses are applied to the gates of SCRs through the control switch *S*. When

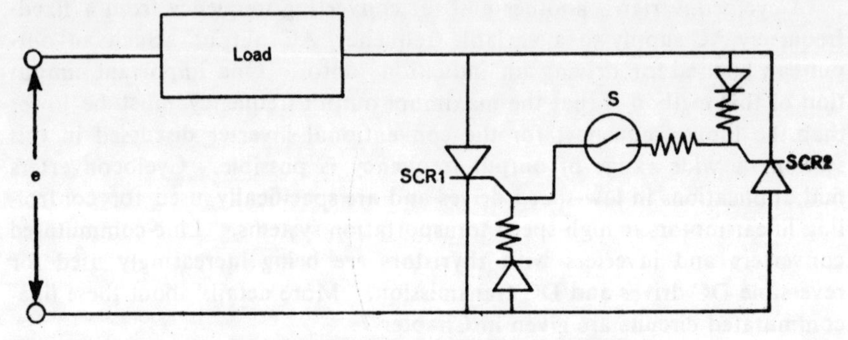

Fig. 5.2 Static AC circuit breaker.

switch *S* is closed, SCR1 will fire at the beginning of the positive half-cycle (the gate trigger current is assumed to be very small). It will turn off when the current goes through the zero value. As soon as SCR1 is turned off, SCR2 will fire since the voltage polarity is already reversed and it gets the proper gate current. When any of the SCRs is triggered, the gate current will be negligible. To break the circuit, switch *S* is opened. Since the current through this switch is small, opening the gate circuit poses no problem. As no further gate signal will be applied to the SCRs when switch *S* is open, the SCRs will not be triggered and the load current will be zero. The maximum time delay for breaking the circuit is one half-cycle.

Figure 5.3 shows a static DC circuit breaker. Here, capacitor *C* pro-

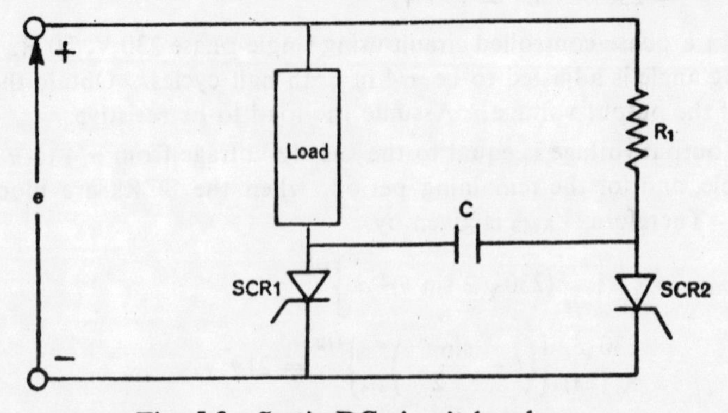

Fig. 5.3 Static DC circuit breaker.

vides the required commutation of the main SCR since the current does not have a natural zero value in a DC circuit. When SCR1 is conducting, the load voltage will be equal to the supply voltage and capacitor C will get charged through R_l. The breaking of the circuit is achieved by turning off SCR1. This is done by firing SCR2, which is called the auxiliary SCR. Then, capacitor C will discharge through SCR2 and SCR1. This current will oppose the load current flowing through SCR1 and when these two currents become equal, the net current will be zero and SCR1 will be turned off. Thereafter, capacitor C will get charged through the load and during this time a reverse potential across SCR1 will be applied. This method of turning off is called forced commutation. There are several methods of forced commutation in DC circuits; their classification and design details will be given in Chapter 8. When capacitor C is fully charged, SCR2 will be turned off because the current through the load is zero and the current through R_l is below the level of the holding current of SCR2. This is the criterion for obtaining the required value for R_l.

5.2.1 Example

Obtain proper values for C and R_l in Fig. 5.3 if the supply voltage is 100 V, load current is 10 A, and SCR1 has a turn-off time of 20 μsec. SCR2 has a holding current of 3 mA.

The value of R_l will be

$$R_l = \frac{100}{3} \times 10^3 = 33.33\ \text{k}\Omega.$$

The voltage across the capacitor after SCR1 is turned off will be

$$V_C = 100(1 - e^{-t/(R_L C)}),$$

where R_L is the load resistance. The duration t_q for which the SCR is reverse-biased must be equal to the device turn-off time. Thus,

$$t_q = R_L C \ln 2 = 20\ \mu\text{sec}.$$

The load resistance is

$$R_L = 100/10 = 10\ \Omega.$$

Therefore,

$$C = 2.9\ \mu\text{F}.$$

5.3 OVERVOLTAGE PROTECTION

Silicon-controlled rectifiers can be used for protecting other equipment from overvoltages because of their fast switching action. These overvoltages may be caused by the bad regulation of supply voltage, or by any switching action. The SCR used for protection is connected in parallel with the load. Whenever the voltage exceeds a specified limit, the gate of the SCR will get energised and trigger the SCR. A large current will be drawn from the supply mains, which will reduce the overvoltage at the load. Since the applied voltage is AC, two SCRs are used—one for the positive half-cycle and the other for the negative half-cycle—as shown in Fig. 5.4. Resistance R_1 limits the short-circuit current when the SCRs are fired. This large current produces enough voltage drop in the source impedance so that the terminal voltage is within safe limits. A zener diode D_5 in series with R_1 and R_2 constitutes a voltage-sensing circuit. When the line voltage is above

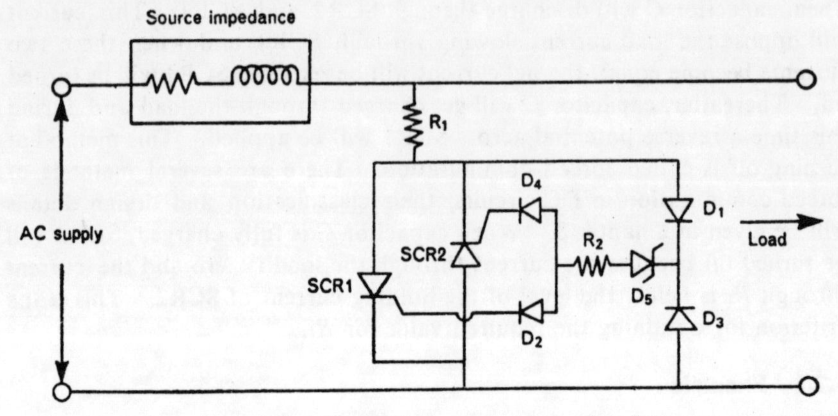

Fig. 5.4 Overvoltage protection circuit.

the specified limit, diode D_5 will break. Then, during the positive half-cycle, the gate of SCR1 will get energised through $R_1D_1R_2D_2$, and trigger it. In the negative half-cycle, the gate of SCR2 will get energised through $D_3R_2D_4R_1$, and SCR2 will turn on if the overvoltage persists. As soon as the load voltage returns to the safe value, the zener diode D_5 will recover and the current through it will be very low so that the SCRs will not fire. Thus, the response time of the protection circuit will be quite fast and protection will be provided in both positive and negative half-cycles.

5.4 ZERO VOLTAGE SWITCH

In some AC circuits it is necessary to apply the voltage to the load when the instantaneous value of this voltage is going through the zero value. This is to avoid a high rate of rise of current if the load is purely resistive (e.g., furnace and lighting loads) and thereby reduce the generation of radio noise and hot-spot temperatures in the device carrying the load current. The circuit shown in Fig. 5.5, which achieves this, is called the *zero voltage switch*. Only half-way control is used here. The portion of the circuit shown by the dashed lines relates to the negative half-cycle. During the positive half-cycle, both the base of transistor Q_1 and the gate of SCR1 get positive currents. Before SCR1 is triggered, the base drive for Q_1 is enough to saturate it, and so the current through R_4 will be bypassed. Therefore, SCR1 will not turn on. If switch S is closed, capacitor C_1 will not get any significant charge and no negative bias will be applied to the base-emitter junction of Q_1, and during every positive half-cycle the SCR1 gate current will be bypassed by the saturated transistor Q_1. So, SCR1 will not be triggered and no voltage will be applied to the load. If switch S is now opened, capacitor C_1 will get charged to the peak of the applied voltage during the negative half-cycle through R_1 and D_1, and after that it will begin to discharge through

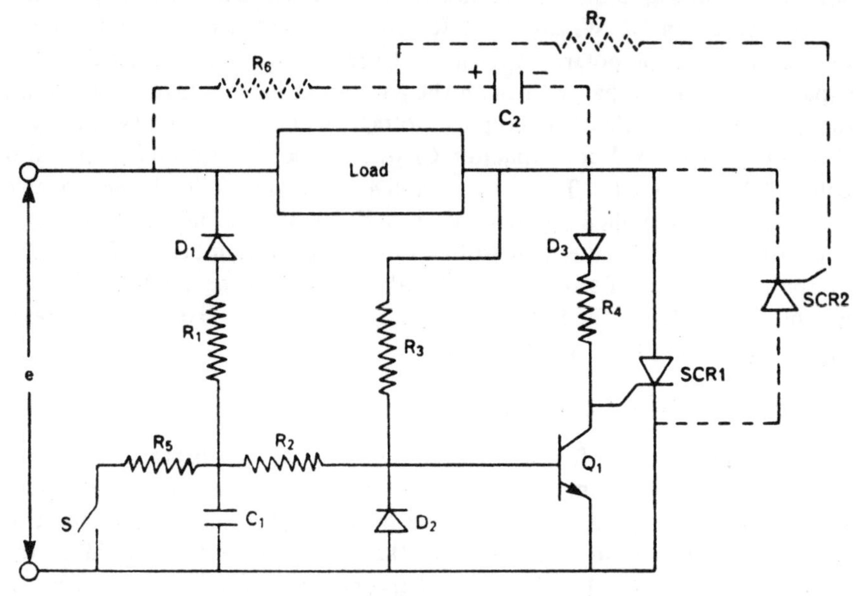

Fig. 5.5 Zero voltage switch.

D_2 and R_2. The time constant for the discharge path is so chosen that D_2 will conduct for a part of the following positive half-cycle. Thus, the base-emitter junction of Q_1 is reverse-biased in the beginning of the positive half-cycle, and during this period SCR1 will be triggered because all the current through R_4 will now flow through the gate. Whatever may be the instant of time when switch S is open (either during the positive or the negative half-cycle), only at the beginning of the following positive half-cycle of the applied voltage will SCR1 be triggered. Similarly, when switch S is closed, SCR1 will stop conducting at the end of the present or previous positive half-cycle and will not get triggered again. Resistors R_3 and R_4 are designed on the basis of minimum base and gate currents required for transistor Q_1 and SCR1. Resistors R_1 and R_2 decide the charging and discharging rates of capacitor C_1. Resistor R_5 is used for preventing large discharge currents when switch S is closed.

5.5 INTEGRAL-CYCLE TRIGGERING

In some circuits it may be required to have the total number of positive half-cycles of load current equal to the number of negative half-cycles during the period the voltage is applied to the load. In such cases, the duration of on-time will be an integral multiple of the period of the input voltage. This is known as *integral-cycle triggering* and is essential to avoid magnetic saturation of inductive loads—a specific requirement when on-off control is used for the control of AC motors. An exactly equal number of positive and negative half-cycles of conduction is provided by what is called *negative slave triggering*. The portion of the circuit shown by the dashed lines in Fig. 5.5 is the required circuit. SCR2 is not triggered by any external gate signal. Whenever SCR1 conducts, capacitor C_2 will charge through R_6 with the polarity shown. At the end of the positive half-cycle, capacitor C_2 will discharge through the gate of SCR2 and turn it on. Then, the negative half-cycle of the applied voltage will appear across the load. If SCR1 is not turned on, capacitor C_2 will not be charged and SCR2 also will not be triggered. Thus, the negative half-cycle follows the positive half-cycle of the applied voltage only if SCR1 is turned on, or if switch S is opened. Therefore, the load will carry an integral number of full cycles of applied voltage. If switch S is closed, the load current will stop at the end of the present or the following negative half-cycle. Hence, the maximum interrupting time will be one cycle.

5.6 TIME DELAY CIRCUIT

Time delay circuits are frequently used in industrial controls to apply power to or remove power from a load at a predetermined time after the initiating signal is applied. SCRs in conjunction with a UJT can be used for this purpose. A relaxation oscillator with the UJT is used for providing the gate pulse for an SCR to turn on, and this results in application or removal of power. The relaxation period of the UJT can be changed by varying the charging rate of the capacitor (see Fig. 4.4), thereby changing the delay time; the timing is initiated either by applying the supply voltage V_b to the oscillator or by opening a shorting contact across capacitor C.

5.7 SOFT START CIRCUIT

In many power control circuits it is desirable to apply power gradually. That is, if phase control is employed for power modulation, the firing angle α must start initially from nearly 180° (zero power) and gradually decrease at some specified rate to any preselected firing angle. This type of variation in the firing angle is necessary for illumination control because the cold resistance of lamp filaments is low and there will be a very high-current surge if full voltage is applied. The SCR used must have a rating which is sufficient to withstand such high-current surges. There will nevertheless be the danger of damage to or a reduction in the life of the filament because of high-current surges.

The soft start scheme can be implemented easily by using a UJT for phase control as shown in Fig. 2.12b. In this method, the charging circuit RC is energised by an exponentially increasing voltage. This voltage is obtained from another capacitor which is charged from a fixed voltage supply V_b through a large resistance. The time constant for the voltage applied to RC can be controlled. Since this voltage gradually increases from zero, the firing angle for the initial few cycles will be nearly 180°. The firing angle will finally move to its steady-state value which is determined by the charging rate of capacitor C. The steady-state voltage across RC is also equal to V_b.

5.8 LOGIC AND DIGITAL CIRCUITS

Figure 5.6 shows a few logic circuits (NAND, NOR, NOT, and threshold) using SCS or GTO devices. Using these as building blocks, any logic function can be realised. In the NAND circuit the string conducts only

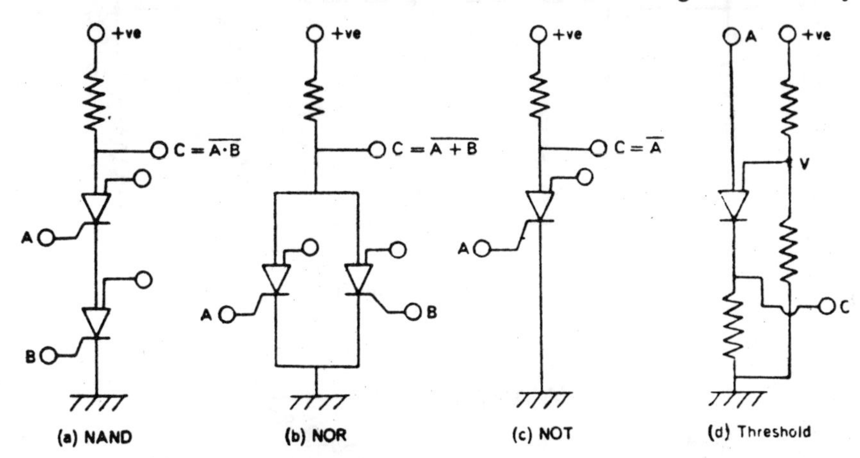

Fig. 5.6 Logic circuits.

when both gates are energised. This is known as *coincident logic*. The operation of other logic circuits can be easily understood from the figure. The functional details of these logic circuits can be obtained from books on digital electronics.

The circuit shown in Fig. 5.3 can be operated as a bistable flip-flop if resistance R_1 is also made equal to the load resistance. In such a case SCR2 will continue to conduct through R_2 after turning off SCR1. When SCR1 is fired, SCR2 will be turned off by capacitor C. Thus, by alternately firing SCRs 1 and 2, a square wave is obtained at the output points connected to the anodes of the two SCRs. The principle of the flip-flop circuit discussed here can be extended to obtain a *ring counter*. The applications and basic operation of this circuit will be discussed in Chapter 8.

Figure 5.7a shows an astable circuit using an SCS. Capacitor C provides forced commutation of the conducting SCS. The time constant RC determines the output period. The output frequency can be changed either by varying R, or by changing the bias voltage applied to the anode gates. The cathode gates are used for synchronising the output to an incoming signal. Figure 5.7b shows a monostable circuit used as a pulse stretcher. Here also an SCS is used. The device is turned on by applying the input pulse to the cathode gate. Capacitor C will discharge through R and provide enough forward current to keep the device conducting. The current from the supply is kept below the level of the holding current by the output resistance R_o. When the discharge current of the capacitor falls below I_h, the device will go into the blocking state. The duration for

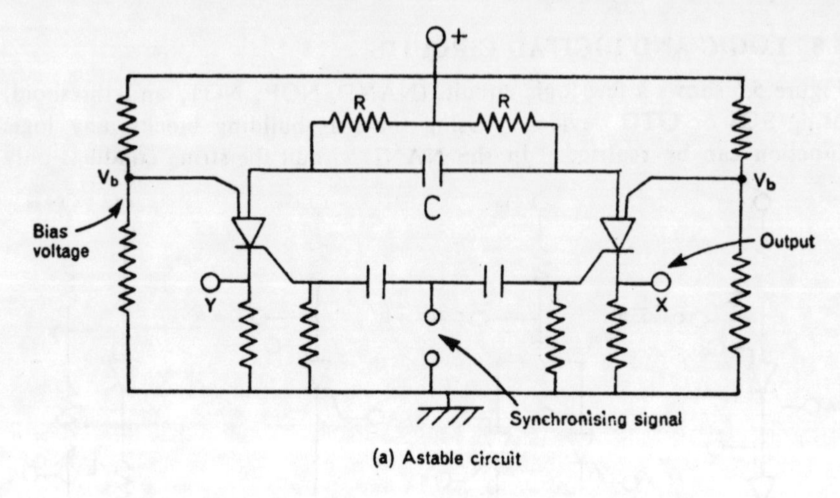

(a) Astable circuit

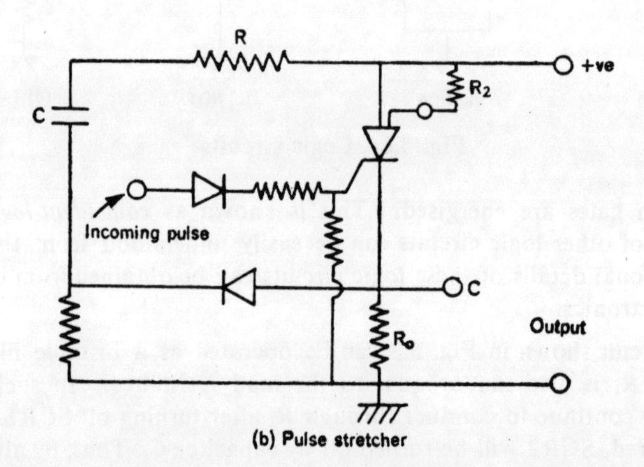

(b) Pulse stretcher

Fig. 5.7 Digital circuits.

which the device is on is equal to the width of the output pulse, and this is controlled by the time constant *RC*. Resistance R_2 is used for reducing the rate effects (dv/dt) when the device goes into the off-state.

5.9 PULSE CIRCUITS

Figure 5.8 shows an important application of SCRs for producing high voltage/current pulses of required waveform and duration. High voltage/current pulses can be used for spot welding, electronic ignition in automobiles, generation of large magnetic fields of short duration, and insulation testing. The capacitor is charged during the positive half-cycle of the input supply and the SCR is triggered during the negative half-cycle. The capacitor will discharge through the output circuit, and when the SCR forward current becomes zero, it will turn off. The output circuit is so

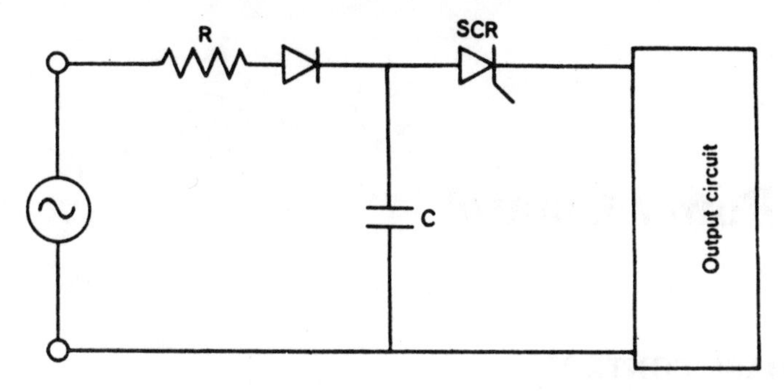

Fig. 5.8 Pulse circuit.

designed that the duration of the discharge current is less than a millisecond. The capacitor will again get charged in the following positive half-cycle and will be triggered again in the negative half-cycle. Thus, the frequency of the output pulse will be equal to the frequency of the input supply. Resistance R is used for limiting the charging current.

There are many more applications of SCRs and of the other devices of the thyristor family. As it is not possible to discuss all these here, the interested reader may consult the references that follow.

REFERENCES

Al Pshaenich, Interface techniques between industrial logic and power devices, Application Note AN712, Motorola Company, Phoenix, USA.

Balenovich, J. D., Thyristor selection for pulse applications, Tech. Tips 1.1, Westinghouse, Youngwood, USA.

Galloway, J. H., Using the triac for control of AC power, Application Note 200.35, G. E. Company, Syracuse, USA.

Grafham, D. R., Using low current SCRs, Application Note 200.19, G. E. Company, Auburn, USA.

Ramamoorty, M., All static point-on-wave selector, *JIE* (India), 1971, p. 442.

Ramamoorty, M., Pop, A. S., A soft start circuit for illumination control, *IEE-IERE Proc.* (India), 1975, p. 128.

Ramamoorty, M., Seth, A., Solid state inverse time over-current relay, *The Electrical Research* (India), 1971, p. 20.

6

AC Power Control

6.1 PHASE CONTROL

In Chapter 2, we considered various methods for triggering SCRs. Of these, the most efficient method for power modulation with thyristors uses gate control. In AC circuits, the SCR can be turned on by the gate at any angle α with respect to the applied voltage. This angle α is the ***firing angle***. Power control is obtained by varying the firing angle, and this is known as *phase control*.

Figure 6.1a shows a simple half-wave circuit. The principle of phase

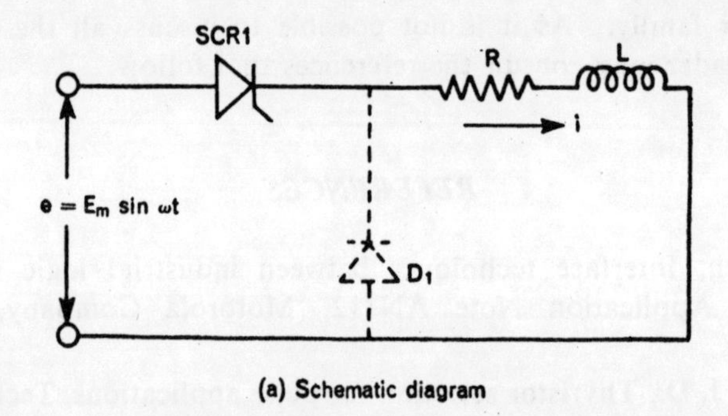

(a) Schematic diagram

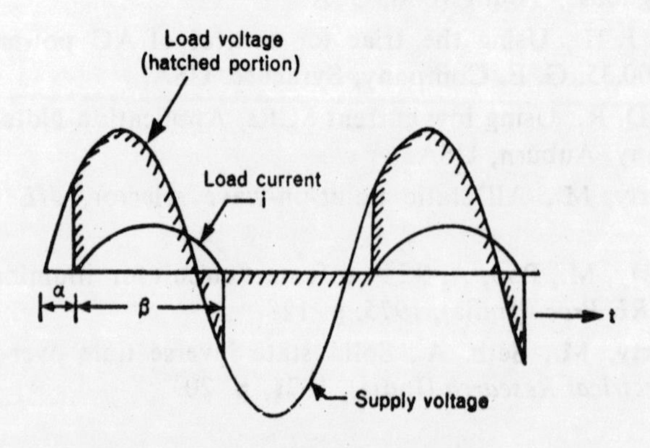

(b) Load current and voltage waveforms

Fig. 6.1 Phase control (*cont.*).

control for an inductive load can be explained by this circuit. Let the SCR be fired at an angle α. The load current, load voltage, and supply voltage waveforms are shown in Fig. 6.1b. The SCR will turn off by natural commutation when the current becomes zero. Angle β is known as the *conduction angle*. By varying the firing angle α, the RMS value of the load voltage can be changed. The firing circuits discussed in Chapter 2 can be used to control the firing angle. Current $i(t)$ in the circuit shown in Fig. 6.1 is given by

$$i(t) = \frac{E_m}{\sqrt{(R^2 + \omega^2 L^2)}} \sin(\omega t + \alpha - \phi) + \frac{E_m}{\sqrt{(R^2 + \omega^2 L^2)}} \sin(\phi - \alpha)\, e^{-Rt/L}, \quad \text{for } 0 \leqslant t \leqslant \beta/\omega, \qquad (6.1)$$

where ϕ is $\tan^{-1}(\omega L/R)$ and E_m is the amplitude of the input voltage. The average value of the load current can be obtained from the equation for $i(t)$. The power consumed by the load decreases as angle α is increased. The reactive power input from the supply increases with the firing angle. The load current waveform can be improved by connecting a free-wheeling diode D_1 as shown by the dashed line in Fig. 6.1a. With the diode, SCR1 will be turned off as soon as the input voltage polarity

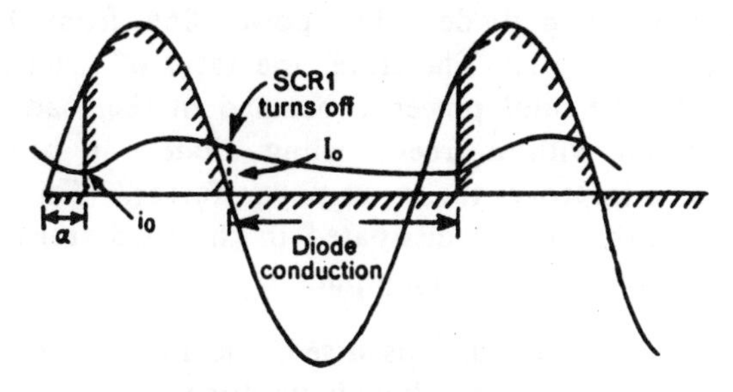

(c) Effect of free-wheeling diode

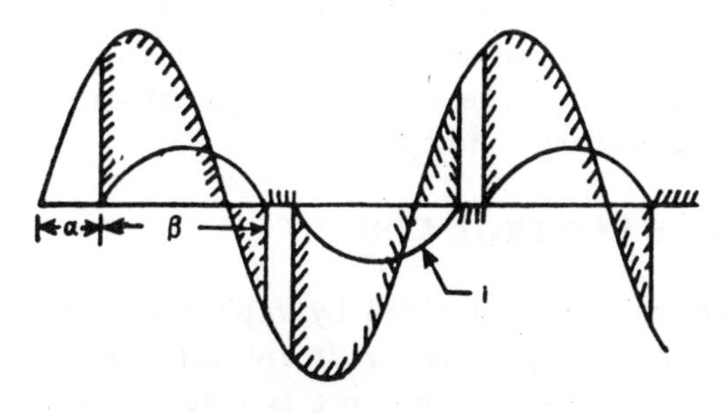

(d) Full-wave control

Fig. 6.1 Phase control.

reverses. After that, the load current will free-wheel through the diode and a reverse voltage will appear across the SCR. Figure 6.1c shows the load current and load voltage waveforms with a free-wheeling diode.

There are two modes of operation for this circuit. In the first mode, diode D_1 will be reverse-biased and SCR1 will conduct. This mode will exist from the instant of firing to the time when the voltage polarity reverses. The duration is given by $(\pi - \alpha)/\omega$. The current waveform is

$$i(t) = \frac{E_m}{\sqrt{(R^2 + \omega^2 L^2)}} \sin(\omega t + \alpha - \phi) + \left\{\frac{E_m}{\sqrt{(R^2 + \omega^2 L^2)}} \sin(\phi - \alpha) + i_0\right\} e^{-Rt/L}, \quad \text{for } 0 \leqslant t \leqslant (\pi - \alpha)/\omega, \tag{6.2}$$

where i_0 is the current in the load when the SCR is fired. Let the current at the end of mode 1 $(t = (\pi - \alpha)/\omega)$ be I_0. During mode 2, SCR1 will be reverse-biased and diode D_1 will conduct. It is assumed that the load is sufficiently inductive to maintain the current in the load circuit until the next instant of firing. Current $i(t)$ in this period is given by

$$i(t) = I_0 e^{-Rt/L}, \quad \text{for } \frac{\pi - \alpha}{\omega} \leqslant t \leqslant \frac{2\pi}{\omega}. \tag{6.3}$$

In the steady state, the current at the end of mode 2 must be equal to i_0. Comparing the load current waveforms in Figs. 6.1b and 6.1c, it will be observed that for the same firing angle the load power consumption is more with a free-wheeling diode. The power flow from the input takes place only during mode 1. Therefore, the ratio of reactive power flow from the input to the total power consumed in the load is less for the phase-control circuit with a free-wheeling diode. In other words, the free-wheeling diode improves the input power factor. This is because the inductive energy of the load is dissipated in the load resistance R during mode 2 instead of returning to the input.

In the half-wave circuits just discussed, the input current has a large DC component since the current flow is unidirectional. This asymmetrical current produces magnetic saturation in the input transformers, if any are used. However, this problem is remedied in three-phase half-wave controlled circuits by the special winding connections on the secondary side of the input transformer. The input power factor can also be improved by connecting the primary windings in delta.

6.2 FULL-WAVE CONTROL CIRCUIT

Full-wave power control is obtained by replacing SCR1 in Fig. 6.1a by a triac or two SCRs in antiparallel. A free-wheeling diode cannot be used for such a circuit. Symmetrical firing is used in each half-cycle. The resulting current waveform is shown in Fig. 6.1d. Any one of the SCRs can be fired only after the other conducting SCR is turned off due to natural commutation. This limits the minimum firing angle α to the impedance angle ϕ of the load circuit at the input frequency. Phase control is also used for obtaining a variable DC voltage. The half-wave control circuit shown in Fig. 6.1a can be employed for this purpose. LC filters are utilised for reducing the ripple in the output and the load is connected on the DC side. The filter size can be reduced by using full-wave control circuits (with polyphase input) because these circuits produce higher ripple frequency in the output as compared, with half-wave control circuits. There are two basic configurations for full-wave control circuits. One configuration requires an input transformer with two identical windings on the secondary side for each phase. The two windings have a common terminal. This is known as the *midpoint configuration.* Single-phase circuits require two SCRs (M-2 connection) and three-phase circuits need six SCRs (M-6 connection) as shown in Figs. 6.2a and 6.2b. These are known also as two-pulse and six-pulse converters. The number of pulses is equal to the order of the lowest harmonic in the output. Variable DC voltage is obtained by controlling the firing angle. Hence, these circuits are also referred to as *controlled rectifiers* or *converters.*

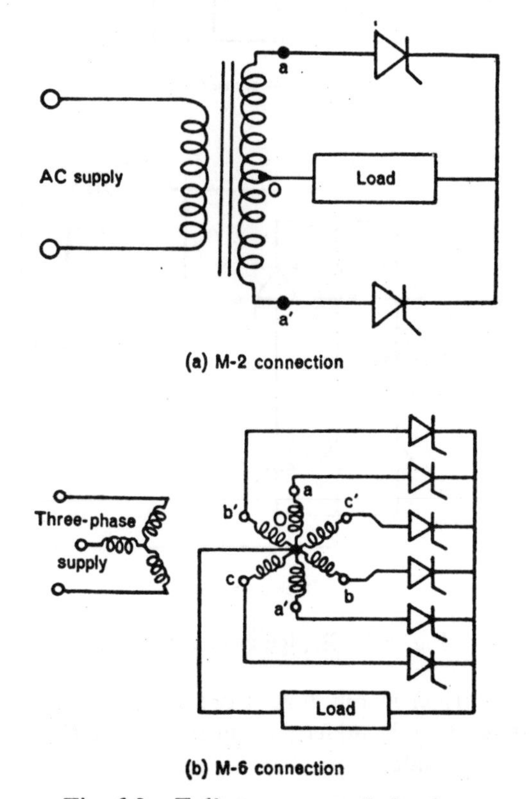

Fig. 6.2 Full-wave control circuits.

The second configuration, called the *bridge circuit,* is shown in Fig. 6.3. Here, no input transformer is required. The single-phase circuit (Fig. 6.3a) is known as the B-2 connection and the three-phase circuit (Fig. 6.3b) is known as the B-6 connection. The numerals in these notations correspond to the number of pulses in the output during one period of the input wave. For a given voltage rating of the SCRs, the load voltage for the M-2 connection is one-half that for the B-2 connection. The volt-ampere

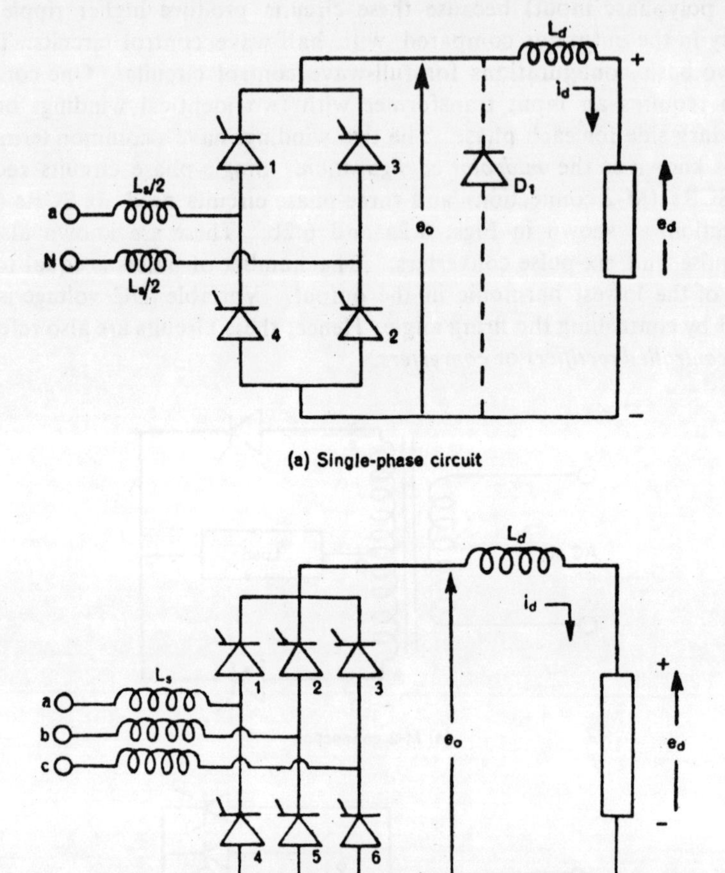

Fig. 6.3 Bridge configurations.

rating of the transformer in Fig. 6.2a is twice that of the load. Therefore, the bridge configuration is preferable unless one of the terminals on the DC side has to be grounded.

6.2.1 Analysis of a Bridge Circuit

The single-phase circuit of Fig. 6.3a is considered for the analysis. Inductance L_d is used in the DC circuit to reduce the ripple. A large value of L_d will result in a continuous steady current in the load. A small L_d will produce a discontinuous load current for large firing angles. The source is assumed to have an internal inductance L_s. Let SCRs 1 and 2 be fired at an angle α in the positive half-cycle. The direction of load current i_d is as shown in the figure. The current waveforms for the two extreme values of L_d are given in Figs. 6.4a and 6.4b. The effect of source inductance is neglected here. It can be seen that for this circuit, if the DC-side negative

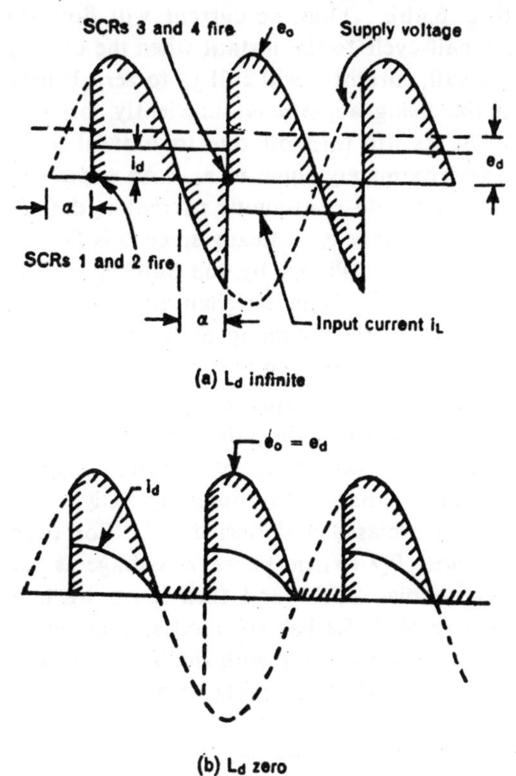

Fig. 6.4 Current and voltage waveforms for bridge circuits (*cont.*).

terminal is grounded (connected to the supply neutral), SCRs 2 and 3 will not fire and there will be a short-circuit on the supply during the negative half-cycle when SCR4 is triggered. The average DC voltage e_d is $(2E_m/\pi)\cos\alpha$.

In Fig. 6.4a, since the smoothing inductor is very large, current i_d in the steady state will be pure DC. Therefore, even when the input voltage polarity is reversed, the current will continue to flow through SCRs 1 and 2 till the other pair, SCRs 3 and 4, is fired symmetrically at an angle α in the negative half-cycle. Since the polarity of the input voltage is already reversed, the firing of SCRs 3 and 4 will reverse-bias SCRs 1 and 2, and turn them off. The load current will then shift from the pair of SCRs 1 and 2 to the pair of SCRs 3 and 4. This is referred to as *class F type forced commutation* (see Section 8.2), or *line commutation*. Current i_d will maintain the same direction of flow in the load. However, the current on the input side will flow in the reverse direction when SCRs 3 and 4 conduct, that is, the input current i_L will be a rectangular AC wave. The output voltage e_o is shown by the hatched lines in Fig. 6.4a; the dashed line corresponds to the average value e of this output voltage. When L_d is zero and source inductance is neglected, current i_d will go to zero at the end of every half-cycle (Fig. 6.4b). Thus, no current will flow through the load from the end of the half-cycle to the instant when the other pair of SCRs is fired. When L_d is small, current i_d will still go to zero before the other pair of SCRs is fired if the firing angle α is sufficiently large. Here also, the conducting pair of SCRs will turn off due to natural commutation. The load voltage e is approximately equal to e_o. An expression for the minimum value of L_d to provide continuous current is derived in Chapter 7. For large values of L_d, voltage e_d is steady since i_d is DC. This will be the average value of voltage e_o (shown by the hatched portion in Fig. 6.4a). The firing angle α for the SCRs can be changed from zero to π. During this period, the potential of the incoming SCRs will be more than that of the conducting SCRs, and proper commutation can take place.

Figure 6.4c shows the waveforms of a three-phase fully-controlled bridge. As in the single-phase bridge, the changeover of conduction from one SCR to the other will take place only if the phase voltage of the incoming SCR is more than that of the outgoing SCR. Then only will the outgoing SCR be reverse-biased and turned off. For example, the firing of SCR1 during period PQ (when phase a voltage is more than phase c voltage) will reverse-bias SCR3 and turn it off by line commutation. The conduction pattern of SCRs has six modes, each mode extending for $2\pi/3$ radians. The six modes, along with the firing sequence of the SCRs, are listed in Table 6.1. Neglecting overlap due to source inductance, only

Table 6.1 Firing sequence of SCRs

Mode	1	2	3	4	5	6
Conducting SCRs	5, 3	1, 5	6, 1	2, 6	4, 2	3, 4
SCR to be fired	1	6	2	4	3	5
Outgoing SCR	3	5	1	6	2	4

two SCRs will be conducting at any time. The load inductance is assumed to be very large so as to produce a steady load current i_d, and the effect of source inductance is neglected. To maintain a symmetrical waveform at the input current, it is necessary to adhere to the sequence of firing given in Table 6.1. The firing frequency will be six times the input frequency. Control circuits to achieve this will be discussed in Chapter 9. The firing angle is measured from point O shown in Fig. 6.4c. This angle can be changed from zero to π. The average DC voltage e, shown by the dashed lines in this figure, is given by $(3\sqrt{3}/\pi)E_m \cos \alpha$ where E_m is the peak value of the line-to-neutral voltage. The variation of e_d as a function of α is shown in

Fig. 6.4d. The input phase current is a rectangular pulse of width $2\pi/3$ and amplitude i_d.

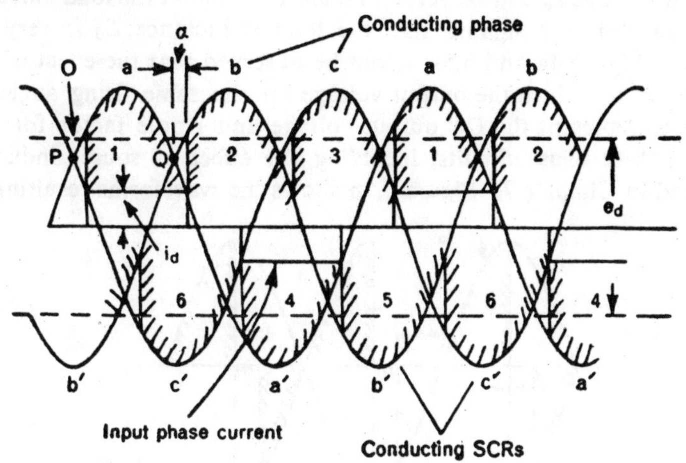

(c) Three-phase bridge circuit

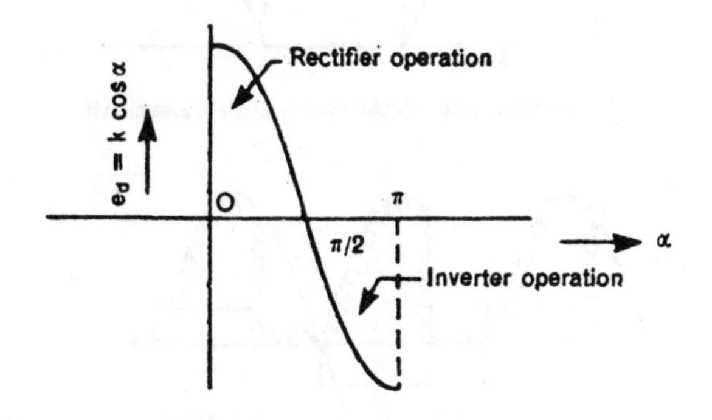

(d) Variation of e_d with firing angle α

Fig. 6.4 Current and voltage waveforms for bridge circuits.

If the bottom SCRs are replaced by diodes in Fig. 6.3, the circuit will become a half-controlled bridge. As will be explained in the following sections, such a circuit cannot produce negative output voltage. If the firing angle of the SCRs in the bridge circuits shown in Fig. 6.3 is made zero, or if all the SCRs are replaced by diodes, an uncontrolled bridge circuit will be obtained.

6.2.2 Effect of Source Inductance

When source inductance L_s is present, SCRs 1 and 2 in Fig. 6.3a will not turn off immediately after SCRs 3 and 4 are fired. Inductance L_s will maintain the current flow through SCRs 1 and 2 for some time more even though the supply voltage polarity has been reversed. Therefore, the current will shift gradually from the pair of SCRs 1 and 2 to the pair of SCRs

3 and 4. This duration is known as *overlap period* μ, and during this period the output voltage e_o will be zero. Figure 6.5a shows the load current and voltage waveforms assuming that the load inductance L_d is very large. Comparing Figs. 6.4a and 6.5a, it can be observed that the effect of source inductance is to reduce the output voltage for the same firing angle. Expressions to represent the DC output voltage and ripple factor for single-phase and three-phase circuits, including the effect of source inductance, are derived in Chapter 7. Figure 6.5b shows the waveforms resulting from

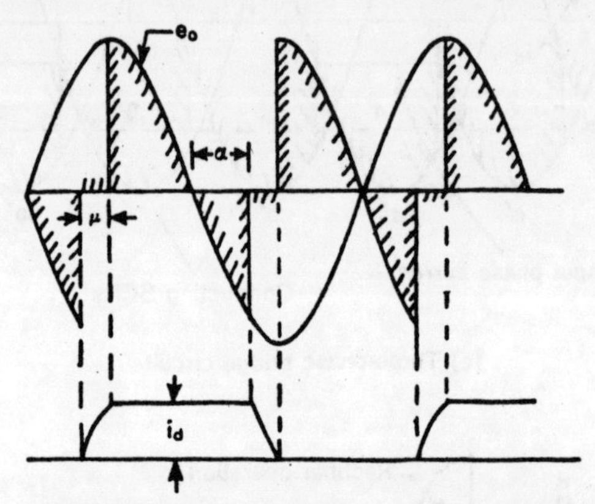

(a) With a large source inductance in a single-phase circuit

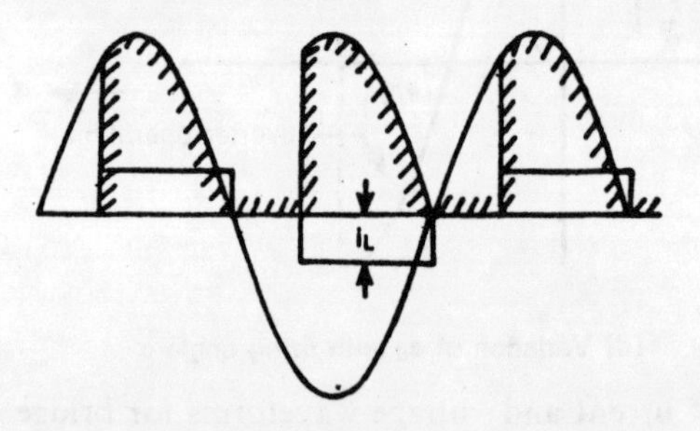

(b) With a free-wheeling diode in a single-phase circuit

Fig. 6.5 Modified load voltage and load current waveforms for bridge circuits (*cont.*).

the use of a free-wheeling diode in the single-phase circuit shown in Fig. 6.3a. This will be discussed in Section 6.2.4. Figure 6.5c shows the change in the output voltage waveform brought about by source inductance for a three-phase bridge circuit. When two phases conduct during the overlap period, the output voltage will be the average of the voltages of the conducting phases.

6.2.3 Operation as a Two-Quadrant Converter

The circuits shown in Fig. 6.3 can also be operated with power flow in the reverse direction, that is, DC power is converted into AC power. This mode of operation is called *inversion.* Details of inverter operation will be given in Chapter 7; in this section, only the basic principle of such an operation will be discussed. With reference to the voltage and current waveforms given in Fig. 6.4a, if the output current is continuous and steady, the average output voltage e_o will be negative when the firing angle α is greater than $\pi/2$. The current direction in the load and in the SCRs will be the

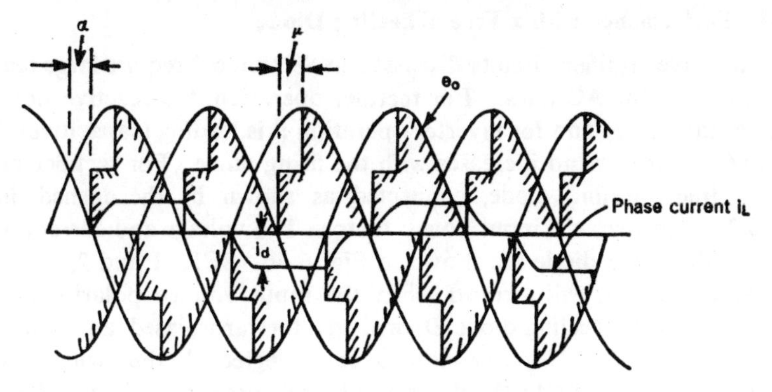

(c) With source inductance in a three-phase bridge circuit

Fig. 6.5 Modified load voltage and load current waveforms for bridge circuits.

same as before. Therefore, the operation of the circuit with firing angles greater than $\pi/2$ will be possible only if a DC source with voltage of opposite polarity is connected at the output. Then, the DC source will supply power to the circuit, and the circuit in turn will feed this power into the AC system. Consider the circuit in Fig. 6.3a. SCRs 1 and 2 can be fired at any instant during the positive half-cycle of the input when their anodes are positive with respect to the cathodes. On being fired, this pair of SCRs will reverse-bias the other pair and turn it off. For firing angles from zero to $\pi/2$, the DC output voltage e_d (given by $(2\sqrt{2}E/\pi)\cos\alpha$, where E is the RMS value of the input voltage) and the DC power output $e_d i_d$ are positive. In other words, for this range of firing angle, power will flow from AC input to DC output. This is the rectifier operation. The firing angle α gives the approximate phase angle between the applied AC voltage and the fundamental component of the input line current i_L (see Fig. 6.4a). The input power factor for this condition will be lagging. For firing angles greater than $\pi/2$, the DC output voltage e_d will be negative and the input power factor angle will be more than $\pi/2$. Thus, the DC power output $e_d i_d$ will be negative and the AC power input $EI_L \cos\alpha$ will also be negative (I_L is the fundamental RMS value of the input current i_L). This means that power will flow from the DC side to AC side through the bridge circuit. In this mode, the bridge circuit will operate as an inverter and the output power factor will be leading. In both modes of operation (rectifier and inverter), the available AC line voltage is used for commutation. Hence, the bridge circuit can also be referred to as a line-commutated rectifier and inverter.

The conversion of AC to DC or DC to AC described here is also known as the *two-quadrant operation* of the converter circuit. A similar operation is possible with the three-phase bridge shown in Fig. 6.3b. Figure 6.4d shows the variation of output voltage e_d with the firing angle α for the two-quadrant converter.

6.2.4 Performance with a Free-Wheeling Diode

The full-wave rectifier circuits discussed in Section 6.2 require large reactive power flow in the AC lines. For rectifier operation, the reactive power is lagging (inductive) and for inverter operation it is leading (capacitive). This reactive power demand increases with the firing angle. For rectifier operation, a free-wheeling diode, connected as shown by the dashed line in Fig. 6.3a, improves the input power factor. The voltage and current waveforms with such a diode are shown in Fig. 6.5b. SCRs 1 and 2, which are fired at an angle α, will turn off when the input voltage polarity reverses because the free-wheeling diode D_l will get forward-biased and take over the load current i_d. If source inductance is neglected, the switch-over of current from the SCRs to the diode will be instantaneous. Since inductance L_d is considered to be large, the current will continue to flow through D_l until the other pair of SCRs is fired. During this period the output voltage e_o will be zero. It can be observed that for the same firing angle the average DC output voltage e_d will be higher when the free-wheeling diode is used. The AC line current i_L flows in any one direction when the angle varies from α to π and in the opposite direction when the angle varies from $(\pi + \alpha)$ to 2π. Therefore, the phase angle of the fundamental component of input current will be less than α, and the reactive power input will be reduced. The phase angle θ is given by

$$\begin{aligned} \theta &= \alpha \quad \text{(without the free-wheeling diode)}, \\ \theta &= \alpha/2 \quad \text{(with the free-wheeling diode)}. \end{aligned} \tag{6.4}$$

The reactive power flow is calculated as follows. Let E be the RMS value of the DC input voltage and i_d the direct current output. The average DC output voltage (neglecting source inductance and assuming a large L_d) is given by

$$\begin{aligned} e_d &= \frac{1}{\pi}\int_{\alpha}^{\pi+\alpha} \sqrt{2}E \sin\theta \, d\theta = \frac{2\sqrt{2}}{\pi} E \cos\alpha \quad \text{(without } D_l\text{)}, \\ e_d &= \frac{1}{\pi}\int_{\alpha}^{\pi} \sqrt{2}E \sin\theta \, d\theta = \frac{\sqrt{2}E}{\pi}(1+\cos\alpha) \quad \text{(with } D_l\text{)}. \end{aligned} \tag{6.5}$$

The fundamental RMS value of the alternating line current I_L is

$$\begin{aligned} I_L &= \frac{2\sqrt{2}}{\pi} i_d \quad \text{(without } D_l\text{)}, \\ I_L &= \frac{2\sqrt{2}}{\pi} i_d \cos(\alpha/2) \quad \text{(with } D_l\text{)}. \end{aligned} \tag{6.6}$$

Thus, the reactive power input is

$$Q_i = E\frac{2\sqrt{2}}{\pi}i_d \sin\alpha \qquad \text{(without } D_l\text{)},$$

$$Q_i = E\frac{2\sqrt{2}}{\pi}i_d \cos(\alpha/2)\sin(\alpha/2) \tag{6.7}$$

$$= E\frac{\sqrt{2}}{\pi}i_d \sin\alpha \qquad \text{(with } D_l\text{)}.$$

Therefore, diode D_l will reduce the reactive power input for any given firing angle with the same direct current by 50 per cent.

6.2.5 Example

A single-phase bridge circuit shown in Fig. 6.3a is used for obtaining a regulated DC output voltage. The RMS value of the AC input voltage is 230 V, and the firing angle is maintained at $\pi/4$ so that the load current is 5 A.

(a) Calculate the DC output voltage, and the active and reactive power input.

(b) Assuming that the load resistance remains the same, calculate the quantities in (a) if a free-wheeling diode is used at the output. The firing angle is maintained at $\pi/4$.

(c) If SCR3 is damaged and gets open-circuited, calculate the average DC output voltage and the average direct current output. For this case, a free-wheeling diode D_l is connected as shown in Fig. 6.3a. The firing angle is $\pi/4$.

(a) The DC output voltage will be

$$e_d = \frac{2\sqrt{2}E}{\pi}\cos\alpha = \frac{2\times\sqrt{2}\times 230}{\pi}\cos(\pi/4)$$

$$= 460/\pi = 146.5 \text{ V}.$$

The active power input is given by

$$P_i = EI_L \cos\alpha,$$

where I_L, the fundamental RMS value of the input current, is

$$I_L = \frac{4}{\sqrt{2}\pi}i_d = \frac{20}{\sqrt{2}\pi} = 4.5 \text{ A}.$$

Therefore,

$$P_i = 230\times 4.5\times\frac{1}{\sqrt{2}} = 730 \text{ W}.$$

The reactive power input Q_i from Eq. (6.7), without diode D_l, for $\alpha = \pi/4$ will be 730 vars.

(b) The load resistance is $146.5/5 = 29.3$ ohms. The firing angle is

$\pi/4$. Therefore, with a free-wheeling diode connected to the controlled circuit, the DC output voltage will be

$$e_d = \frac{\sqrt{2}E}{\pi}(1 + \cos\alpha)$$

$$= \frac{1.707 \times \sqrt{2} \times 230}{\pi} = 177 \text{ V}.$$

The load current i_d is 177/29.3 = 6.05 A. From Eq. (6.6), the fundamental RMS value of the input current will be

$$I_L = \frac{2.828}{\pi} \times 6.05 \times 0.924 = 5.03 \text{ A}.$$

From Eq. (6.4), the input power factor is equal to $\cos(\alpha/2) = 0.924$. Therefore,

$$P_i = EI_L \cos(\alpha/2) = 230 \times 5.03 \times 0.924$$

$$= 1070 \text{ W}.$$

From Eq. (6.7),

$$Q_i = \frac{230 \times 6.05}{\pi} = 450 \text{ vars}.$$

(c) If SCR3 is open-circuited, the circuit will behave as a half-wave controlled circuit. Assuming that there is large inductance on the DC side to make the current continuous in the load, the average output voltage e_d for a firing angle α is given by

$$e_d = \frac{1}{2\pi}\int_\alpha^\pi \sqrt{2}E \sin\theta \, d\theta$$

$$= \frac{E}{\sqrt{2}\pi}(1 + \cos\alpha).$$

Therefore, for $\alpha = \pi/4$, we have

$$e_d = \frac{230}{\sqrt{2}\pi} \times 1.707 = 88.05 \text{ V}.$$

The load current i_d will be 88.05/29.3 = 3.02 A.

6.3 HALF-CONTROLLED BRIDGE CIRCUITS

The improvement in the input power factor obtained by connecting the free-wheeling diode to the circuit shown in Fig. 6.3a can also be achieved if one pair of SCRs is replaced by diodes. Such a circuit is called a half-controlled circuit, whereas the circuits in Fig. 6.3 are referred to as fully-controlled circuits. The main difference between a fully-controlled circuit and a half-controlled circuit is that the former can operate as an inverter when the firing angle is between $\pi/2$ and π, and the latter can operate only in the rectifying mode as the firing angle changes from zero to π. Figure 6.6 shows two versions of the half-controlled circuit. The output waveforms for the two versions are identical. Since the cathodes of the two SCRs are at the same potential in the symmetrical configuration (Fig. 6.6a), their gates can be connected and a single gate pulse can be used for triggering either SCR. The SCR which has the forward-bias at the instant of firing will turn on. In the asymmetrical configuration (Fig. 6.6b), separate triggering circuits have to be used.

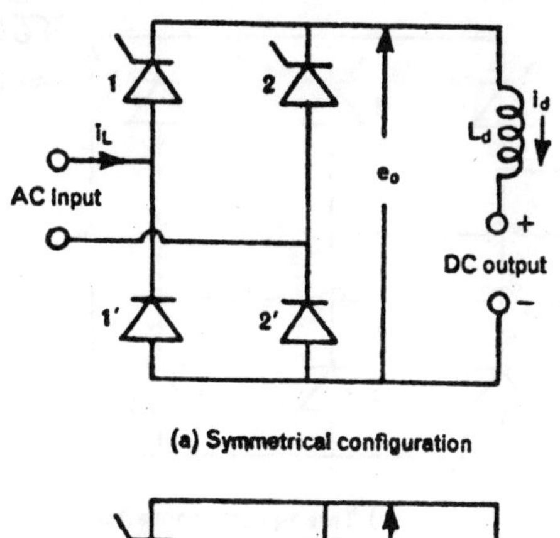

(a) Symmetrical configuration

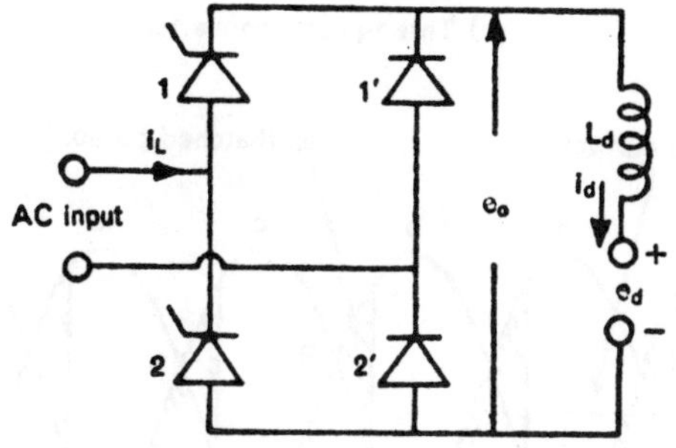

(b) Asymmetrical configuration

Fig. 6.6 Half-controlled circuit (*cont.*).

In Fig. 6.6a the free-wheeling action will take place through 1 and 1′ or 2 and 2′ when the input voltage polarity is reversed. Therefore, SCR1 will not be turned off even though the input current i_L and the output voltage e_o go to zero. For the asymmetrical circuit, the free-wheeling action will take place through 2′ and 1′, and so the SCR current will be zero and the SCR will be turned off as soon as the input voltage polarity reverses. Thus, the average current rating of the SCR will be more for the symmetrical circuit than for the other configuration. The output voltage and current waveforms are as shown in Fig. 6.5b. The expressions with the free-wheeling diode for the output voltage, line current, and reactive power input given, respectively, by Eqs. (6.5), (6.6), and (6.7) will also apply to the half-controlled circuit. Comparing the expressions for DC output voltage in Eq. (6.5), it will be observed that for a half-controlled circuit the output voltage e_d will always be positive as the firing angle changes from zero to π. In a similar manner, it can be shown that the performance of a three-phase half-controlled circuit differs from that of the fully-controlled circuit of Fig. 6.3b. However, in this case either the top three or bottom three SCRs can be replaced by the diodes for the half-control operation. Figure 6.6c shows a three-phase half-controlled bridge. The asymmetrical configuration is not used as it introduces imbalance in line

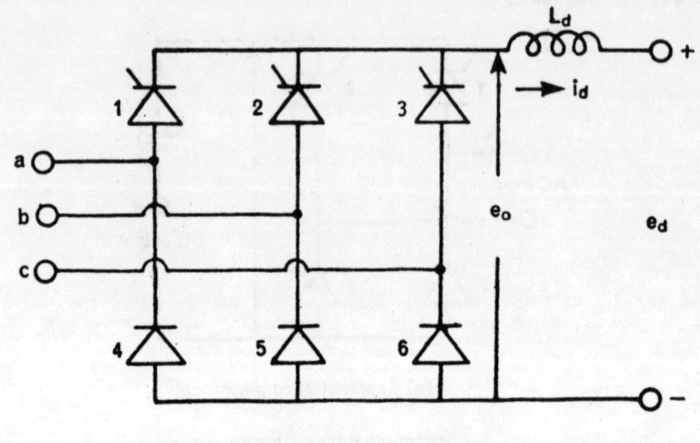

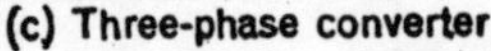
(c) Three-phase converter

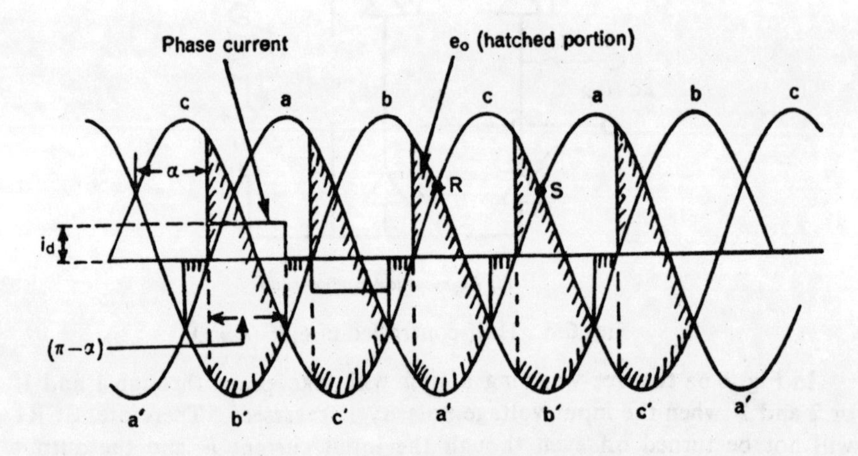

(d) Voltage and current waveforms

Fig. 6.6 Half-controlled circuit.

currents on the AC side. Referring to the waveforms shown in Fig. 6.4c for the three-phase fully-controlled bridge circuit, it will be seen that free-wheeling action for the three-phase half-controlled circuit will take place when the firing angle α is more than $\pi/3$, and the output voltage e_o will become zero during this period $(\alpha - \pi/3)$. If the load inductance L_d is large, the output current i_d will be maintained continuous. For $\alpha < \pi/3$, the input phase current waveform will be the same as that shown in Fig. 6.4c. When $\alpha > \pi/3$, the pulse width of the line current will become

$(\pi - \alpha)$. This will increase the harmonic content. The input phase current waveform is shown in Fig. 6.6d.

For the fully- and half-controlled circuits discussed here, the input power factor is poor when the DC power output is low. For example, when the firing angle approaches $\pi/2$ in a fully-controlled circuit, the average DC output voltage e_d will be low and therefore the active power input will also be low. However, the reactive power input will be high because of the large phase angle. For a half-controlled circuit, the reactive power input will be lower than that of a fully-controlled circuit, but the input power factor will still be poor over a wide range of operation of the circuit. Therefore, to control the DC output voltage and power, it is necessary to devise a method by which the phase angle of the line current i_L with respect to the input voltage can be kept reasonably small as the firing angle of the SCRs is varied. This is achieved by connecting a fully-controlled bridge in series with an uncontrolled bridge as shown in Fig. 6.7a. Inductance L_d will maintain a constant current i_d through the two bridges.

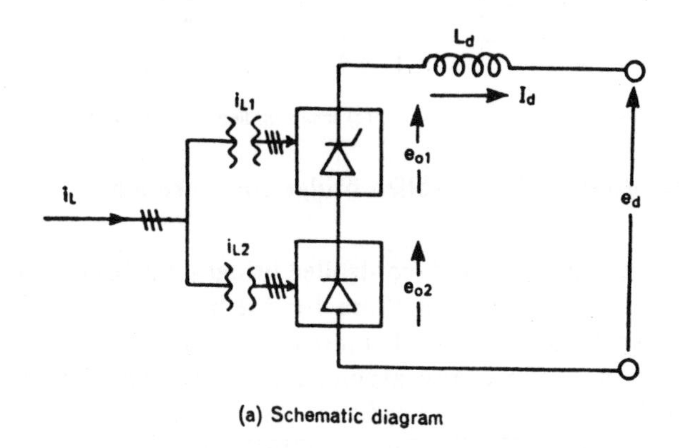

(a) Schematic diagram

Fig. 6.7 Improved half-controlled bridge configuration (*cont.*).

The output voltage e_{o2} of the uncontrolled bridge will be constant, whereas that of the fully-controlled bridge e_{o1} can change from a positive maximum to a negative minimum by varying the firing angle. Thus, the resultant output voltage will change from zero to $2e_{o1}$ because the input voltage to the two bridges is the same. This circuit can operate only when the power flow is from the AC input to the DC output. In other words, the net output voltage cannot become negative.

For any value of output voltage, the line currents i_{L1} and i_{L2} will be identical in magnitude but displaced by the firing angle of the controlled bridge. Figure 6.7b shows these current waveforms when the controlled bridge operates in the inverting mode with a firing angle of about 150°, resulting in a small value of output voltage and power. The total input current i_L, which is the vector sum of i_{L1} and i_{L2}, will be in the form of pulses, placed symmetrically with respect to the phase voltage waveform. The fundamental component of current i_L is shown by the dashed lines in the figure. It is observed that the input phase angle will be small. Thus, although the output power can be varied by controlling the firing angle of the SCR bridge, the overall input power factor will remain high as compared with that obtained from a controlled or half-controlled bridge circuit.

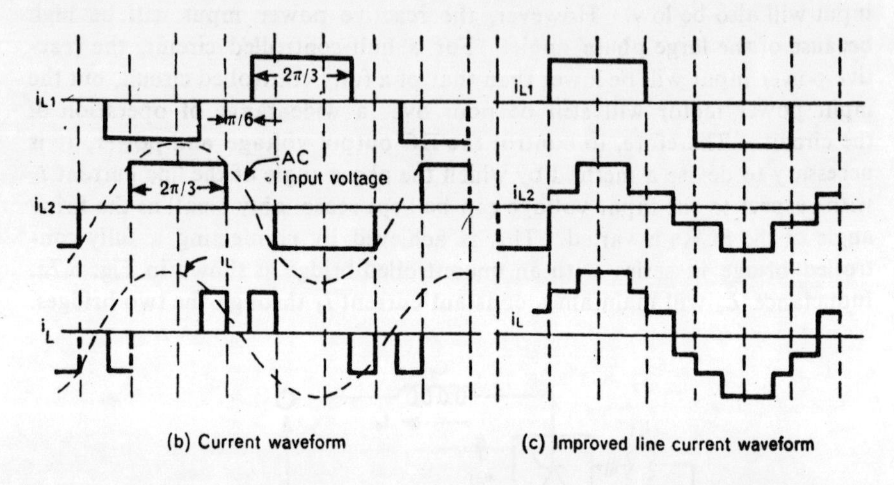

Fig. 6.7 Improved half-controlled bridge configuration.

By using asymmetrical triggering in fully-controlled bridges (the firing angles for the top SCRs are made greater than those for the bottom SCRs), the input power factor can be improved. It must be remembered, however, that this control will not produce any asymmetry in the line current i_L since all the SCRs will conduct for $2\pi/3$ radians. A similar type of circuit can also be used for improving the line current waveform. Here, both the bridges are controlled and the input to one of the bridges is obtained through a phase-shifting transformer. Normally, an input transformer with a star-star or star-delta connection is used for each bridge (if the transformer for one bridge has a star-star connection, the transformer for the second bridge should have a star-delta connection). This produces a phase difference of 30° between the two input phase voltages. If the firing angles are the same for both the bridges, then the primary currents i_{L1} and i_{L2} will be as shown in Fig. 6.7c. The resultant current will have a stepped waveform (see Fig. 6.7c), which is an improvement over the current waveform shown in Fig. 6.4c. Another method of reducing the reactive power requirement of a three-pulse converter is by *pulse-skipping*. This results in a one-and-a-half-pulse operation, and each SCR will conduct for a duration of $4\pi/3$ radians. The firing sequence of the SCRs will be opposite to that for normal three-pulse operation. For more details on pulse-skipping and asymmetrical triggering, see Davis (1971), and Möltgen (1962), listed in General References.

6.4 DUAL CONVERTERS

As the name indicates, a dual converter consists of two converters, both either fully-controlled or half-controlled, connected to the same load. The purpose of a dual converter is to provide a reversible DC voltage to the load. It is needed for DC motor drives where speed reversal is required. Figure 6.8 shows the schematic arrangement of a dual converter. The two modes of its operation are the noncirculating-current mode and the circulating-current mode. In the former, only one of the bridges is triggered.

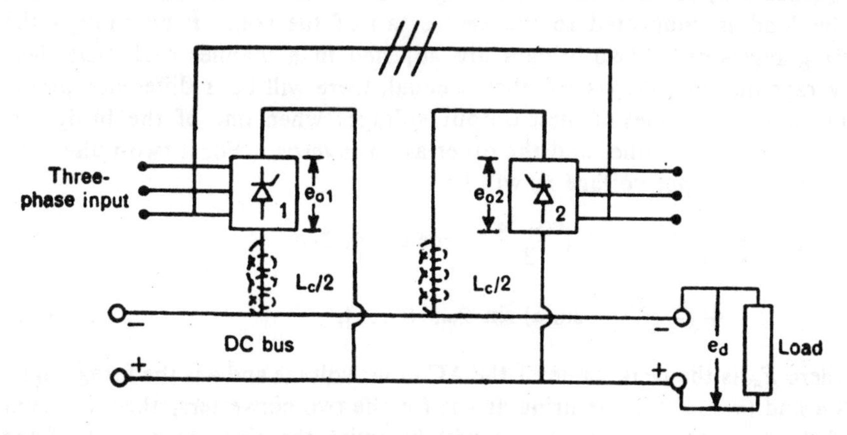

Fig. 6.8 Dual converter.

When reversal of output voltage is required, the firing pulses for the conducting bridge are stopped and the second bridge is gated. Since the conducting SCRs in the first bridge will turn off only when the current goes to zero, a small dead time must be allowed before the second bridge is gated. Otherwise, the AC input will be shorted through the two bridges. For this, a current sensor is required which ensures that all the SCRs in bridge 1 will be turned off before the firing pulses are applied to SCRs of the second bridge. In the circulating-current mode, both bridges are gated simultaneously, one operating in the rectifying mode and the other in the inverting mode to avoid short-circuits. This scheme requires fully-controlled bridges. For the polarity shown in Fig. 6.8, bridge 1 is a rectifier and bridge 2 is an inverter. The internal voltage of the rectifier is higher and that of the inverter is lower than the output voltage. If the firing angle for bridge 1 is α, then its internal voltage $(3\sqrt{3}/\pi)E_m \cos \alpha$ (where E_m is the peak value of the input phase-to-neutral voltage) is made slightly more than the DC output voltage e_d; the firing angle for bridge 2 is made slightly less than $(\pi - \alpha)$ so that its internal voltage is lower than e_d. If the reversal of output voltage polarity is required, then the firing angles of the two bridges are changed simultaneously such that bridge 2 will operate as a rectifier and bridge 1 as an inverter. The polarity of the output voltage will be the same as that of the rectifier. The internal voltage of the inverter must be close to but smaller than the DC output voltage. This is to ensure that the circulating currents between the two bridges are minimised. The main advantage of the circulating-current scheme is the rapidity with which the phase reversal of the output current can be obtained. However, this scheme will produce a continuous flow of circulating current between the two bridges, resulting in increased power losses. A similar scheme is also used for cycloconverters for AC-to-AC conversion. This method of frequency conversion and firing angle control will be explained in Chapter 7.

To reduce the circulating current, it is necessary to include inductance L_c (shown by the dashed lines in Fig. 6.8) in the circulating-current path. The load is connected to the centre tap of the coil. Even though the firing angles of the two bridges are adjusted in a manner such that their average output voltages are almost equal, there will be a difference in the instantaneous values of these output voltages when one of the bridges is operating as a rectifier and the other as an inverter. For a two-pulse converter, the output voltage e_o will be

$$e_o = \frac{2E_m}{\pi}\Big[\cos\alpha + \Big(\frac{\cos 3\alpha}{3} - \cos\alpha\Big)\cos 2\omega t + \Big(\frac{\sin 3\alpha}{3} - \sin\alpha\Big)\sin 2\omega t + \ldots\Big], \tag{6.8}$$

where E_m is the peak value of the AC input voltage and α is the firing angle. If α and $(\pi - \alpha)$ are the firing angles for the two converters, then the sum of the two voltages $(e_{o1} + e_{o2})$ will be twice the sine component of the second harmonic (neglecting all other higher harmonics) which drives the circulating current. This current will have a maximum peak value when $\alpha = \pi/2$, and the corresponding average load voltage and current will be zero. Assuming that the maximum peak circulating current is one-fifth of the maximum load current I_d, the required value of inductance L_c can be obtained from

$$\frac{4E_m}{\pi} \times \frac{4}{3} \times \frac{1}{2\omega L_c} = 0.2I_d. \tag{6.9}$$

In this derivation for L_c it is assumed that the load current is steady. If there are AC components in the load, there will be an additional self-inducted circulating current component which is independent of L_c (see Chapter 7).

If the pulse number of each converter is increased, the required value of L_c to satisfy the constraint imposed in Eq. (6.9) on the maximum peak circulating current will decrease and the dynamic performance of the system will improve.

6.4.1 Example

Two fully-controlled three-phase bridges are connected in antiparallel across a load to provide reversible DC voltage to the load. The bridges operate in the circulating-current mode. The input is three-phase, 400 V, 50 Hz, AC supply, and the maximum load current is 30 A. Obtain a suitable value of L_c to limit the circulating current.

The output voltage e_o for a six-pulse converter is given by

$$e_o = \frac{3\sqrt{3}E_m}{\pi}[\cos\alpha + (\frac{\cos 7\alpha}{7} - \frac{\cos 5\alpha}{5})\cos 6\omega t$$

$$+ (\frac{\sin 7\alpha}{7} - \frac{\sin 5\alpha}{5})\sin 6\omega t + \ldots],$$

where α is the firing angle and E_m is the peak phase-to-neutral voltage. The peak value of the circulating current is taken to be 6 A. This will occur when $\alpha = \pi/2$. Therefore,

$$6 = \frac{3\sqrt{3}E_m}{\pi} \times 2\,[\frac{1}{7} + \frac{1}{5}] \times \frac{1}{6\omega L_c},$$

$$L_c = 32.8 \text{ millihenrys (mH)}.$$

6.5 PHASE-CONTROL CIRCUIT

For the half-wave and full-wave circuits discussed in the preceding sections, the SCRs have to be fired synchronously in each half-cycle of the input AC supply. For single-phase circuits, no additional logic circuit is required to choose the particular SCR to be fired each time as is the case for three-phase circuits. One SCR or two SCRs conduct depending on whether the single-phase bridge is half-controlled or fully-controlled. The other SCRs will have a reverse-bias across them, and so will not turn on even if gated. The phase of the firing pulses has to be changed with respect to the input supply to vary the firing angle, and thereby to control the DC power.

For three-phase circuits there will be three SCRs for a half-controlled bridge and six SCRs for a fully-controlled bridge which in each case have to be triggered in a proper sequence to produce balanced currents in the input lines and low ripple on the DC output. There will be more than one SCR with proper forward-bias, and so the firing pulses cannot be applied to all the SCRs at the same time as in single-phase circuits. For a half-controlled circuit if the firing angle variation is restricted, the control circuit can be simplified very much. In this case, all the top SCRs have a common cathode (Fig. 6.6c) and so their gates can be connected and a common gate signal can be used for triggering. Only one SCR (that with the highest anode voltage in all the phases) will conduct and the remaining two SCRs will be reverse-biased. For example, if a gating pulse is applied anywhere in the interval *RS* (Fig. 6.6d), only SCR3 will be triggered since phase *c* voltage will be the highest in this interval. Thus, the maximum variation of the firing angle can be from *R* to *S* (i.e., $2\pi/3$), and the frequency of the gating pulse will be three times the supply frequency. But, when the circuit is required to operate both as a rectifier and an inverter, large variations in the firing angle are generally possible only if there is a logic circuit to guide each firing pulse to the proper SCR. Since each of the six SCRs for a fully-controlled bridge has to be gated once during one period of the input, the firing frequency will be six times the supply frequency. The required control and logic circuits are given in Chapter 9. Here, the firing angle can be changed from zero to π.

The R and RC firing circuits for the phase control of SCRs have been discussed in Chapter 2. The UJT relaxation oscillator, described in Chapter 4, is frequently used for firing SCRs because with this oscillator the firing angle can be controlled over a wide range. The method of synchronising the UJT pulses with the input is shown in Fig. 6.9. A full-wave phase-control circuit will be discussed here.

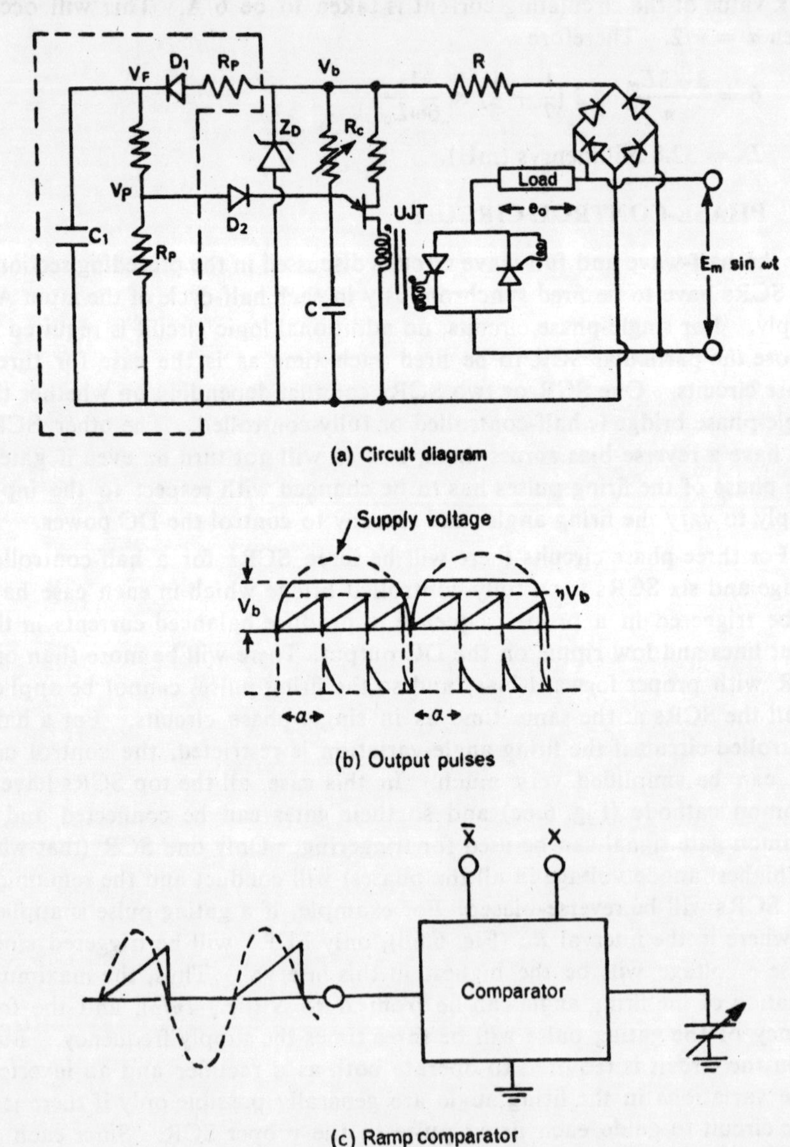

Fig. 6.9 UJT phase-control circuit (*cont.*).

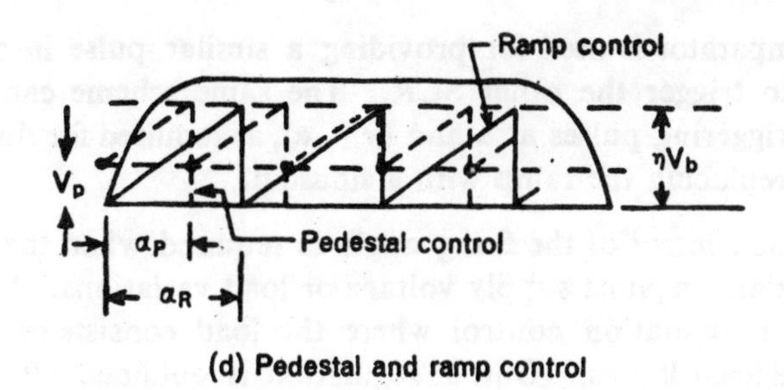

(d) Pedestal and ramp control

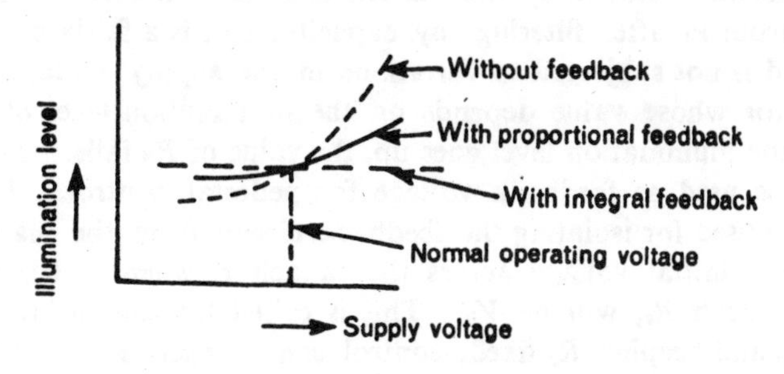

(e) Input-output characteristics

Fig. 6.9 UJT phase-control circuit.

The input voltage is rectified and clipped to V_b by the zener diode Z_D as shown in Fig. 6.9a. This is the input to the UJT relaxation oscillator. At the end of every half-cycle, voltage V_b will become zero and capacitor C will be discharged. Thus, time t $(=\alpha/\omega)$, when the first firing pulse is applied to the SCRs, will remain constant. Two SCRs in antiparallel connection are shown in the figure. A triac can also be used in this circuit. The firing pulses are applied to both SCRs. Only one of these will conduct depending on the input polarity. Even though the UJT gives out a train of pulses as shown in Fig. 6.9b, only the first pulse will determine the firing angle. By varying the charging resistance R_c we can bring about a change in the oscillator frequency and thereby in the firing angle which, in turn, controls the output power. This control of output power by R_c is known as *ramp control*; it is also referred to as *open loop control* or *manual control*. Normally, R_c is used in this manner to adjust the output power to the desired level.

Figure 6.9c shows another circuit for obtaining synchronised firing pulse to gate the SCRs. Here, ramp voltage is derived from the AC input voltage. This voltage is obtained by charging a capacitor by a constant current during the positive half-cycle, and is reset to zero during the negative half-cycle of the input. The comparator will compare the ramp voltage with the DC voltage. The voltage at X will become high whenever the ramp voltage is more than the DC reference voltage. By varying the reference voltage, a control can be exercised on the instant the voltage at X becomes high. The voltage at X is used for triggering the SCR.

Another comparator is used for providing a similar pulse in the negative half-cycles to trigger the other SCR. The same scheme can be used for generating triggering pulses at α and $(\pi - \alpha)$, as required for dual converter control, by replacing the ramp with a sinusoid.

Automatic control of the firing angle is required when the load power is to be regulated against supply voltage or load variations. For example, consider an illumination control where the load consists of a number of parallel-connected lamps. Output regulation is obtained by the pedestal control, shown enclosed by the dashed lines in Fig. 6.9a. Voltage V_F, obtained from V_b after filtering (by capacitor C_1), is a fairly constant DC voltage and is not subjected to variations in the supply voltage. R_P is a photoresistor whose value depends on the illumination level of the lamp load. If the illumination level goes up, the value of R_P falls. The voltage across R_P is used as feedback voltage for pedestal control. Diodes D_1 and D_2 are used for isolating the feedback circuit from the main control circuit. The initial voltage across the capacitor, when it begins to get charged through R_c, will be V_P. This is called the *pedestal voltage*. By varying V_P and keeping R_c fixed, control can be exercised on the instant the capacitor voltage becomes equal to ηV_b and discharges through the UJT. This is shown in Fig. 6.9d. For the same value of R_c, the firing angle will be α_R without the pedestal voltage V_P and it will decrease to α_P when V_P is applied. These firing angles are given by

$$\alpha_R = \omega R_c C \ln \frac{1}{1-\eta},$$
$$\alpha_P = \omega R_c C \ln \frac{V_b - V_P}{V_b(1-\eta)}, \tag{6.10}$$

where ω is the input frequency and η is the stand-off ratio of the UJT. Thus, by controlling the pedestal voltage, the output power or the illumination level can be controlled. The same circuit can also be used for triggering SCRs in a bridge configuration (Fig. 6.3).

If, for a given setting of R_c with pedestal control, the illumination level increases due to a rise in input voltage, then R_P will decrease and the pedestal voltage will fall as V_F is maintained constant. This will increase the firing angle and reduce the output power. Here, the feedback is proportional to the output. This is a first-order system, and hence has a nonzero static error, that is, the output deviates from the reference value. However, the error can be minimised by increasing the gain in the feedback circuit, or by employing the error integrator (second-order control). Figure 6.9e shows the input-output characteristics of the controlled system. Similar circuits can be used for temperature regulators, regulators for DC power supplies, and speed controllers for motors.

6.5.1 Modified Ramp Control

In Fig. 6.9a the half-cycle average E_o of the output voltage e_o is given by

$$E_o = \frac{2E_m}{\pi} \cos \alpha, \tag{6.11}$$

where α is the firing angle and E_m is the peak of the input sinusoidal voltage. For the control circuit described here, the variation of α with ramp control, i.e., by varying R_c, or pedestal control, i.e., by varying R_P, is approximately linear. Therefore, the output voltage E_o will change in a cosinusoidal manner. The amplification factor $\Delta E_o/\Delta\alpha$ is small for low values of α. For a feedback control system where it is necessary to maintain E_o constant, it is not very important how ΔE_o changes with $\Delta\alpha$ as long as the static error is maintained low. In some cases it is desirable to have a linear relationship between input and output. If E_o is to vary linearly with R_c, then the firing angle should change in a cosinusoidal manner with respect to R_c. This is achieved by connecting the charging circuit R_cC to a sinusoidal supply as shown in Fig. 6.10a. The voltage across C will then be

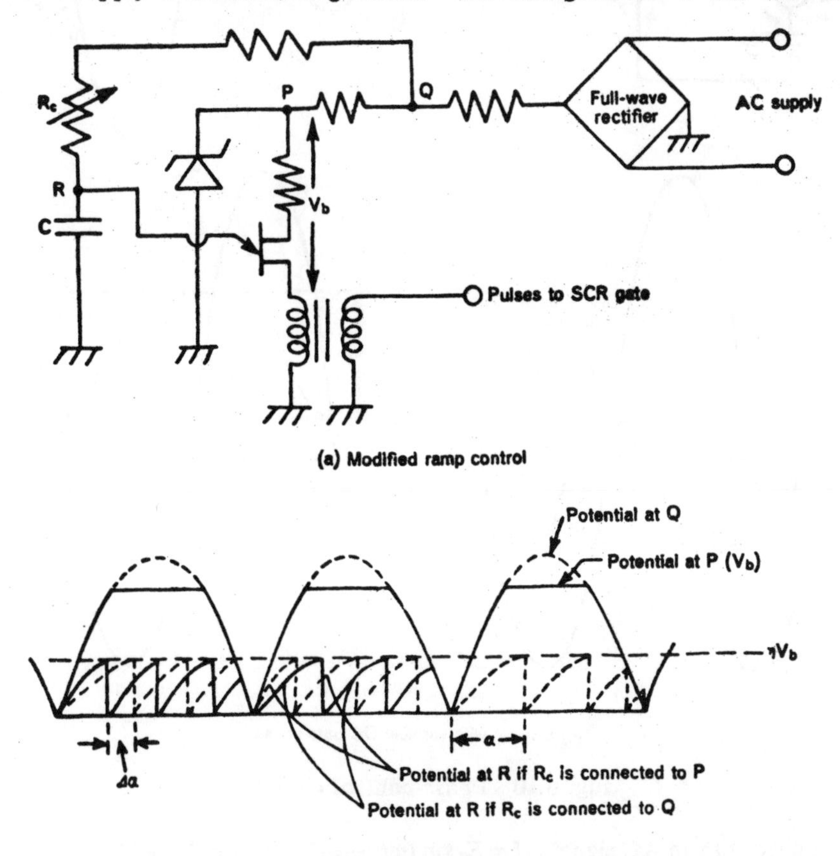

Fig. 6.10 Phase-control circuits (*cont.*).

a ramp, with a sinusoidal signal superimposed. The variation of α with R_c is shown in Fig. 6.10b.

Another simple method of producing a nonlinear change in the firing angle with respect to the control signal is shown in Fig. 6.10c. The gate is

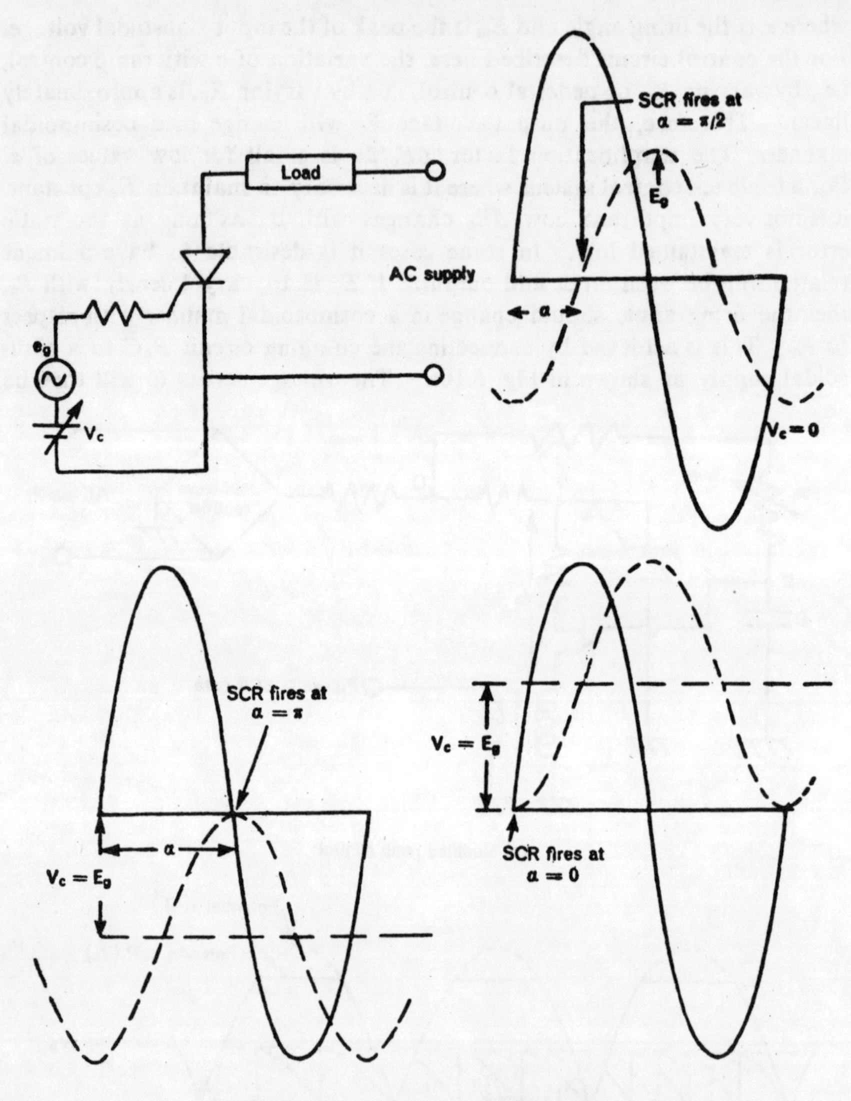

Fig. 6.10 Phase-control circuits.

supplied with an AC signal $e_g [= E_g \sin(\omega t - \pi/2)]$, phase-shifted by $\pi/2$ with respect to the main supply. The control signal V_c (DC voltage) is connected in series with the gate-cathode circuit. The phase angle of the resultant voltage is changed by varying V_c. Figure 6.10c also shows the firing angle for three different values of V_c. The maximum value of V_c is limited to the peak value of the AC voltage applied to the gate. The firing angle α is given by $\cos^{-1}(V_c/E_s)$.

6.5.2 Control by a Voltage-Controlled Oscillator (VCO)

In a phase-control circuit, feedback is used for the automatic adjustment of the firing angle to keep the output voltage/current constant. The control circuits described in Sections 6.5 and 6.5.1 produce inherent regulation, that is, the output voltages/currents deviate from their reference values in the steady state, resulting in nonzero static error. To make this point clear, let us consider the illumination control scheme shown in Fig. 6.9a. Let the intensity of illumination be reduced by a fall in the supply voltage. Then, resistance R_P will increase and so also pedestal voltage V_P. Due to this, the firing angle α will decrease and the load voltage will rise, resulting in an increase in the illumination level. However, if the illumination intensity returns to the specified value, then R_P will decrease, bringing down the pedestal voltage. Therefore, the firing angle will advance. In the steady state, the illumination will be slightly lower than the reference level and this error will produce a suitable pedestal voltage to keep the firing angle at the new value. This is known as *proportional feedback* and the error is made small by increasing the feedback gain. Figure 6.9e shows the input-output characteristics with this control scheme. One method of bringing the static error to zero is by integrating the error signal. This is known as *integral feedback* and can be obtained by using a voltage-controlled oscillator (VCO) to provide triggering pulses to the SCRs. In Fig. 6.9a the UJT is replaced by the VCO. When the AC input voltage is normal, the free-running frequency of the VCO is adjusted to be 100 Hz (twice the input frequency) so that a firing pulse will appear in each half-cycle and the SCRs will be triggered symmetrically in successive half-cycles. Adjustment of the firing angle is made by producing temporal changes in the VCO output frequency.

When the input voltage is reduced, the error signal raises the VCO frequency and the next firing pulse will appear prematurely, producing a smaller firing angle. The firing angle will decrease as long as there is an error in the output, and the VCO frequency will return to the normal value (100 Hz) when the error becomes zero. The new firing angle will be maintained in the steady state. Thus, the static error with VCO control is zero. This is because there is inherent integration of the error signal. Similar operations take place when the supply voltage is increased. Here, the error will be of polarity opposite to that which occurs when the input voltage is reduced, and the VCO frequency will be lowered. The firing of the SCRs will be delayed and in the steady state when the illumination is brought back to the reference level, the firing angle will remain constant at a higher value. The astable circuit discussed in Chapter 5 can be used as a VCO by providing a variable-bias voltage to the anode gates. VCOs are available as IC chips and are increasingly used in such power control circuits as discussed here. The main drawback of this VCO scheme is that if the supply frequency changes the output will also drift from the desired level.

6.5.3 Derating of SCRs in Phase Control

It was mentioned in Chapter 2 that the permissible average current rating of SCRs decreases as the firing angle is increased. The SCR rating curves shown in Fig. 2.13 can be used only when the input voltage to the phase-controlled circuit is sinusoidal and the output is resistive. Even for inductive loads, the form factor of the load current will increase as the conduction angle for the SCR decreases, and hence a derating of the device will be necessary to maintain the internal power dissipation within limits. For triacs, ratings are specified in terms of the RMS value and, therefore, there will be no such derating problem. Triacs can carry the same RMS value of the current at all conduction angles, provided the peak amplitude ratings are not exceeded.

6.5.4 Example

For the circuit shown in Fig. 6.9a if the breakdown voltage for the zener diode Z_D is 12 V, η for the UJT is 0.7, and the supply voltage is 325 sin 314t, calculate the firing angle α: (a) when R_c is adjusted to give an output frequency of 1 kHz and zero pedestal voltage V_P, and (b) with R_c kept constant but with $V_P = 5$ V.

(a) Since the pulse frequency is 1 kHz and the UJT output is synchronised with the input frequency (capacitor C is reset to zero voltage at the beginning of every half-cycle), the first pulse will always fire the SCRs, and hence will determine the firing angle. Therefore,

$$\alpha_R = \frac{1}{10^3} \times 314 \times \frac{180}{\pi} = 18°.$$

(b) With $V_P = 5$ V, Eq. (6.10) gives the new firing angle α_P. Thus,

$$\frac{\alpha_P}{\alpha_R} = \frac{\ln\left[\frac{1 - \frac{5}{12}}{1 - 0.7}\right]}{\ln\left[\frac{1}{1 - 0.7}\right]} = 0.56.$$

Therefore,

$$\alpha_P = 18 \times 0.56 = 10.08°.$$

6.6 APPLICATION TO SPEED CONTROL OF MOTORS

Phase control can be very conveniently used for the speed control of AC and DC motors; this is achieved by applying a variable voltage to the motor. As the speed of synchronous motors does not change when the input voltage is varied, this method is useful only for commutator or induction motors. For AC motors, full-wave phase-control circuits are required. Figure 6.11 shows the schematic arrangement for the speed control of single-phase and three-phase induction motors. By varying the firing angle, the RMS input voltage can be changed. In the case of single-phase motors, an additional starting winding is required. Two SCRs connected in anti-parallel are preferred to one triac since the motor is an inductive load.

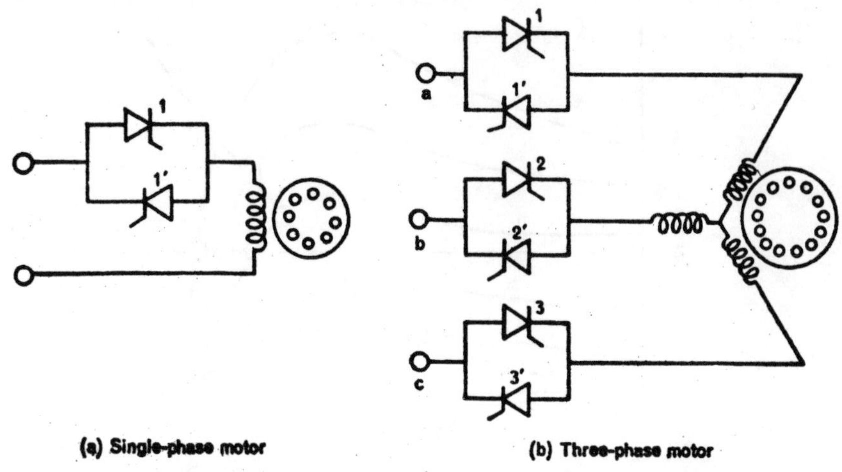

Fig. 6.11 Phase control of an AC motor.

For single-phase circuits, the UJT relaxation oscillator shown in Fig. 6.9 can be used. The input current waveform is shown in Fig. 6.12a. The motor winding will experience open-circuits in every half-cycle if the angle of conduction β is less than π. During this period, the rotor currents will

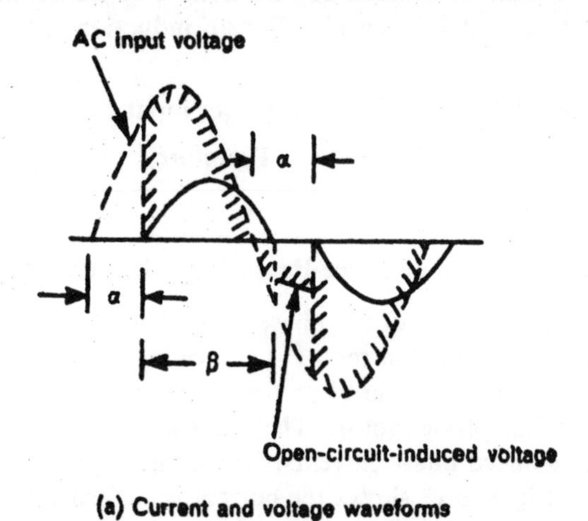

Fig. 6.12 Characteristics of a phase-controlled AC motor (*cont.*).

induce a voltage in the stator phase winding. The area shown by hatched lines in Fig. 6.12a is the motor phase voltage. As the firing angle increases, the RMS value of this voltage will decrease. The characteristics of an induction motor with variable applied voltage are shown in Fig. 6.12b. This method of control is very simple and economical. It provides a wide range of speed control if the load torque increases with speed as shown by curve 1 in Fig. 6.12b. The points indicated by circles on this curve show

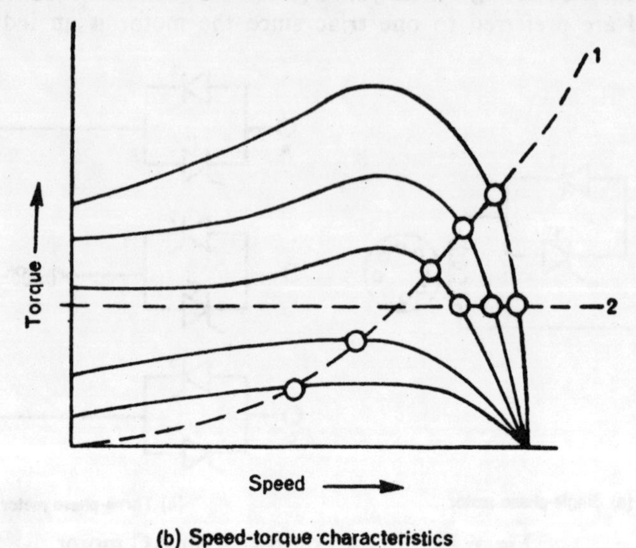

(b) Speed-torque characteristics

Fig. 6.12 Characteristics of a phase-controlled AC motor.

the several speeds that are possible by varying the voltage. If the load torque is constant, the speed variation is very much limited as shown by curve 2. Another drawback of this method of control is that the efficiency falls off with decrease in speed. For an induction motor, the power output is given by

$$\text{mechanical power output} = \text{power input to rotor} \times (1 - S),$$

where S is the slip of the motor. Therefore,

$$\text{efficiency} = (1 - S). \tag{6.12}$$

6.6.1 Phase Control of Three-Phase Induction Motors

The schematic diagram of the phase-control circuit for controlling the speed of a three-phase induction motor is shown in Fig. 6.11b. The speed-torque characteristics and the overall performance are similar to those for a phase-controlled single-phase motor. The SCRs have to be triggered in a sequence to obtain balanced phase currents. The firing angle must be the same for all the phases. Figure 6.13 shows the current waveforms for three firing angles. It will be observed that the motor exhibits three different modes of operation. When the conduction angle β is more than $2\pi/3$, the number of SCRs conducting at any time will be either 3 or 2. This is called the 3/2 mode. The phases get open-circuited for a very short time. If the conduction angle is between $\pi/3$ and $2\pi/3$, the number of SCRs that conduct at any one time will be 2 or 1. This is the 2/1 mode. When angle β is less than $\pi/3$, only one SCR or none will conduct at any time. This is the

1/0 mode. The last two modes, i.e., the 2/1 mode and the 1/0 mode, are possible only when the stator winding neutral is connected to the supply

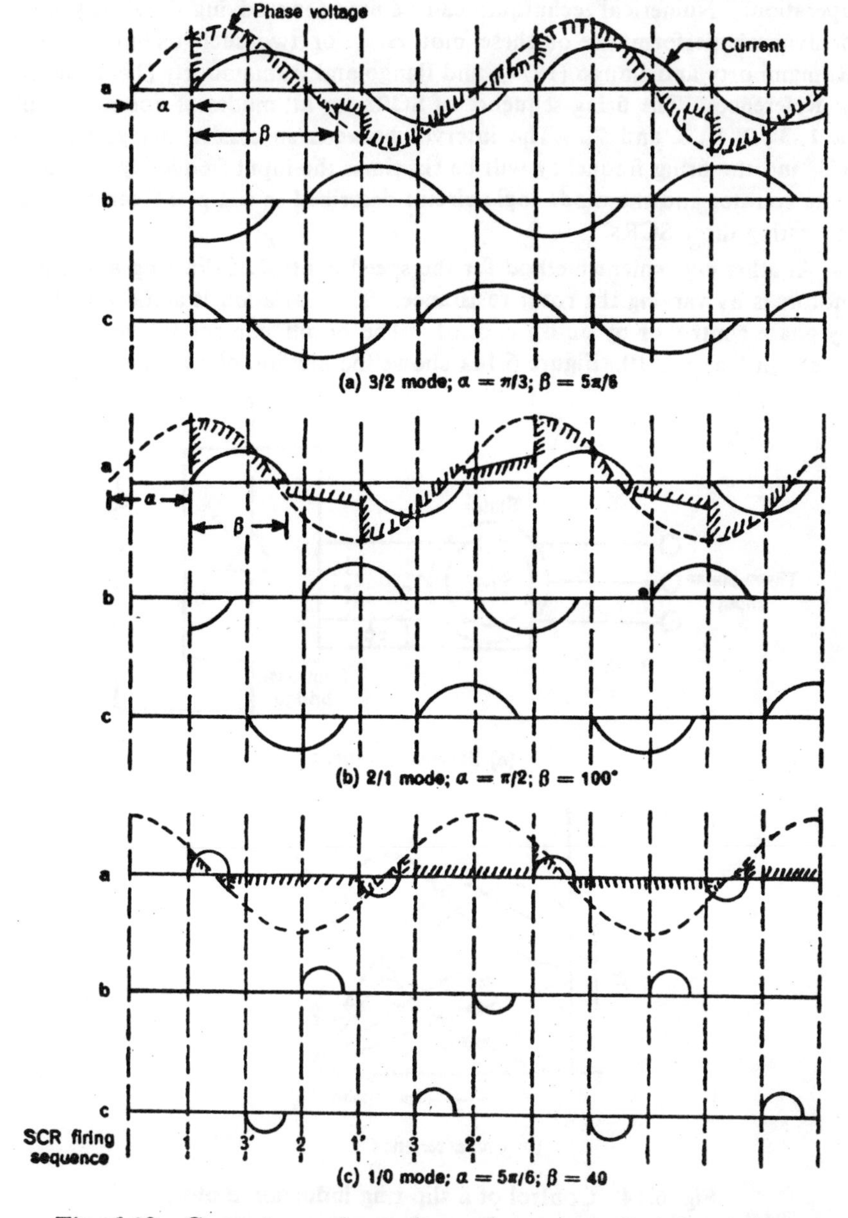

Fig. 6.13 Current waveforms for a phase-controlled three-phase AC induction motor.

neutral. As long as the SCRs conduct, the corresponding phase voltage will be known. When the phase gets open-circuited the corresponding voltage across the phase will be the induced voltage due to currents flowing in the other stator and rotor windings. The procedure for obtaining the steady-state current waveforms requires the solution of the motor differential equations; such equations are not the same for every mode of operation. Numerical techniques can be used for studying the steady-state or dynamic performance of these motors. For two such techniques, see Ramamoorty and Ilango (1974), and Ilango and Ramamoorty (1971), listed in References. The firing sequence of SCRs for all modes of operation will be 1, 3′, 2, 1′, 3, and 2′. The interval between successive firings will be $\pi/3$, and the firing frequency will be six times the input frequency. A six-state counter and the diode logic circuit described in Chapter 9 can be used for gating these SCRs.

Another convenient method for the speed control of slip-ring induction motors is by varying the rotor resistance. This variation is achieved either by phase control or by on-off control. The on-off control scheme is discussed in Chapter 10. Figure 6.14a shows the method of varying the rotor

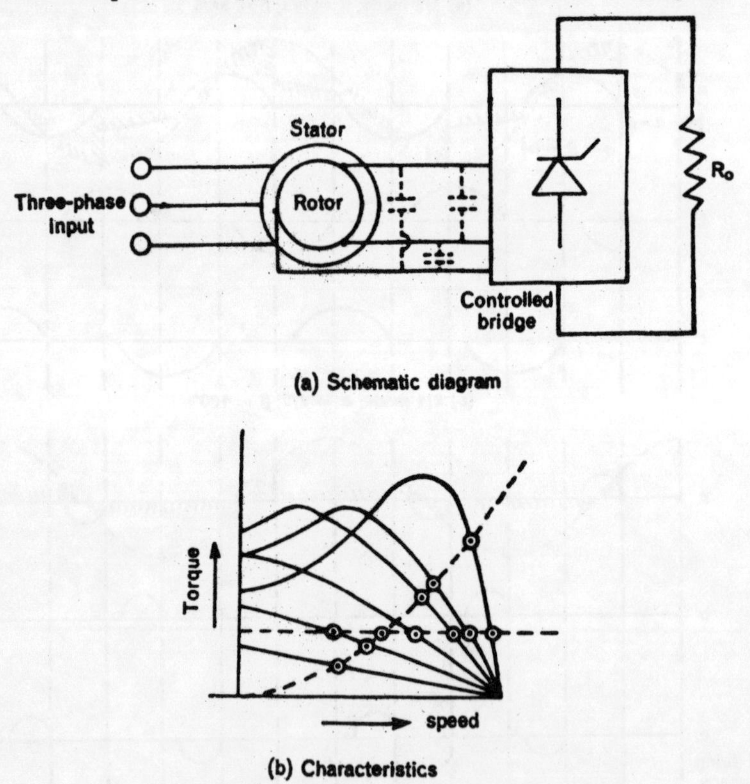

Fig. 6.14 Control of a slip-ring induction motor.

resistance using phase control. The corresponding speed-torque characteristics are shown in Fig. 6.14b. By increasing the firing angle of the SCRs in the bridge, the power consumed by resistance R_o is reduced. The main drawback of this method is the low input power factor for the bridge circuit. Resistance R_o appears as a variable inductive load to the input terminals. The effect of the lagging power factor load in the rotor circuit is to reduce the pull-out torque. Capacitor C shown by the dashed lines (Fig. 6.14a) is used for improving the rotor power factor. This method of control can be applied for both constant torque and variable torque loads and will produce a wide range of speed and a good starting torque (Fig. 6.14b). Another difficulty with this control scheme is that the frequency of the input voltage to the bridge varies with the speed of the motor. Thus, the firing frequency and the firing angle must be synchronised with the rotor frequency. A half-controlled bridge can be used. This will give a wide range of control for the firing angle and produce symmetric current waveforms on the AC side of the rotor. Because of the resistive load, the input harmonic content will be high and result in additional rotor heating.

Yet another method of speed control of slip-ring induction motors, called the *slip power recovery scheme*, is explained in Chapter 7. Details of phase control of DC motors are given in Section 6.8.

6.7 REGULATED DC POWER SUPPLIES

An important application of the controlled bridge circuits, discussed in Sections 6.1 and 6.2, is for regulated DC power supplies. A large series inductor and a shunt capacitor are used for reducing the ripple in the output. A feedback arrangement similar to that shown in Fig. 6.9 can be used for automatic adjustment of the firing angle of SCRs to control the output DC voltage. Figure 6.15 shows a commonly-used rectifier with an inductor input filter. A single-phase uncontrolled full-wave rectifier is considered

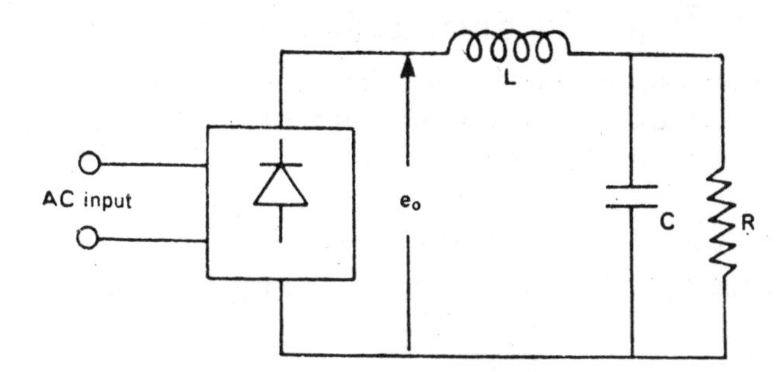

Fig. 6.15 Single-phase full-wave rectified power supply.

for the filter design. If E_s is the effective value of the AC input, the Fourier series expansion for the rectified output voltage waveform is given by

$$e_o = \sqrt{2}E_s \left[\frac{2}{\pi} - \frac{4}{3\pi}\cos 2\omega t - \frac{4}{15\pi}\cos 4\omega t - \ldots\right]. \tag{6.13}$$

The ripple factor RF for the filtered output is defined as

$$\text{RF} = \frac{\text{RMS value of the harmonic voltages at the output}}{\text{DC voltage output}}.$$

For a single-phase full-wave bridge, we have

$$\mathrm{RF} = \frac{1}{6\sqrt{2}\omega^2 LC}, \tag{6.14}$$

where ω is the frequency of the input. Harmonics of order higher than the second are neglected.

To maintain continuous conduction, the average direct current I_d must be at least equal to the peak amplitude of the second harmonic current through the inductor. If this condition is maintained by imposing a limit on the maximum value for load resistance R, then the DC output voltage will be approximately constant for all load currents. This criterion will be satisfied by

$$R_{max} = 3\omega L. \tag{6.15}$$

Equations (6.14) and (6.15) can be used for designing the required values of L and C for given RF and R_{max}. As the load resistance is increased beyond R_{max}, the load current will become discontinuous and the output voltage will rise above the average value $2\sqrt{2}E_s/\pi$ and reach $\sqrt{2}E_s$ when the load current is zero. A similar operation will take place when a controlled bridge is used.

6.7.1 Example

Design an LC filter for a DC power supply obtained from a three-phase full-wave bridge circuit if the ripple factor is required to be not more than 1 per cent and the maximum load resistance is 10 ohms.

The Fourier series expansion for the output voltage e_o is given by

$$e_o = 2\sqrt{2}E_s\,[0.828 - 0.0472\cos 6\omega t - 0.0116\cos 12\omega t - \ldots],$$

where E_s is the effective value of the input line-to-neutral voltage. With an LC filter as shown in Fig. 6.15, the ripple factor at the output (considering only the sixth harmonic) will be

$$\mathrm{RF} = \frac{0.0472}{\sqrt{2} \times 0.828} \times \frac{1}{36\omega^2 LC},$$

where ω is the input frequency (50 Hz). The ripple factor being 0.01, we have

$$LC = \frac{4.72}{\sqrt{2} \times 0.828 \times 36 \times 314 \times 314} = 1.13 \times 10^{-6}.$$

The condition given by Eq. (6.15) is modified for a three-phase full-wave circuit as

$$R_{max} = \frac{6\omega L \times 0.828}{0.0472} = (33 \times 10^3)L.$$

Therefore,

$$L = \frac{1}{3.3 \times 10^3} = 0.11\ \mathrm{mH},$$

$$C = \frac{1.13 \times 10^{-6} \times 10^{3}}{0.11} = 10^{4}\ \mu\text{F}.$$

6.8 DC MOTOR CONTROL

The phase-controlled circuits discussed in this chapter can also be used for the variable speed control of DC motors. The output filter will not be required here as the input to the DC motor need not be steady though it must be unidirectional. The speed of DC motors can be varied by controlling either the field current or the armature voltage. For a given armature voltage, a decrease in field current will increase the motor speed. Since the air gap flux changes in the field current, the full load torque will decrease with an increase in speed. Hence, this is known as *constant power control.* For controlling the armature voltage, the field is separately excited. This is called *constant torque control,* because the motor full load torque remains the same irrespective of the speed. The motor response with armature control is faster than with field control since the time constant of the field is very much larger than that of the armature. Generally, field control is used for speeds above and armature control for speeds below the rated speed. A schematic diagram for DC motor control with separate field excitation is shown in Fig. 6.16a. The field is supplied from a full-

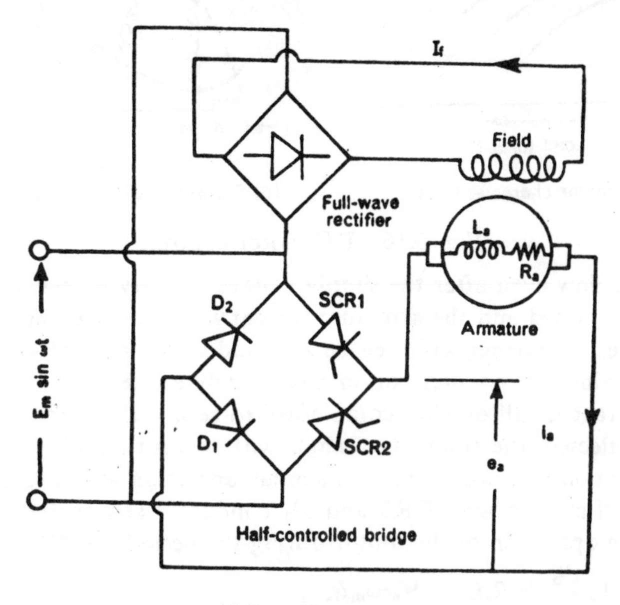

(a) Schematic diagram

Fig. 6.16 DC motor control (*cont.*).

wave rectifier bridge. All DC shunt motors are controlled in this manner. A symmetrical half-controlled bridge is used for varying the armature voltage.

SCRs 1 and 2 (Fig. 6.16a) are fired in each half-cycle. In the positive half-cycle, SCR1 and D_1 will conduct from α to $(\alpha + \beta)$, where α is the firing angle and β is the conduction angle. If angle β is less than $(\pi - \alpha)$, then the armature current will be discontinuous as shown in Fig. 6.16b. However, if the motor speed is low and armature inductance high, current i_a will

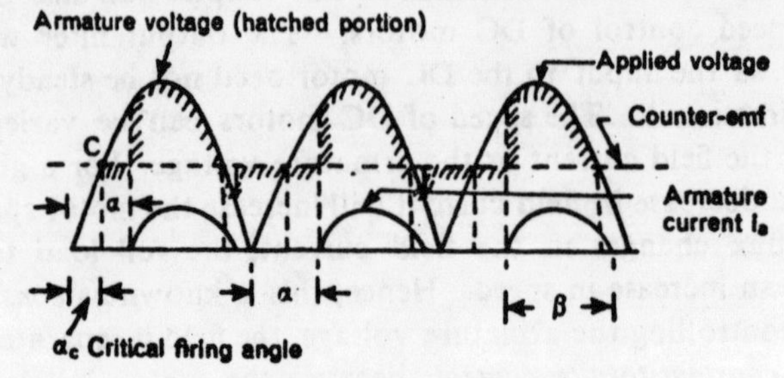

(b) Current and voltage waveforms

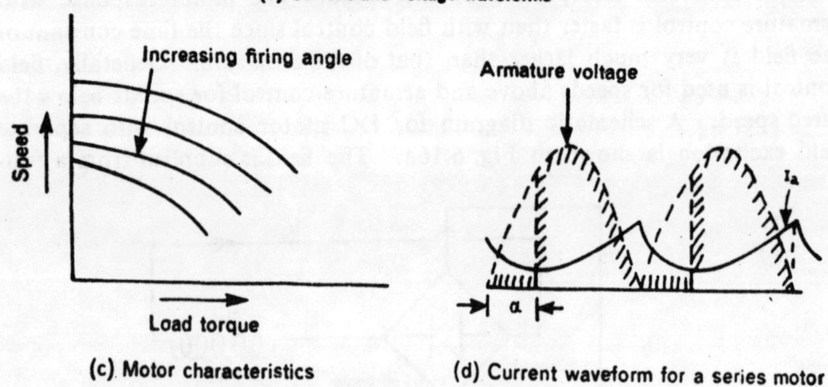

(c) Motor characteristics

(d) Current waveform for a series motor

Fig. 6.16 DC motor control.

continue to flow even after the supply voltage polarity is reversed. D_2 will get forward-biased and the armature current will free-wheel through D_2 and SCR1. The line current will then be zero and so too the voltage across the armature. Since the counter-emf in a separately-excited DC motor is fairly steady, current i_a will quickly come down to zero. If the armature inductance is neglected, the conduction angle will be less than $(\pi - \alpha)$ and free-wheeling action will not occur. A similar operation will take place in the negative half-cycle when SCR2 and D_2 conduct. Differential Eqs. (6.16) describe the operation of the motor during the period the SCRs conduct:

$$e_a = L_a \frac{di_a}{dt} + R_a i_a + M_{af} \omega_m I_f,$$

$$T_d = M_{af} I_f i_a = J \frac{d\omega}{dt} + B\omega_m + T_L, \tag{6.16}$$

where e_a is the applied voltage, R_a is the armature resistance, L_a is the armature inductance, I_f is the field current, T_d is the developed torque, M_{af} is the mutual inductance between the field and armature, ω_m is the motor speed, B is the damping constant, J is the moment of inertia of the rotating parts, and T_L is the load torque. Because of the large inertia of the motor, the ripple in speed is small and is therefore neglected. If each term in Eq. (6.16) for e_a is integrated from α to $(\alpha + \beta)$ and divided by π, the instantaneous voltage, current, and speed will be converted to their respective average values. The field current I_f is assumed to be constant. Thus,

$$\frac{\omega}{\pi}\int_{\alpha/\omega}^{(\alpha+\beta)/\omega} E_m \sin \omega t\, dt = \frac{\omega L_a}{\pi}\int_{\alpha/\omega}^{(\alpha+\beta)/\omega} \frac{di_a}{dt}\, dt + \frac{R_a\omega}{\pi}\int_{\alpha/\omega}^{(\alpha+\beta)/\omega} i_a\, dt + M_{af}\frac{\omega_m I_f}{\pi}\int_{\alpha/\omega}^{(\alpha+\beta)/\omega} dt, \tag{6.17}$$

$$E_{av} = R_a I_{av} + M_{af\,av} I_f \frac{\beta}{\pi},$$

where

$$E_{av} = \frac{E_m}{\pi}\left[\cos \alpha - \cos (\alpha + \beta)\right].$$

The average value of voltage across L_a is zero. Similarly,

$$T_{av} = M_{af} I_{av} I = \beta\omega_{av} + T_L. \tag{6.18}$$

For given values of the firing angle and load torque, the average values of armature current and motor speed can be calculated from Eqs. (6.17) and (6.18) if the value of β is known. For a separately-excited DC motor, the armature inductance is small and can be neglected; in this case the armature current will fall to zero at the instant the counter-emf is equal to the supply voltage. Then,

$$\beta = \pi - \sin^{-1}\left(\frac{E_b}{E_m}\right), \tag{6.19}$$

where E_b is the back-emf and E_m is the peak value of the input AC voltage. Thus, there are three equations, viz., (6.17), (6.18), and (6.19) and three unknowns, namely, I_{av}, ω_{av}, and β. These can be solved by using any iterative method and the characteristics of the motor can be plotted for several firing angles to give the relation of load torque and speed as shown in Fig. 6.16c. The control circuit shown in Fig. 6.16a can also be used for the variable speed operation of a DC series motor. Here, the motor field will be in series with the armature. Therefore, the total circuit inductance will be high and the armature current will become continuous as shown in Fig. 6.16d. Irrespective of the firing angle α, the current will free-wheel through D_2 or D_1 when the supply voltage polarity is reversed. The hatched portion in Fig. 6.16d shows the voltage across the armature. For a series motor, the armature counter-emf will not be steady as in a shunt motor even though the speed fluctuations can be neglected. This will also make the armature current continuous. There are two modes of operation for the circuit. In one mode, SCR1 and D_1 or SCR2 and D_2 will conduct and the supply voltage will appear across the motor. In the second mode, SCR1 and D_2 or SCR2 and D_1 will conduct and the applied voltage to the motor will be zero. The differential equations relating the voltage to the current of the motor are

$$e_a = E_m \sin \omega t = (L_a + L_f)\frac{di_a}{dt} + (R_a + R_f)i_a + M_{af}i_a\omega_m,$$
$$\text{for} \quad \alpha/\omega \leqslant t \leqslant \pi/\omega, \tag{6.20}$$
$$e_a = 0 = (L_a + L_f)\frac{di_a}{dt} + (R_a + R_f)i_a + M_{af}i_a\omega_m,$$
$$\text{for} \quad \pi/\omega \leqslant t \leqslant (\pi + \alpha)/\omega,$$

where L_f and R_f are the inductance and resistance, respectively, of the series field. The developed torque is given by

$$T_d = M_{af}i_a^2 = J\frac{d\omega_m}{dt} + B\omega_m + T_L. \tag{6.21}$$

Equations (6.20) and (6.21) are integrated over one half-cycle and then divided by π to obtain the following relationships for the average values:

$$E_{av} = \frac{E_m}{\pi}(1 + \cos \alpha) = (R_a + R_f)I_{av} + M_{af}I_{av}\omega_{av},$$
$$T_{av} = M_{af}k^2I_{av}^2 = B\omega_{av} + T_L, \tag{6.22}$$

where k is the form factor for the armature current. Since the armature current is continuous for all firing angles and its waveform is as shown in Fig. 6.16d, the form factor can be taken to be unity for all values of α. Then, Eqs. (6.22) can be solved for ω_{av} and I_{av} for given values of α and T_L.

6.8.1 Stability in DC Motor Drives

In a series motor, since the armature counter-emf becomes zero if the SCRs do not conduct, the SCRs can be fired at any instant during the positive half-cycle. But, for a separately-excited motor the counter-emf remains constant as shown in Fig. 6.16b. Let the firing angle initially be close to 180° when the motor speed and counter-emf are low, and assume that the SCRs can be gated by a single sharp pulse in each half-cycle. If the firing angle is gradually decreased for a given load, then both the speed and the counter-emf will increase, and at a particular firing instant, indicated by point C in Fig. 6.16b, the instantaneous value of the AC input voltage will be equal to the counter-emf. If the firing angle is decreased still further, the SCR will be reverse-biased at the time of gating. Thus, the SCR will not turn on, and there will be no voltage applied to the armature. The motor speed will fall as the average developed torque is reduced. If α is maintained constant, then the SCR will turn on only when the counter-emf magnitude falls below the value of the AC input voltage at the instant of gating. When this happens, the SCR will fire, voltage will be applied to the armature, and the motor speed will begin to rise. Thus, the SCRs will be subjected to intermittent firing which results in speed fluctuations. This is known as *instability* in DC motor drives, and the firing angle α_c at which it takes place is known as the *critical firing angle*. Angle α_c is a function of the load and speed of the motor. This intermittent firing of SCRs, and consequent speed fluctuations, can be avoided if the SCR gating pulse has a width of $\pi/2\omega$, or some inductance is included in the armature circuit to maintain continuous armature current. Series motors are not subjected to this instability. However, since the motor circuit is highly inductive, the gating pulses for the SCRs must be of sufficient duration to properly turn them on.

6.8.2 Example

A series motor is supplied from a rectified single-phase supply of RMS voltage 230 V and frequency 50 Hz. The armature and field resistance together equal 2 ohms. The torque constant M_{af} is 0.23 H and the load torque is 20 newton-metres (N-m). Neglecting damping, calculate the average armature current and speed.

The average applied voltage to the motor E_{av} is 207 V. Using Eq. (6.22), we have

$$207 = 2I_{av} + 0.23\,\omega_{av}I_{av},$$

$$20 = 0.23I_{av}^2.$$

Therefore,

$$I_{av} = 9.35 \text{ A},$$

$$\omega_{av} = 122 \text{ rad/sec},$$

$$N_{av} = 1165 \text{ rpm}.$$

6.8.3 Other Control Circuits for DC Drives

Figure 6.17a shows a simple circuit for speed control of DC motors, based on full-wave rectified supply to the armature through an SCR. This control circuit is applicable only to shunt or separately-excited motors where the counter-emf in the armature is steady and the circuit inductance is small, so that the current is discontinuous and does not spill into the next half-cycle. The voltage and current waveforms are given in Fig. 6.17b. The SCR is fired at an angle α (point P). At Q, the input voltage will be equal in magnitude to the counter-emf. However, as the inductance of the motor armature is small, the current will go to zero a little beyond Q and the SCR will be turned off at R. Reverse bias will appear on the SCR during the period from R to S. This period must be longer than the turn-off time of the SCR. It is for this reason that such a circuit with large firing angles may not work properly as the counter-emf will be low and the period from R to S will be shorter than it would be for smaller angles. If the current goes to zero beyond S because of large armature inductance, then the SCR will not turn off and control will be lost. Series motors therefore cannot be controlled in this manner.

The dual converter schemes discussed in Section 6.4 can be used for reversible-speed DC motor drives. Here, the field polarity is maintained constant and the polarity of the armature voltage is changed. If converter 1 (Fig. 6.8) produces a clockwise rotation, then converter 2 will drive the motor in the opposite direction. It is simpler and more convenient to use the noncirculating-type dual converter scheme. When speed reversal is

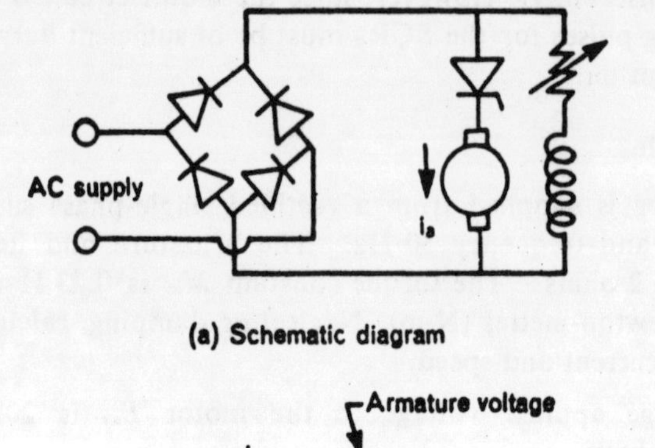

(a) Schematic diagram

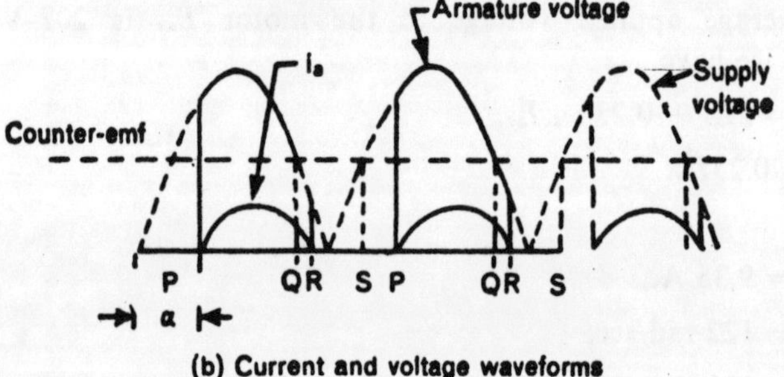

(b) Current and voltage waveforms

Fig. 6.17 Simple circuit for speed control of a DC motor.

desired, converter 2 is gated only after all the SCRs in converter 1 have been properly turned off. Since, due to inertia, the motor will still be running in the same direction as before, the counter-emf in the armature will be in the same direction as the rectified output of converter 2, and hence the armature current will be very high. This will produce a large braking torque which will stop the motor and eventually rotate it in the opposite direction. The firing angle of the second bridge must be controlled so that current rating is not exceeded. This method of braking the motor is called *plugging*.

Another control scheme for DC motors requires that after the gating pulses to converter 1 have been stopped, converter 2 be operated in the inverting mode. For this operation, fully-controlled bridges must he used. The counter-emf in the armature will then drive the current through the second converter and power will be fed back into the AC mains. Thus, the kinetic energy of the motor will be converted into electrical power. This will slow down the motor. When the motor stops, the firing angle of the converter will decrease to a value below $\pi/2$, and the circuit will operate as a rectifier, applying a DC voltage across the armature which will drive the motor in the opposite direction. This braking scheme is known as *regenerative braking*. The high armature current experienced in plugging can be avoided in this scheme. To facilitate the operation of the bridge in the rectifying and inverting modes, a suitable external inductance must be connected in the armature circuit. This will also result in continuous current in the armature.

With a dual converter, and using regenerative braking, the DC motor will be subjected to what is generally known as *four-quadrant operation*. The various operations are: motor running in the forward direction with (a) DC power flowing into the armature (converter 1 operating as rectifier) or (b) DC power flowing out of the armature (converter 2 operating as inverter); and motor running in the reverse direction with (c) DC power flowing into the armature (converter 2 operating as rectifier) or (d) DC power flowing out of the armature (converter 1 operating as inverter). Reversible-drive control is also possible with one converter using the reversal of armature connections by magnetic contactors.

The extra features usually required for DC motor drives are the *inching* and *jogging* operations. In the inching operation, the motor moves over a fraction of a revolution at a time. It requires that the SCR be fired only once. In the jogging operation, the SCRs are fired in each cycle but the firing angle is made sufficiently large so that the torque developed will be only slightly greater than the total load torque. Therefore, the motor will rotate at a very slow speed. For both operations, the armature counter-emf is low and the current high. The feedback circuits discussed in Section 6.5 can be used for automatic adjustment of the firing angle to maintain constant speed or constant torque. To protect the SCRs, it is necessary to provide means for limiting the current. This is done by a comparator which short-circuits the gating pulses applied to the SCRs when the armature current exceeds the permissible value. At the time of starting the motor, the firing angle must be made large and then gradually decreased at a suitable rate so that the current-limiting device does not go into operation. The control circuits for the various operations discussed here can be obtained from the references listed at the end of this chapter.

6.8.4 Simple Speed Control Circuit for DC Motors

The automatic adjustment of the firing angle to keep the speed constant at various loads is obtained by a feedback signal which is proportional to the actual speed of the motor. If the speed goes above the reference speed, the firing angle is advanced, and vice versa. The signal can be obtained from a tacho-generator or from the armature counter-emf. Figure 6.18 shows a very simple half-wave circuit for the speed control of a DC series motor. Such a motor is also known as a *universal motor* since it can run with both AC and DC inputs. Half-rectified AC is used in this circuit for driving the motor. SCR1 is gated during the positive half-cycle. This will apply a positive average voltage across the armature. Resistor R will control the rate of rise of voltage across the capacitor. When this voltage e_C becomes equal to the armature voltage e_a plus the breakdown voltage

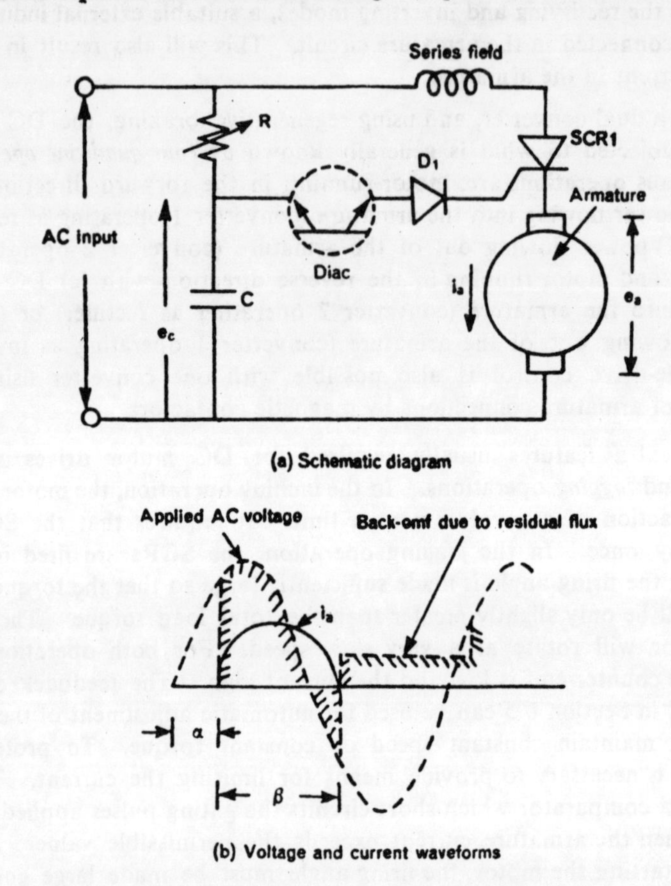

Fig. 6.18 Control for a DC series motor.

V_{BR} of the diac, SCR1 will get triggered. When SCR1 is not conducting, voltage e_a across the armature will be due to the residual flux. This is approximately proportional to the speed. Thus, the firing angle α can be changed by varying resistance R, which in turn will change the speed of the motor. For a given value of R, let the speed fall below the set speed due to the addition of load on the motor. Then, voltage e_a across the armature (after the SCR is turned off) will be low because of the reduction in speed. Therefore, in the next positive half-cycle, voltage e_C across the capacitor will become equal to $(e_a + V_{BR})$ much earlier than before and the SCR will get triggered. That is, the firing angle will be reduced and, because of the increased average voltage across the armature, the motor speed will increase. Similarly, if the motor speed rises because of load throw-off, the firing angle will automatically advance and bring down the speed. Thus, the regulation of speed will improve.

In this simple control scheme using proportional feedback, the speed will not be constant for all loads. For maintaining constant speed, the error in the value of speed, obtained from a tacho-generator, is integrated and the resulting signal is used for controlling the firing angle. The effect of integration is to reduce the static error in the motor speed to zero. A typical control scheme for DC drives is described in Section 6.8.5. This can be used for both series- and separately-excited motors. Speed reversal is obtained either by using dual converters or by armature reversal. A similar scheme can also be used for the AC drives discussed in Section 6.6.

6.8.5 Improved Speed Control Scheme for DC Motors

Figure 6.19a shows the block diagram of the scheme for improved control of speed. The error voltage signal proportional to $(\omega_0 - \omega_m)$, where ω_0 is the reference speed, is passed through a PI (proportional-integral) controller block. This provides a reference signal e_1 for the motor current and the difference between e_1 and the motor current i_a is passed through a second PI block. The first PI controller ensures that the steady-state error in speed will be zero. The second controller provides the current limit. The maximum value of the motor current will be limited by the output voltage e_1 of the first PI controller. Figure 6.19b shows a method of obtaining a proportional and integral value of the input signal using an operational amplifier. The proportional part is used to improve the response time of the controlled system. The integration will result in a nonzero output even when the input goes to zero. This will produce zero static error with step inputs to the control system. Diode D_1 is used for avoiding negative excursions of voltage e_1. The maximum possible value of this voltage can be changed by varying resistance R_1. For details of the operational amplifier circuits, see Graeme and Tobey (1971) under References.

The output voltage e_2 of the second PI block is compared with a ramp to obtain the instant of firing the SCRs, as explained in Section 6.5 (see Fig. 6.9c). If a two-pulse converter is used, the output pulses from the comparator can be used for directly firing the SCRs in the converter. For multiple-pulse converters, the SCRs have to be triggered in a sequence; this is done by the firing controller block. Details of this controller will be given in Chapter 9. For a two-pulse converter supplying variable DC voltage to the motor armature, the variation of the output voltage is given by

$$\Delta e_a = -\frac{2E_m}{\pi} \sin \alpha_0 \Delta\alpha, \tag{6.23}$$

where Δe_a is the change in armature voltage, $\Delta\alpha$ is the change in the firing angle produced by the comparator, and α_0 is the initial firing angle. The transfer function for the converter can be approximated by a gain K $[=(-2E_m/\pi) \sin \alpha_0]$, and a time constant T_1 $(=T/4$, where T is the period of the AC input voltage). Similar transfer functions can be developed for

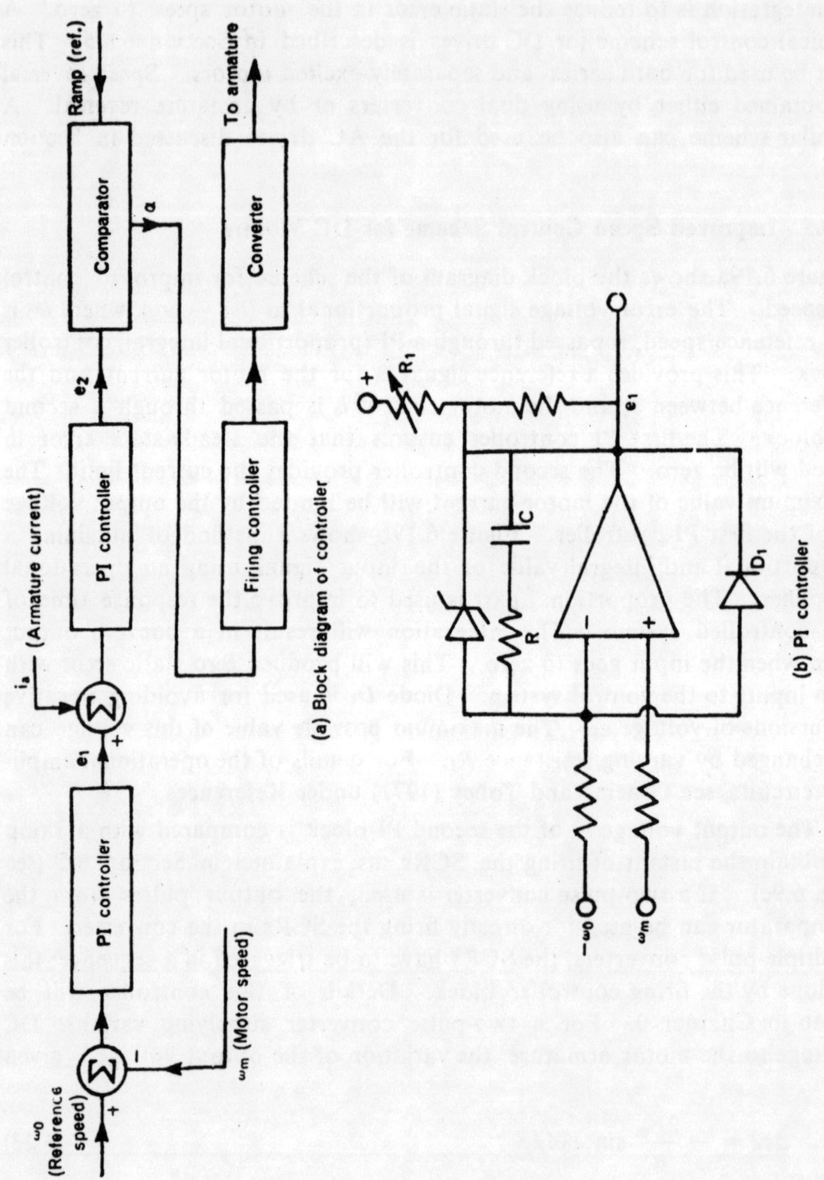

Fig. 6.19 Improved speed control scheme (*cont.*).

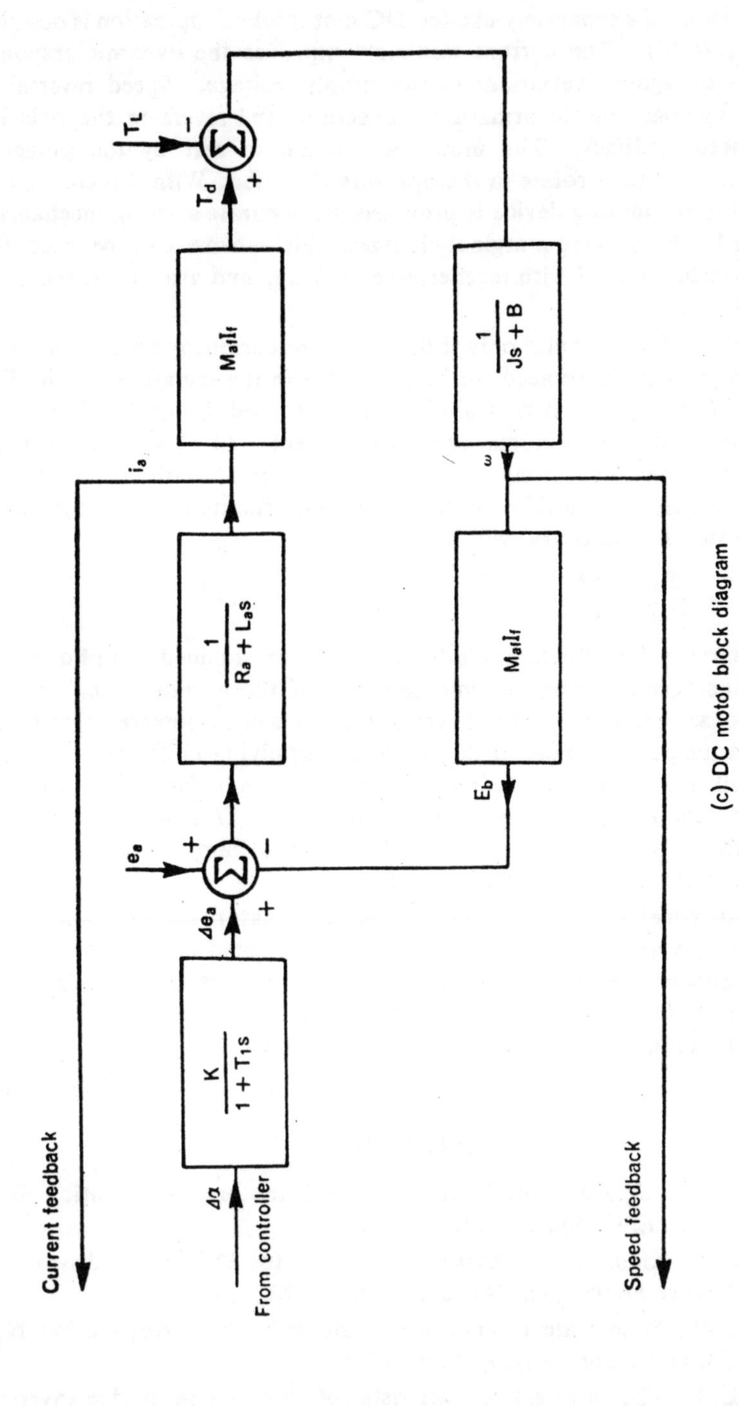

(c) DC motor block diagram

Fig. 6.19 Improved speed control scheme.

three- and six-pulse converters. Figure 6.19c shows the block diagram representation of a separately-excited DC motor whose operation is described by Eq. (6.16). The current feedback improves the dynamic response of the system against variations in the supply voltage. Speed reversal is obtained by changing the armature connections and reversing the polarity of the speed feedback. The motor will come to rest by the plugging action, and will then rotate in the opposite direction. With this controller, since a current-limiting device is provided, no separate starting mechanism is required. By making marginal changes, this scheme can be used for dual converter control with regenerative braking, and also for speed control of AC motors.

Equation (6.23) is valid only if the armature current is continuous. For this, a small inductance needs to be connected in the armature circuit. For a two-pulse bridge, a criterion similar to that used in Eq. (6.9) can be derived to produce continuous armature current. In this case, the peak value of the second harmonic current is made one-fifth of the motor full load current I_M. With this criterion, the total armature circuit inductance L_e (expressed in millihenry) will be

$$L_e = \frac{4E_m \times 10^3}{0.6\pi\omega I_M}. \tag{6.24}$$

The required value of external inductance can be obtained from Eq. (6.24). This inductance increases the time constant of the armature and thereby the response time of the control system too. It will be observed that an increase in the pulse number of the converter supplying DC to the armature will result in a decrease in the required value of inductance L_e. Further, time constant T_1 for the six-pulse converter will be $T/12$, which is one-third of that for a two-pulse converter. Thus, the dynamic response with a multiple-pulse converter will be better.

Another method of obtaining continuous armature current (and thereby reducing torque pulsations) for a separately-excited motor is by employing multiple-chopping, using forced commutation in each half-cycle. Forced commutation methods will be explained in Chapter 8 and the performance of choppers will be discussed in Chapter 10.

REFERENCES

Adem, A. A., Speed controls for universal motors, Application Note 200.47, G. E. Company, Auburn, USA.

Davis, J. A., Kidd, A. C., Thyristor converter for DC motor drives, IEE Conference Publication, No. 53, 1969, p. 472.

Fox, R. W., Solid state incandescent lighting control, Application Note 200.53, G. E. Company, Auburn, USA.

Freris, L. L., The universal characteristic of three-phase bridge inverters, *Direct Current* (UK), 1961, p. 198.

Graeme, J. G., Tobey, G. E., Operational Amplifiers—Design and Applications, McGraw-Hill, New York, 1971.

Gutzwiller, F. W., An all solid state phase controlled rectifier system, AIEE Paper No. 217, IEEE Summer Power Meeting, 1959.

Ilango, B., Ramamoorty, M., Digital algorithms for direct steady state solution of thyristor controlled single-phase induction motors, *Proceedings of Computer Society of India*, 1971, p. 239.

Krishnan, T., Ramaswami, B., A fast response DC motor speed control system, *IEEE Trans.* (IA), 1974, p. 643.

Ramamoorty, M., Bril, J. M., Reversible drive control for elevator doors, Parts I and II, *IEEE Trans.* (IECI), 1975, p. 19.

Ramamoorty, M., Ilango, B., The transient response of a thyristor controlled series motor, *IEEE Trans.* (PAS), 1971, p. 289.

Ramamoorty, M., Ilango, B., A new approach to the steady-state analysis of single pulse fired thyristor controlled DC motors, *JIE* (India), 1972, p. 205.

Ramamoorty, M., Ilango, B., Steady state analysis of thyristor controlled three-phase induction motors, *IEEE Trans.* (PAS), 1974, p. 1165.

Rayworth, G., Variable phase SCR triggering, IEE Conference Publication, No. 17, 1965, p. 121.

Read, J. C., The calculation of rectifier and inverter performance characteristics, *Proc. IEE*, 1945, p. 495.

Rice, J. B., Nickels, L. E., Commutation *dv/dt* effect in thyristor 3 phase bridge converters, *IEEE Trans.* (IGA), 1968, p. 665.

Shepherd, W., Steady state analysis of the series *R-L* circuit controlled by silicon controlled rectifiers, *IEEE Trans.* (IGA), 1965, p. 259.

Shepherd, W., On the analysis of the three-phase induction motor with voltage control by thyristor switches, *IEEE Trans.* (IGA), 1968, p. 304.

7

Line-Commutated Converters and Inverters

7.1 LINE-COMMUTATED CIRCUITS

The method by which AC supply voltage is used for the commutation of conducting SCRs has been discussed in Chapter 6; it is very convenient for both rectification and inversion in phase-controlled bridge circuits and mid-point-connection circuits. This method of commutation is referred to as class F type of commutation (see Chapter 8), and is also known as line commutation. All types of AC to variable voltage DC converters used for motor control and regulated power supplies, in both of which the AC input current is made continuous by a large reactor on the DC side, make use of line commutation without any external commutating components. The input-output characteristics of such line-commutated converters will be considered in detail in this chapter. The effect of source impedance of various types (*R*, *L*, or *C*) on the output voltage, and the minimum value of inductance required on the DC side to provide continuous load current for both active and passive loads will also be discussed.

Line commutation can also be used for AC-to-AC frequency conversion, that is, when the AC input at one frequency is to be transformed to AC output at another frequency. Such a frequency conversion takes place directly without an intermediate DC stage such as that required for forced-commutated circuits (Chapters 8 and 9). The operating principle and performance of AC-to-AC frequency converters will also be covered in this chapter. Another important application of line commutation is in high-voltage DC power transmission, which will also be discussed here.

7.1.1 Input-Output Characteristics of Bridge Circuits

The schematic diagrams of single-phase and three-phase fully-controlled bridge circuits used for variable voltage DC output are shown in Fig. 7.1a. The unfiltered output voltage waveform e_o is shown in Fig. 7.1b. It is convenient to assume that the output current is constant and to neglect source impedance. The waveforms for the current through one SCR and the average output voltage e_d for single-phase and three-phase circuits are shown in Fig. 7.1b. These waveforms are for rectifier operation with a firing angle α. The operation of bridge circuits and the required firing sequence have been explained in Chapter 6. The output voltage and input current waveforms will be analysed here for various operating conditions. A similar analysis can

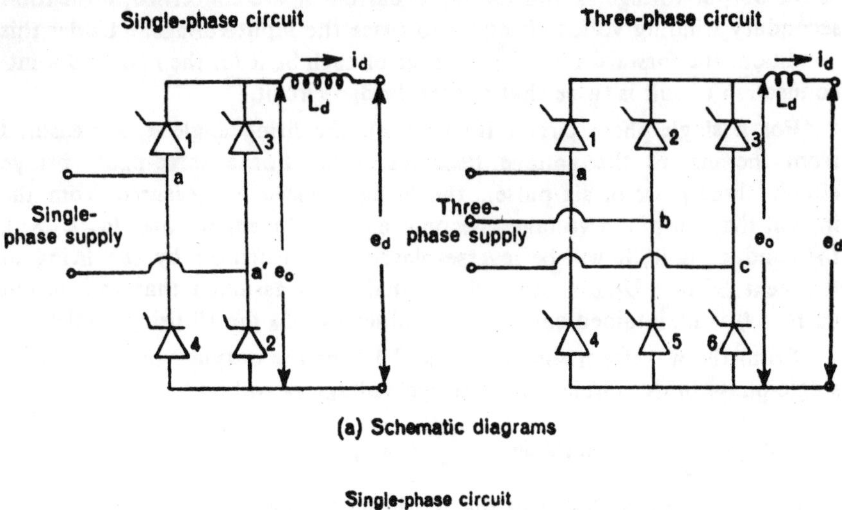

(a) Schematic diagrams

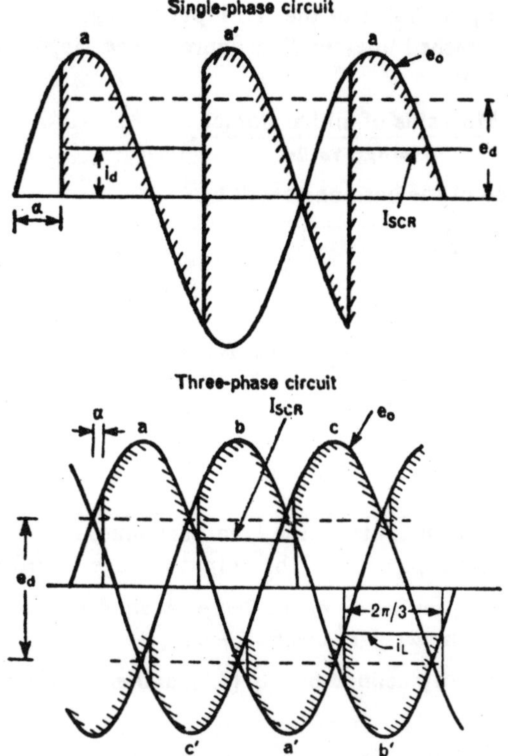

(b) Voltage and current waveforms

Fig. 7.1 Controlled rectifiers.

also be made for circuits of the other configuration, that is, for circuits employing a midpoint connection. In fact, the M-2 midpoint-connection circuit and the single-phase fully-controlled bridge are identical in so far as the output voltage e_d and the input current i_L are concerned, if the total secondary winding voltage is equal to twice the input voltage. Under this condition, the forward blocking voltage of each SCR for the M-2 midpoint-connection circuit is twice that for the bridge circuit.

For a single-phase circuit (two-pulse), the firing angle α is measured from the instant the voltage becomes zero. For a three-phase bridge circuit (three-pulse or six-pulse), the firing angle α is measured from the instant the two phase voltages become equal. In either case, for $\alpha > 0$, the conducting SCR will be reverse-biased and turned off by the firing of the next SCR. Unless otherwise stated, it is assumed that the output current i_d is maintained constant by inductance L_d for all firing angles.

From the waveform shown in Fig. 7.1b for the output voltage e_o of a single-phase bridge circuit, the average voltage e_d will be

$$e_d = \frac{1}{\pi}\int_{\alpha}^{\pi+\alpha} E_m \sin\theta \, d\theta = \frac{2E_m}{\pi}\cos\alpha, \tag{7.1}$$

where E_m is the peak value of the AC input voltage. This average voltage is shown by the dashed lines in this figure. The ripple factor RF in the output is given by

$$\text{RF} = \frac{\text{RMS value of all harmonics}}{\text{average value}}. \tag{7.2}$$

The RMS value of the harmonics will be

$$I_{h\,\text{RMS}} = \sqrt{e^2_{\text{RMS (total)}} - e_d^2}, \tag{7.3}$$

where

$$e_{\text{RMS (total)}} = \sqrt{\left(\int_{\alpha}^{\pi+\alpha} E_m^2 \sin^2\theta \, d\theta\right)\Big/\pi} = E_m/\sqrt{2}.$$

Therefore,

$$\text{RF} = \frac{\pi}{2}\sqrt{\tfrac{1}{2}\sec^2\alpha - 4/\pi^2} \tag{7.4}$$

The output waveform contains the even harmonics of the input frequency. The Fourier series expansion of the output voltage waveform is given by

$$e_o = e_d + e_2 \sin 2\omega t + e_2' \cos 2\omega t + e_4 \sin 4\omega t + e_4' \cos 4\omega t + e_6 \sin 6\omega t + e_6' \cos 6\omega t + \ldots, \tag{7.5}$$

where e_d is as already defined by Eq. (7.1), and e_m, e_m' (for $m = 2, 4, 6, \ldots$) are

$$e_m = \frac{2E_m}{\pi}\left[\frac{\sin(m+1)\alpha}{m+1} - \frac{\sin(m-1)\alpha}{m-1}\right],$$

$$e_m' = \frac{2E_m}{\pi}\left[\frac{\cos(m+1)\alpha}{m+1} - \frac{\cos(m-1)\alpha}{m-1}\right].$$

The input current waveform is rectangular with an amplitude equal to I_d. This can be expressed as

$$i_L(t) = \frac{4}{\pi} I_d [\sin(\omega t - \alpha) + \tfrac{1}{3}\sin 3(\omega t - \alpha) + \tfrac{1}{5}\sin 5(\omega t - \alpha) + \ldots]. \tag{7.6}$$

The fundamental input power factor angle is equal to the firing angle α. Therefore, the fundamental power input is obtained as

$$\begin{aligned} \text{active power} &= \frac{4}{2\pi} I_d E_m \cos\alpha, \\ \text{reactive power} &= \frac{4}{2\pi} I_d E_m \sin\alpha. \end{aligned} \tag{7.7}$$

There will be additional flow of reactive power due to the presence of harmonics in the line current. Assuming no losses in the bridge circuit and in the DC reactor, the output power given by $e_d I_d$ will be equal to the active power input given by Eq. (7.7); this can be verified by substitution from Eq. (7.1) for e_d. As mentioned in Chapter 6, the fully-controlled bridge will operate at a low input power factor for large firing angles, as can be seen from Eq. (7.7). If the single-phase bridge is half-controlled (SCRs 2 and 4 in Fig. 7.1a replaced by diodes), the output voltage and input current will be zero for the period from π to $(\pi + \alpha)$. The voltage and current waveforms for this circuit are shown in Fig. 7.2a. An analysis similar to that for fully-controlled circuits can be performed for the harmonics in the current and output voltage waveforms of half-controlled circuits. Figure 7.2b shows the relation of active power input to the fundamental reactive power for fully-controlled and half-controlled circuits. For both, the output current i_d is assumed to be steady. The maximum fundamental reactive power input for the half-controlled circuit is seen to be one-half of that for a fully-controlled circuit.

For the three-phase bridge circuit shown in Fig. 7.1a, waveforms for the output voltage e_o and input current i_L are given in Fig. 7.1b. The average output voltage e_d is

$$\begin{aligned} e_d &= 2\left\{\frac{3}{2\pi}\int_{-\pi/3+\alpha}^{\pi/3+\alpha} E_m \cos\theta \, d\theta\right\} \\ &= \frac{3\sqrt{3}}{\pi} E_m \cos\alpha, \end{aligned} \tag{7.8}$$

where E_m is the peak value of input phase-to-neutral voltage. The multiplying factor 2 appears in the foregoing equation because the positive and negative portions of the output voltage are identical. If the circuit is half-controlled (SCRs 4, 5, and 6 in Fig. 7.1a replaced by diodes), then the firing angle for the lower half of the bridge (diodes) will be zero and the total output voltage will be

$$e_d = \frac{3\sqrt{3}}{2\pi} E_m(1 + \cos\alpha). \tag{7.9}$$

In a similar manner, the output voltage for a three-phase half-wave

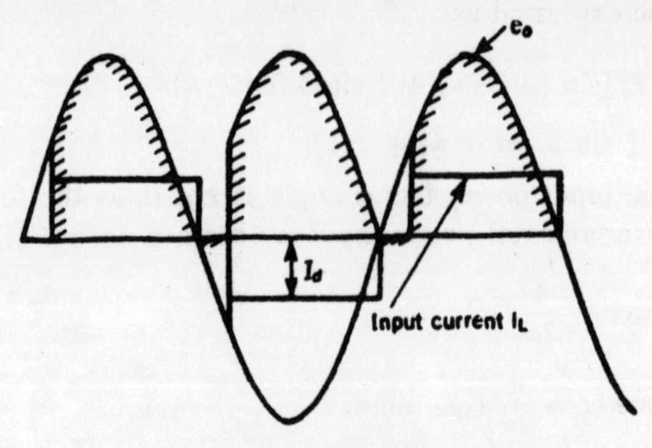

(a) Half-controlled single-phase bridge

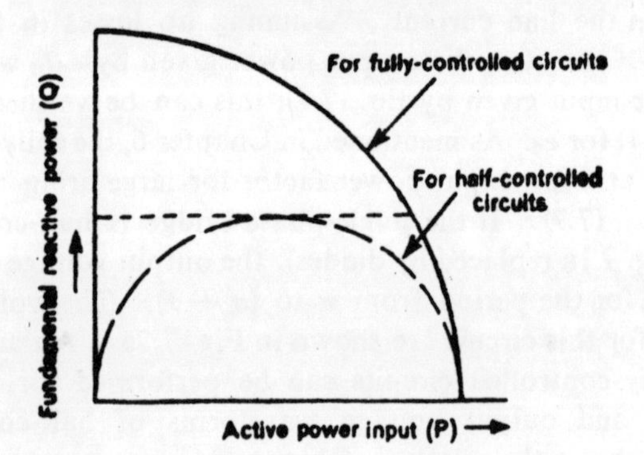

(b) Relation of active power input to fundamental reactive power

Fig. 7.2 Performance of line-commutated circuits (*cont.*).

controlled circuit (also for M-3 connection) is given by

$$e_d = \frac{3\sqrt{3}\ E_m}{2\pi} \cos \alpha. \tag{7.10}$$

The input current waveform in each phase for a fully-controlled three-phase (six-pulse) converter will be rectangular with a pulse width of $2\pi/3$ (Fig. 7.1b). This can be expressed as

$$i_L(t) = \frac{2\sqrt{3}}{\pi} I_d [\sin (\omega t - \alpha) - \tfrac{1}{5} \sin 5 (\omega t - \alpha) - \tfrac{1}{7} \sin 7 (\omega t - \alpha) - \ldots]. \tag{7.11}$$

The fundamental power factor angle will again be equal to α. If the three-phase bridge is half-controlled, two different modes of operation [i.e., with $\alpha < \pi/3$ (mode 1) and $\alpha > \pi/3$ (mode 2)] will take place. If the firing angle is less than $\pi/3$, then the current will switch from one SCR to the other at the instant of firing. The positive half-cycle of the line current will be displaced with respect to the negative half-cycle by the firing angle α. Figure 7.2c shows the voltage and current waveforms for this case. The duration of conduction of current in the positive and negative half-cycles will be $2\pi/3$. The average value of the input current will be zero. When the firing angle is more than $\pi/3$, then the phase voltage of the conducting SCR will be more negative than the negative terminal of

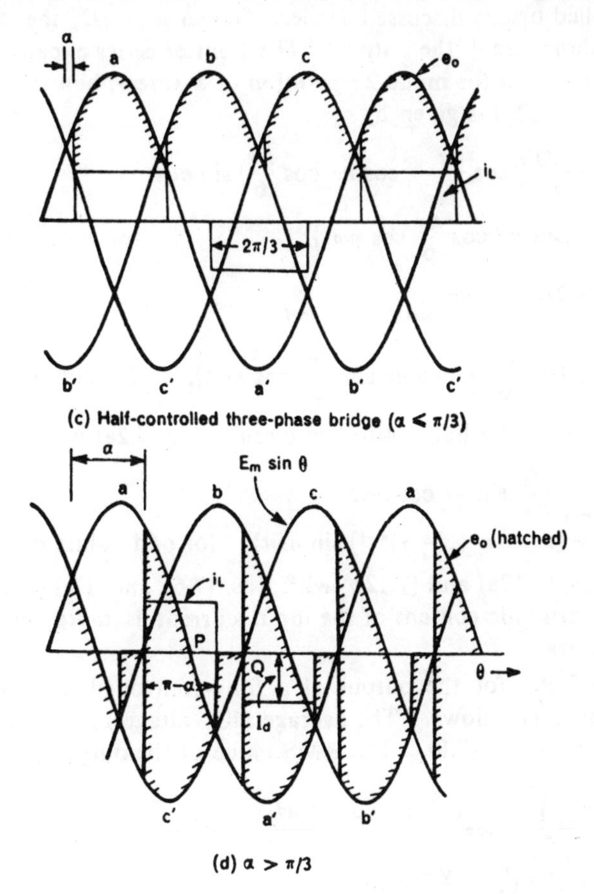

(c) Half-controlled three-phase bridge ($\alpha < \pi/3$)

(d) $\alpha > \pi/3$

Fig. 7.2 Performance of line-commutated circuits.

the DC output, and so free-wheeling action will take place through the conducting SCR and its associated diode (1 and 4 in Fig. 7.1a). The load voltage e_o and the input current i_L will be zero for this period (PQ in Fig. 7.2d). The free-wheeling action will stop when the next SCR is fired. The period during which the line current flows will be $(\pi - \alpha)$ and the average value of the current will be zero. The harmonic content in the input current will be more than that in the input current for mode-1 operation or that in the input current for the fully-controlled circuit, as no free-wheeling action through the bridge can take place. The average output voltage for the half-controlled circuit will be maximum at $\alpha = 0$ and will reduce to zero at $\alpha = \pi$ [see Eq. (7.9)], i.e., the half-controlled bridge will not operate as an inverter since the polarity of the DC output voltage cannot reverse.

Similarly, a fully-controlled bridge will not work satisfactorily if an external free-wheeling diode is connected across the bridge terminals even for a firing angle between $\pi/3$ and $\pi/2$. Free-wheeling action will not take place for firing angles less than $\pi/3$ and the operation will be similar to that for a fully-controlled bridge discussed earlier. When $\alpha > \pi/2$, the diode will permanently short-circuit the output. The Fourier series expansion of the line current waveform for mode-2 operation of a three-phase half-controlled rectifier (Fig. 7.2d) is given by

$$i_L(t) = \frac{-2I_d}{n\pi}\{[\cos\frac{n\pi}{6} - \cos n\alpha \cos\frac{n\pi}{6}]\sin n\omega t + \sin n\alpha \cos\frac{n\pi}{6}\cos n\omega t\}, \quad \text{for even values of } n,$$

$$i_L(t) = \frac{-2I_d}{n}\{\sin\frac{n\pi}{6}\sin n\alpha \sin n\omega t + [\sin\frac{n\pi}{6} + \cos n\alpha \sin\frac{n\pi}{6}]\cos n\omega t\}, \quad \text{for odd values of } n, \tag{7.12a}$$

and that of a two-pulse half-controlled circuit (Fig. 7.2a) is

$$i_L(t) = \sum_{n=1}^{\infty}\frac{2I_d}{n\pi}\{\sin n\alpha \cos n\omega t + [\cos n\alpha + (-1)^{n-1}]\sin n\omega t\}, \quad \text{for odd values of } n. \tag{7.12b}$$

Comparing Eqs. (7.12a) and (7.12b) with Eqs. (7.6) and (7.11), it can be seen that the harmonic content of the input current is more for half-controlled converters.

The ripple factor for the output of a fully-controlled six-pulse bridge can be computed as follows. The average DC voltage e_d for a given firing angle α is given by Eq. (7.8). The RMS value of the output is given by

$$e^2_{o\,RMS} = \frac{3}{\pi}\int_{-\pi/6+\alpha}^{\pi/6+\alpha}(\sqrt{3}E_m\cos\theta)^2\,d\theta$$

$$= \frac{9E_m^2}{2\pi}[\frac{\pi}{3} + \frac{\sqrt{3}}{2}\cos 2\alpha]. \tag{7.13}$$

Therefore,

$$\text{RF} = \frac{\sqrt{e^2_{o\,RMS} - e^2_d}}{e_d} = \frac{\sqrt{\frac{\pi}{2}[\frac{\pi}{3} + \frac{\sqrt{3}}{2}\cos 2\alpha] - 3\cos^2\alpha}}{\sqrt{3}\cos\alpha} \tag{7.14}$$

For a three-phase converter (half-wave), we have

$$e^2_{o\,RMS} = \frac{3}{2\pi}\frac{E_m^2}{2}(\frac{2\pi}{3} + \frac{\sqrt{3}}{2}\cos 2\alpha),$$

$$e_d = \frac{3\sqrt{3}}{2\pi}E_m\cos\alpha, \tag{7.15}$$

from which the ripple factor can be calculated. A similar analysis can be made for the other half-controlled circuits. It is clear that RF will be minimum (i.e., 0.076) for a fully-controlled bridge with zero firing angle.

The harmonic content of the output voltage e_o (Fig. 7.1b) of a fully-controlled three-phase bridge can be obtained as follows. The bridge is assumed to be made up of two half-wave circuits operating at the same firing angle, one producing positive DC voltage $e_d/2$ and the other producing negative voltage $-e_d/2$ with respect to the common neutral, where e_d is the total DC output of the fully-controlled bridge. The harmonic amplitudes of the output voltage of the positive half-wave circuit are obtained first. Then, the negative harmonics are obtained by displacing each harmonic component by $n\pi/3$ (where n is the order of the harmonic) and reversing the sign of the amplitudes. In the output of each half-wave circuit it can be observed from waveform symmetry that only the triplen harmonics of the input frequency are present. Therefore, n will take the values 3, 6, 9, 12, ... The voltage due to the positive half-wave circuit will then be

$$e_{o\,+ve} = \frac{3\sqrt{3}E_m}{2\pi}\cos\alpha + \Sigma[e_n \cos n\omega t + e_n' \sin n\omega t], \tag{7.16}$$

where

$$e_n = \frac{3}{\pi}\int_{-\pi/3+\alpha}^{\pi/3+\alpha} E_m \cos\beta \cos n\beta \, d\beta$$

$$= \frac{3\sqrt{3}\,E_m}{2\pi}(-1)^{n-1}\left[\frac{\cos(n-1)\alpha}{n-1} - \frac{\cos(n+1)\alpha}{n+1}\right],$$

$$e_n' = \frac{3}{\pi}\int_{-\pi/3+\alpha}^{\pi/3+\alpha} E_m \cos\beta \sin n\beta \, d\beta$$

$$= \frac{3\sqrt{3}\,E_m}{2\pi}(-1)^{n-1}\left[\frac{\sin(n-1)\alpha}{n-1} - \frac{\sin(n+1)\alpha}{n+1}\right].$$

The corresponding output voltage waveform for the negative half-wave circuit will be

$$e_{o\,-ve} = \frac{-3\sqrt{3}\,E_m}{2\pi}\cos\alpha - \Sigma[e_n \cos n(\omega t - \pi/3) + e_n' \sin n(\omega t - \pi/3)], \tag{7.17}$$

where e_n and e_n' are as already defined. The total output voltage will be

$$e_o = e_{o\,+ve} - e_{o\,-ve}$$

$$= \frac{3\sqrt{3}E_m}{\pi}\cos\alpha + 2\Sigma[e_m \cos m\omega t + e_m' \sin m\omega t], \tag{7.18}$$

where $m = 2n$, and n takes the values 3, 6, 9, 12, ... Thus, the lowest order harmonic in the output of a three-pulse converter will be 3, and that in a six-pulse converter will be 6. Therefore, the size of the filter required for obtaining smooth DC at the output will be very much reduced for a fully-controlled bridge. If the three-phase bridge is half-controlled, then $e_{o\,+ve}$ can be obtained from Eq. (7.16) and $e_{o\,-ve}$ from Eq. (7.17) by setting $\alpha = 0$. The total output voltage is then obtained by adding e_{o+ve} and $-e_{o-ve}$. However, in this case the cancellation of the third and ninth harmonics will not take place as in Eq. (7.18).

From the foregoing discussion it can be inferred that as the pulse number for the converter increases, the ripple in the DC output voltage and the harmonic content in the input line current are reduced. Also, that for a given AC input voltage and firing angle, the average DC output voltage increases with the pulse number.

7.1.2 Example

(a) A six-pulse fully-controlled converter is connected to three-phase AC supply of 400 V and 50 Hz. It operates with a constant output current of 8 A and a firing angle $\alpha = \pi/4$. Calculate the input current, the harmonic amplitudes of the output voltage e_o, and the RF of the unfiltered output.

Using Eq. (7.11), the input current $i_L(t)$ will be

$$i_L(t) = 8.83 \sin (\omega t - \pi/4) - 1.766 \sin 5(\omega t - \pi/4) - 1.26 \sin 7(\omega t - \pi/4) - \dots,$$

and using Eq. (7.18), we have

$$e_o(t) = 383 + 133.5 \sin (6\omega t + \alpha_6) + 64.3 \sin (12\omega t + \alpha_{12}) + \dots,$$

where

$$\alpha_6 = \pi + \tan^{-1} 6,$$

$$\alpha_{12} = \pi - \tan^{-1} 0.084.$$

From Eqs. (7.2) and (7.3), the ripple factor RF will be

$$\text{RF} = \frac{\sqrt{133.5^2 + 64.3^2}}{\sqrt{2} \times 383} = 0.306.$$

On examination, Eq. (7.14) shows that RF increases rapidly with the firing angle α.

(b) Calculate the average output voltage of a three-phase half-controlled bridge operating with a firing angle of $\pi/2$ and connected to three-phase AC supply of 400 V and 50 Hz. The load current i_d is assumed to be continuous.

Referring to Fig. 7.2d, the average output voltage will be

$$e_d = \frac{3E_m}{2\pi} \int_{2\pi/3}^{\pi+\pi/6} [\sin \theta - \sin (\theta + 2\pi/3)]\, d\theta$$

$$= \frac{3\sqrt{3}}{2\pi} E_m = \frac{3\sqrt{3}}{2\pi} \times \frac{400}{\sqrt{3}} \times \sqrt{2} = \frac{1200}{\sqrt{2}\,\pi} = 270 \text{ V}.$$

Reference may also be made to Eq. (7.9), which gives the output voltage of a three-phase half-controlled circuit for any firing angle.

7.2 EFFECT OF SOURCE IMPEDANCE

For the two bridge circuits shown in Fig. 7.1a, it is assumed that the commutation of current from one SCR to another will take place instantaneously. That is, as soon as SCRs 3 and 4 (in the case of single-phase bridge) are fired, the conducting SCRs 1 and 2 will turn off due to the application of reverse voltage, and the current will shift to SCRs 3 and 4. This is possible only when the source has no internal impedance. The effect of source inductance is to delay the commutation of current from one pair of SCRs to the other. Figure 7.3a shows the input current waveform (including the effect of source impedance) for a single-phase bridge circuit. The output current is assumed to be constant due to the presence

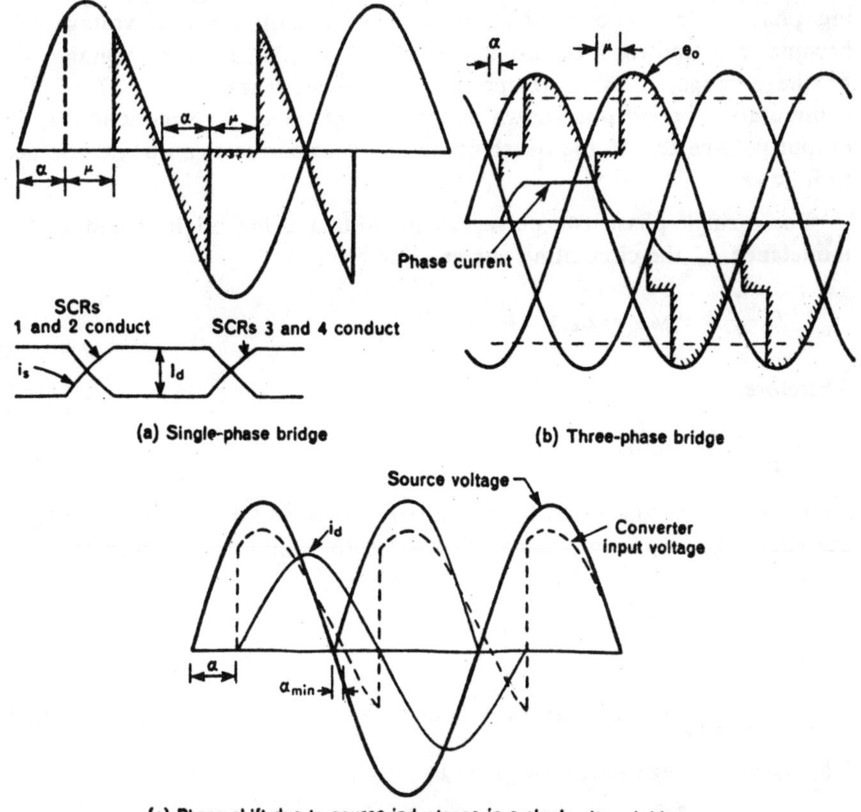

Fig. 7.3 Effect of source impedance.

of a large DC reactor. If the source impedance is resistive, then the commutation will be complete when the circulating current, following the firing of the next pair of SCRs, is equal to load current I_d. This circulating current is equal to e_{input}/r_s for a single-phase circuit, and $(e_b - e_a)/(2r_s)$ for a three-phase bridge, where r_s is the source resistance per phase and $(e_b - e_a)$ is the line voltage between the phases undergoing commutation. Since the source resistance is usually small, it can be assumed that the duration of commutation will also be very small and can be neglected, and that the current transfer will take place immediately after the SCRs are fired. However, the source resistance will produce a constant voltage drop $I_d r_s$ in the DC voltage, and therefore this must be subtracted from the average output voltage e_d obtained by using Eq. (7.1). For three-phase fully-controlled and half-controlled bridges, the corresponding drop will be $2I_d r_s$.

If the source impedance is purely inductive, then the commutation period will be μ and the phase current waveform will be as shown in Fig. 7.3a. The incoming and outgoing SCRs will conduct together, and during the commutation period the output voltage will be the average of the conducting phases. In the case of a single-phase circuit, the load voltage will become zero, and for a three-phase circuit (Fig. 7.3b) the load voltage will be the average of the corresponding phase voltages $[(e_a + e_b)/2]$. The commutation period μ is known as the *overlap angle*. The reduction in the output voltage due to the overlapping of phase currents can be determined as follows.

For a single-phase two-pulse circuit with a firing angle α and source inductance L_s, the circulating current i_s is given by

$$L_s \frac{di_s}{dt} = e_{\text{phase}} = E_m \sin(\omega t + \alpha).$$

Therefore,

$$\frac{di_s}{dt} = \frac{E_m}{L_s} \sin(\omega t + \alpha). \tag{7.19}$$

Commutation is completed by the time $t = \mu/\omega$, and during this period the current in the incoming phase will build up to I_d and in the outgoing phase it will become zero (see Fig. 7.3a). Therefore,

$$I_d = \frac{E_m}{L_s} \int_0^{\mu/\omega} \sin(\omega t + \alpha)\, dt$$

$$= \frac{E_m}{\omega L_s} [\cos \alpha - \cos(\alpha + \mu)]. \tag{7.20}$$

The output voltage e_d is now given by

$$e_d = \frac{1}{\pi} \int_{\mu+\alpha}^{\pi+\alpha} E_m \sin \theta \, d\theta$$

$$= \frac{E_m}{\pi} [\cos \alpha + \cos(\mu + \alpha)]. \tag{7.21}$$

Using Eqs. (7.20) and (7.21), we get

$$e_d = \frac{2E_m}{\pi} \cos \alpha - \frac{\omega L_s}{\pi} I_d. \tag{7.22}$$

Thus, it can be observed that the effect of source impedance is to produce a voltage drop in the average DC output voltage by an amount $(r_s + \omega L_s/\pi)I_d$. This is the equivalent source impedance drop referred to the DC side. It can be similarly shown that the equivalent voltage drop for a three-phase bridge will be $(2r_s + 3\omega L_s/\pi)I_d$. For a purely reactive source impedance, the average DC voltage for a given firing angle α will be

$$e_d = e_{do} - \frac{3\omega L_s}{\pi} I_d, \tag{7.23}$$

where

$$e_{do} = \frac{3\sqrt{3}E_m}{\pi} \cos \alpha, \quad \text{for a fully-controlled bridge (Fig. 7.3b)}$$

$$= \frac{3\sqrt{3}E_m}{2\pi} (1 + \cos \alpha), \quad \text{for a half-controlled bridge}$$

$$= \frac{3\sqrt{3}E_m}{2\pi} \cos \alpha, \quad \text{for a half-wave three-phase circuit.}$$

The term e_{do} is called the *internal voltage of the rectifier*. The unfiltered voltage e_o and the alternating phase currents for a fully-controlled three-phase bridge are shown in Fig. 7.3b. The average voltage e_d can also be written as

$$e_d = \frac{3\sqrt{3}E_m}{\pi} \cos (\alpha + \mu) + \frac{3\omega L_s}{\pi} I_d, \tag{7.24}$$

which is similar to Eq. (7.22) derived for a two-pulse converter.

If the load current is not steady, another effect of source inductance will be a phase-shift in the voltages appearing at the input terminals of the converter. Figure 7.3c shows the source voltage (solid line) and the converter input voltage (dashed line) for a single-phase bridge. Because of the phase-shift produced, the incoming SCR will not be forward-biased immediately after the source voltage reverses its polarity. This determines α_{min} (minimum firing angle) for the converter. In Fig. 7.3c, the load current has been assumed continuous and zero at the instant of commutation so that no overlapping effect is produced because of source inductance.

7.2.1 Example

A three-phase six-pulse fully-controlled converter is connected to three-phase AC supply of 400 V and 50 Hz, and operates with a firing angle $\alpha = \pi/4$. The load current is maintained constant at 10 A and the load voltage is 360 V. Calculate the load resistance, the source inductance, and the overlap angle.

From Eq. (7.23), we get

$$360 = \frac{3\sqrt{3} \times 400\sqrt{2}/\sqrt{3}}{\pi} \times \frac{1}{\sqrt{2}} - \frac{3\omega L_s \times 10}{\pi}.$$

Therefore, L_s, the source inductance per phase, will be 7.3 mH and R_d, the load resistance, will be 360/10 = 36 Ω. From Eq. (7.24), for six-pulse operation, we have

$$e_d = \frac{3\sqrt{3}E_m}{\pi}\cos(\alpha+\mu) + \frac{3\omega L_s}{\pi} I_d.$$

This gives

$$\cos(\alpha+\mu) = 0.63,$$

$$\mu = 6°.$$

7.2.2 Effect of Load Inductance

The characteristics of single-phase and three-phase bridge circuits discussed in Sections 7.2 and 7.2.1 are obtained by assuming that the load current is constant. This assumption will be valid if the value of inductance L_d of the DC reactor on the load side is made high. For converters used as regulated DC power supplies, an output filter is required to reduce the ripple in the direct current and voltage of the load. Inductance L_d is then made reasonably large to act as the filter choke. For such applications, the assumption of constant-current operation of the bridge circuit is valid. But, there are other applications of controlled rectifiers in which the output voltage e_o is not filtered and only the rectified voltage is used (e.g., battery chargers and speed controllers for DC motors). Under this condition, the load current will not be constant. Figures 7.4a and 7.4b show the output voltage and

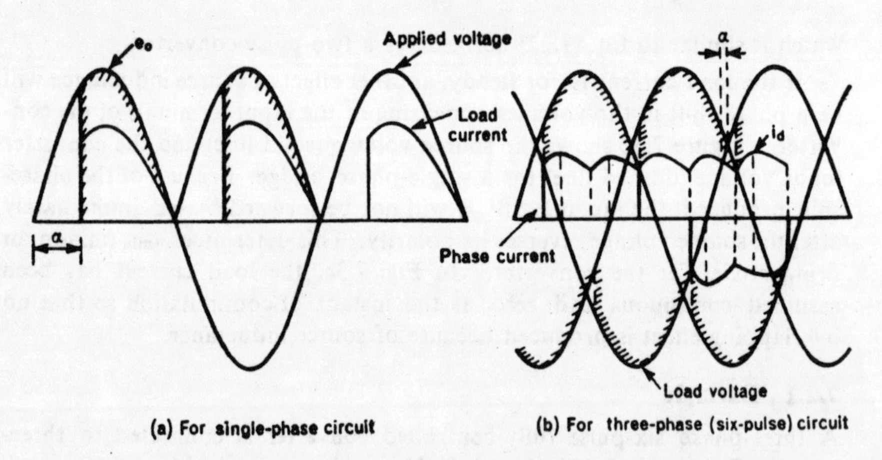

Fig. 7.4 Converter operation when DC inductance is zero (*cont.*).

current waveforms when L_d is zero and the load is resistive. For a single-phase circuit, the load current will be discontinuous since the SCRs will turn off by natural commutation at the end of every half-cycle (neglecting the effect of source inductance). For a three-phase half-controlled circuit, there will be no discontinuity in the load currents if the firing angle is less than $\pi/3$. The alternating current input will have a higher harmonic content than when L_d is large. For a three-phase fully-controlled bridge, the load current will be continuous if the firing angle is less than $\pi/3$ as shown in Fig. 7.4b. The SCRs will not turn on if the firing angle is made larger than $\pi/3$ unless each SCR is continuously gated for a period of $2\pi/3$. Since the load current waveform closely follows the input voltage, the calculation of the overlap angle μ due to source inductance will be very cumbersome. If the DC load is active (e.g., in a battery being charged and a DC motor with counter-emf), the load current waveform will be decided not only by the firing angle but also by the opposing voltage in the load. Figure 7.4c shows the waveforms of output voltage e_o and load current i_d for a single-

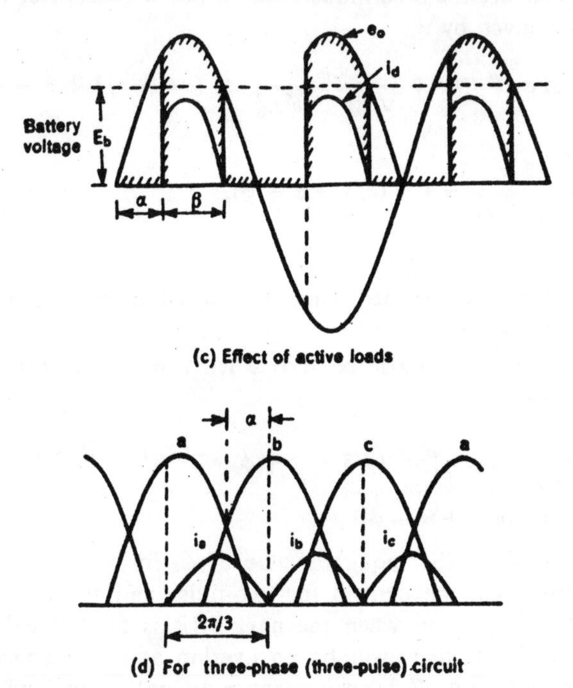

(c) Effect of active loads

(d) For three-phase (three-pulse) circuit

Fig. 7.4 Converter operation when DC inductance is zero.

phase fully-controlled circuit feeding a battery load. The source impedance is neglected. Angle β is known as the *conduction angle*. The average DC voltage will be

$$e_d = \frac{1}{\pi} E_m \int_{\alpha}^{\alpha+\beta} \sin\theta \, d\theta$$

$$= \frac{1}{\pi} E_m [-\cos(\alpha+\beta) + \cos\alpha], \qquad (7.25)$$

where

$$\alpha + \beta = \pi - \sin^{-1}\left(\frac{E_b}{E_m}\right),$$

E_b = battery voltage.

7.2.3 Discontinuous-Current Operation

Figure 7.4d shows the waveforms for a three-phase half-wave bridge using an M-3 connection. The limiting value for load inductance to produce discontinuous input current is obtained as follows. For a given firing angle α, the phase current will become just zero when the next SCR is fired. This is the limit for discontinuous operation. The differential equation relating the phase current i_d and the driving voltage E_m is given by

$$L_d \frac{di_d}{dt} + i_d R_d = E_m \cos(\omega t - \pi/3 + \alpha), \tag{7.26}$$

where L_d and R_d are the load inductance and load resistance, respectively. Current $i_d(t)$ is given by

$$i_d(t) = A\, e^{-R_d t/L_d} + \frac{E_m}{\sqrt{R_d^2 + \omega^2 L_d^2}} \cos(\omega t - \pi/3 + \alpha - \phi),$$

where

$$A = \frac{-E_m}{\sqrt{R_d^2 + \omega^2 L_d^2}} \cos(\pi/3 - \alpha + \phi),$$

$$\tan\phi = \omega L_d / R_d.$$

The value of A can be obtained using the condition that i_d will be zero when t is zero.

The phase current will also be zero when $t = 2\pi/(3\omega)$ for the limiting case. Therefore,

$$0 = \frac{E_m}{\sqrt{R_d^2 + \omega^2 L_d^2}} [\cos(\pi/3 - \alpha + \phi)\, e^{-(R_d/L_d)[2\pi/(3\omega)]} + \cos(\pi/3 + \alpha - \phi)]. \tag{7.27}$$

Equation (7.27) gives the required value of L_d for the given value of α. Similar equations can be derived for two-pulse and six-pulse converters. At the limiting condition when the next SCR is fired, the load current becomes just zero and there will be no overlap angle. Values of L_d less than those given by Eq. (7.27) will produce discontinuous load current. In such a case, the SCRs will undergo natural commutation and the expression for the average output voltage given by Eq. (7.10) will not be valid.

Discontinuous-current operation may also take place for any value of L_d if there is a counter-emf in the load circuit, as is the case when a battery or a DC motor forms the load circuit. If E_c is the counter-emf, then for the same operating conditions as for Eq. (7.26) current $i_d(t)$ will be

$$i_d(t) = [\frac{E_c}{R_d} - \frac{E_m}{R_d^2 + \omega^2 L_d^2} \cos(\frac{\pi}{3} - \alpha + \phi)]\, e^{-R_d t/L} - \frac{E_c}{R_d} + \frac{E_m}{\sqrt{R_d^2 + \omega^2 L^2}} \cos(\omega t - \pi/3 + \alpha - \phi). \tag{7.28}$$

The required conditions for the limiting case are obtained by substituting $i_d[2\pi/(3\omega)] = 0$ in Eq. (7.28), from which the desired value of E_c can be obtained.

If the load resistance R_d is assumed to be zero, then the counter-emf

2 4 6 8 9 10 12 14 16 18 20 22 24 25
1 2 4 5 6 7 8 9 10 11 12 13 14

2 pole 1 pM

15 - 100 Hz o/p f

range of no load speed
for motor

1500 RPM.

ns - nr

$$f = \frac{Np}{}$$

15 - 100 RPM.

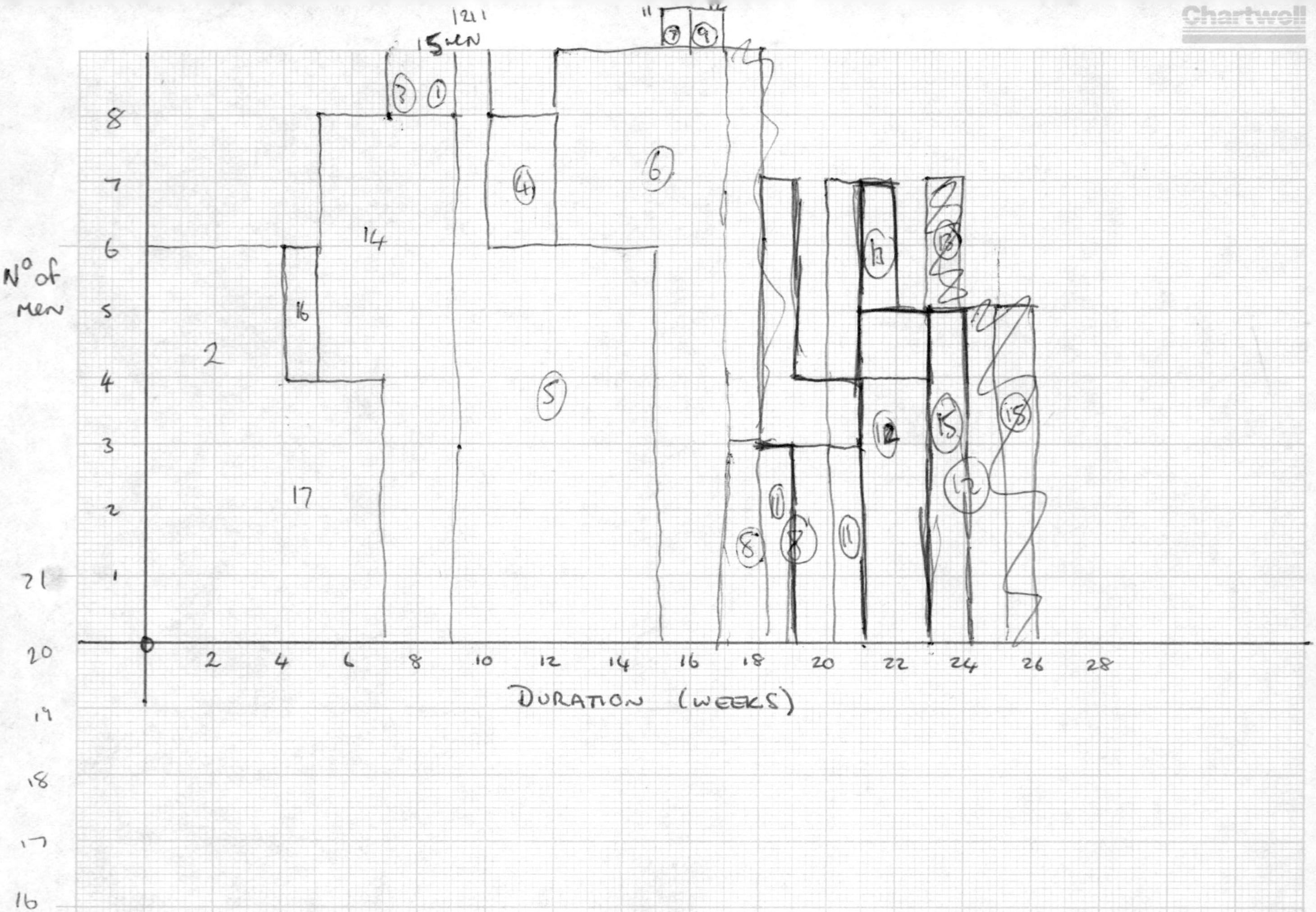

N° of Men
Duration (weeks)
15 men

E_c will be equal to the average output voltage of the converter and given by $(3\sqrt{3}\, E_m \cos\alpha)/(2\pi)$. Then, the required value of L_d can be obtained for the limiting case from Eq. (7.28). The limiting condition for continuous conduction can be obtained similarly for a two-pulse converter. Here, the duration of SCR conduction will be π/ω. Equations (7.29) give the expression for load current $i_d(t)$, and the relationship for the average load current at discontinuous-current limit I_{dis}, the firing angle α, and inductance L_d:

$$i_d(t) = \frac{E_m}{\omega L_d}\left[\cos\alpha - \cos(\omega t + \alpha) - \left(\frac{2}{\pi}\cos\alpha\right)\omega t\right],$$

$$I_{dis} = \frac{2E_m}{\pi\omega L_d}\sin\alpha. \tag{7.29}$$

Figure 7.5a shows the variation of the load voltage E_c as the load current is decreased while the firing angle α of the converter is held constant. It is assumed that the SCRs of the converter are gated by a continuous signal of duration π/ω and its source inductance is zero. As long as the average load current I_d is more than I_{dis}, the average output voltage

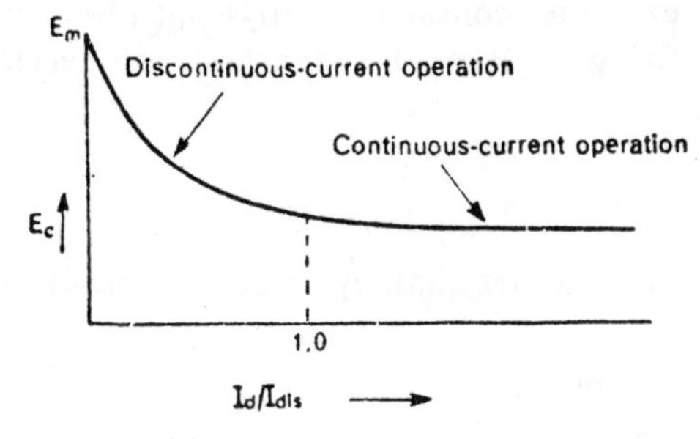

(a) Variation of load voltage

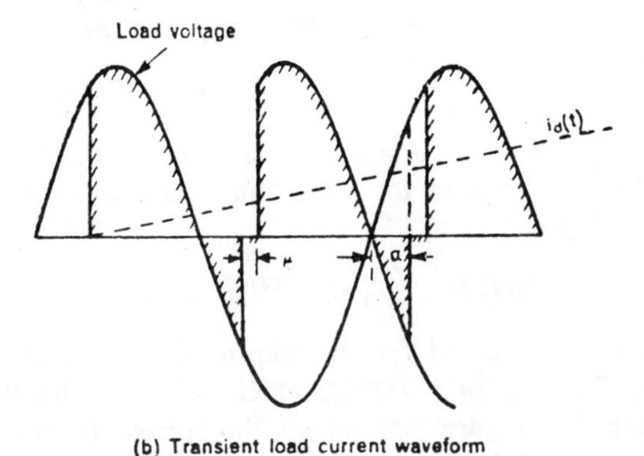

(b) Transient load current waveform

Fig. 7.5 Converter performance.

will be constant [Eq. (7.1)]. When the current becomes discontinuous, the average output voltage will be more than that given by Eq. (7.1) and, finally, when $I_d = 0$, the load voltage will be E_m. For a specified average current, Eq. (7.29) can be used to obtain the minimum inductance required in the load circuit to maintain continuous-current conduction. If load resistance R_d is not negligible, then the criterion developed in Section 6.7 can be conveniently used.

7.2.4 Transient Operation

In a bridge circuit a DC voltage e_d is applied to the load. For an inductive load, current i_d will be initially zero and will gradually build up to a value $I_d = e_d/R_d$, where R_d is the load resistance. The time constant for the load circuit is L_d/R_d. For large values of L_d, the current in the steady state can be assumed to be constant; e_d will then be given by the average value of the output voltage e_o. During the starting stage, the current through the circuit will not be constant, and therefore the overlap angle due to source inductance will be different for each successive commutation. For the sake of simplicity, it can be assumed that the change in load current over each half-cycle is small, and therefore the overlap angle can be calculated on the basis of the current at the instant of firing in every half-cycle. For example, consider a fully-controlled single-phase circuit operating with a firing angle α. For the first half-cycle, the DC voltage will be

$$e_{d1} = \frac{2E_m}{\pi} \cos \alpha.$$

This will, by the time the next half-cycle begins, produce in the circuit a current given by

$$i_d(T/2) = \frac{2E_m \cos \alpha}{\pi R_d} (1 - e^{-R_d(T/2)L_d}). \tag{7.30}$$

Equation (7.30) can be used to calculate the voltage drop due to the overlap angle. The corresponding DC voltage will be

$$e_{d2} = \frac{2E_m}{\pi} \cos \alpha - \frac{\omega L_s}{\pi} \times \frac{2E_m}{\pi} \times \frac{\cos \alpha}{R_d} (1 - e^{-R_d(T/2)L_d}), \tag{7.31}$$

where L_s is the source inductance. The value of the voltage obtained from Eq. (7.31) can be used for calculating current i_d for the next half-cycle, which will be

$$i_d(T) = \frac{e_{d2}}{R_d} + [i_d(T/2) - \frac{e_{d2}}{R_d}] \, e^{-R_d(T/2)L_d}. \tag{7.32}$$

This procedure is continued for the step-by-step calculation of the load current till $i_d(nT/2) = i_d[(n + 1)T/2]$, which will show that the steady state has been reached. Figure 7.5b shows the current build-up in a single-phase fully-controlled bridge. It is assumed that load inductance L_d is large so that the current is continuous for all firing angles upto $\pi/2$.

7.2.5 Example

A fully-controlled single-phase bridge connected to AC supply of 230 V and 50 Hz is used for the speed control of a DC motor with separate field excitation. The full load average armature current is 10 A and the converter operates at a firing angle $\alpha = \pi/4$. Neglecting the inductance and resistance of both armature and source, calculate the minimum value of series inductance L_d required in the armature circuit to provide for continuous-current conduction.

From Eq. (7.29), we have

$$L_d = \frac{2 \times 230 \times \sqrt{2} \times 1/\sqrt{2}}{10 \times 314 \times \pi} \times 1000 = 46.5 \text{ mH}.$$

Using the criterion developed in Chapter 6, the amplitude of the second harmonic current is made equal to the average armature current. From Eq. (7.5), the second harmonic amplitude E_2 of the output voltage e_o of the converter is $\sqrt{(e_2^2 + e_2'^2)}$. Therefore, we have

$$E_2 = \frac{4E_m\sqrt{5}}{3\sqrt{2}\pi} = \frac{4 \times 230 \times \sqrt{5}}{3\pi} = 218 \text{ V}.$$

Hence,

$$2\omega L_d = \frac{218}{10} = 21.8,$$

$$L_d = \frac{21.8}{2 + 314} \times 1000 = 34.7 \text{ mH}.$$

The criterion developed in Chapter 6 gives a lower value of inductance for maintaining continuous current because the effect of higher harmonics is neglected. It can also be seen that the minimum required value of inductance will decrease as the pulse number of the converter is increased.

7.2.6 Effect of Overlap Angle

The effect of source inductance in delaying current commutation has been discussed in Section 7.2. The overlap angle μ (and the corresponding reduction in the average DC output voltage) is a function of the load current and source inductance. For a given value of L_s, the overlap angle μ will increase with load current. As the overlap angle increases (with a constant firing angle α), the circuit will go into various modes of operation. The waveforms of current and voltage shown in Fig. 7.3 are for small values of μ. In single-phase circuits, as long as μ is less than π, Eq. (7.22) for the output voltage will be valid. When $\mu = \pi$, the load will be permanently shorted by the SCRs and the output voltage will be zero because, during the overlap period, all the SCRs will be conducting. Figure 7.6a shows the load characteristic of a two-pulse fully-controlled bridge operating in the rectifier mode.

For a six-pulse three-phase bridge, as the overlap angle increases, the number of conducting SCRs will also increase (see Fig. 7.3b). When angle μ is between zero and $\pi/3$ for a given firing angle α, during commutation only three SCRs will conduct, and two SCRs will conduct after commutation has been completed. Therefore, there will be no overlap of consecutive commutations. This is mode-1 operation and the output voltage will be given by Eq. (7.23). The load characteristic shown in Fig. 7.6b has a negative slope of $\tan^{-1}(3\omega L_s/\pi)$ in mode 1. This mode will exist till $\mu = \pi/3$.

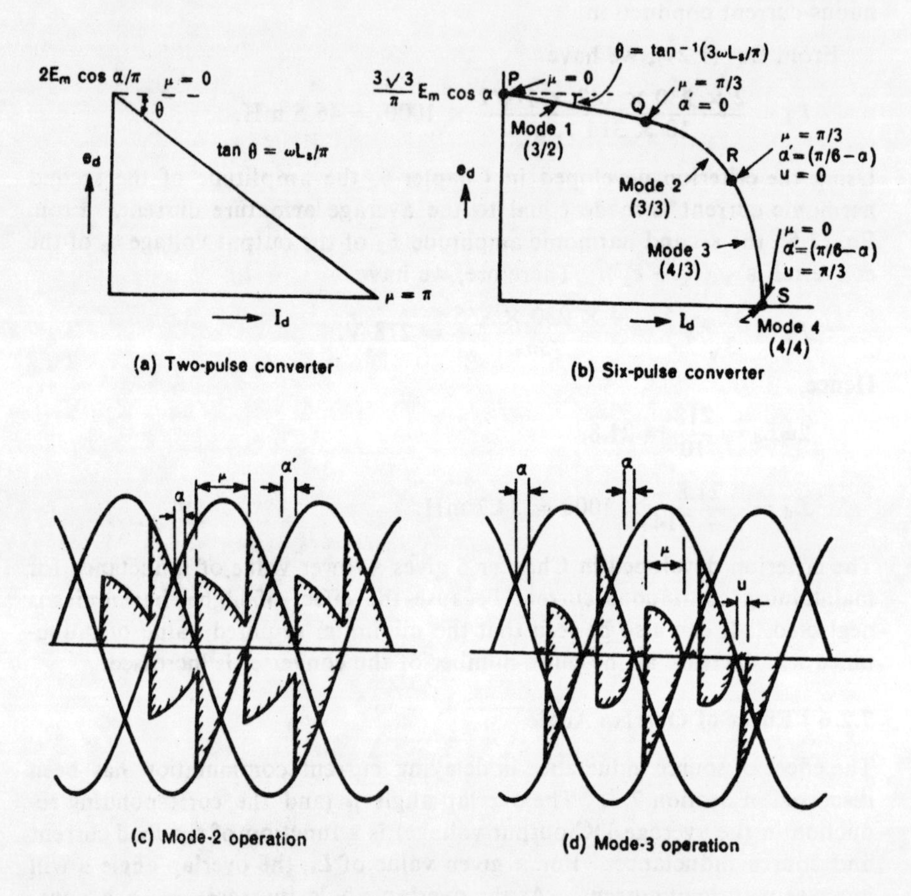

Fig. 7.6 Load characteristics of converters.

When the load current is increased further, consecutive commutations will just overlap and three SCRs will be conducting all the time. The incoming SCR cannot turn on until the previous commutation has been completed. Therefore, there will be an additional firing angle delay α' as shown in Fig. 7.6c, and the overlap period will continue to be $\pi/3$. This is mode-2 operation. The additional angle α' is a function of the load current, and as the load current is increased, the average output voltage will fall off more rapidly (see curve QR in Fig. 7.6b) than in mode 1. To obtain this mode of operation, it is assumed that the SCRs are fired continuously for a minimum period of $\pi/6$ starting from the firing angle α. When $(\alpha + \alpha')$ becomes equal to $\pi/6$ and the current is increased still further, the successive commutations will overlap and four SCRs will conduct for some time. During this period, the output will be short-circuited. This is mode-3 operation; the corresponding output voltage waveform will be as shown in Fig. 7.6d. The output is shorted six times in a cycle, each time for a duration u. Here, the total delay angle $(\alpha + \alpha')$ will remain constant at $\pi/6$ and the total overlapping period $(u + \mu)$ will be equal to $\pi/3$. As the load current is increased still further, the duration u of each short-circuit will increase, and finally when u becomes equal to $\pi/3$, the output will be completely shorted and the load voltage will be zero. The variation of the output voltage with load current for this mode is shown by curve *RS* in Fig. 7.6b. Mode 3 is also known as the 4/3 mode because of the conducting pattern of the SCRs. When $u = \pi/3$, four SCRs will conduct all the time (4/4 mode), resulting in zero output voltage. If the initial firing angle α is more than $\pi/6$, the changeover from mode 1 to mode 3 will take place when μ is $\pi/3$; mode 2 will be absent.

7.2.7 Example

A six-pulse fully-controlled converter operates with a firing angle $\alpha = \pi/6$. The input is three-phase AC supply of 400 V and 50 Hz. The load current on the DC side is such that the output voltage experiences a short-circuit of duration $\pi/6$, six times in each cycle. Calculate the average output voltage. Assume that the load current is steady.

The average load voltage is

$$e_d = 2 \times \frac{3}{2\pi} E_m \left\{ \frac{1}{2} \int_{\pi/3}^{\pi/2} [\sin(\theta + \pi/6) + \sin(\theta + 5\pi/6)]\, d\theta \right.$$

$$\left. + \int_{2\pi/3}^{5\pi/6} \sin(\theta + \pi/6)\, d\theta \right\}$$

$$= \frac{400}{\sqrt{3}} \times \sqrt{2} \times \frac{3}{\pi} [0.2] = 62 \text{ V}.$$

7.2.8 Interphase Reactor Connection

It has been shown in Section 7.1.1 that, for a given AC input and firing angle, a six-pulse converter produces a higher DC voltage and a lower output ripple factor than those produced by a three-pulse (three-phase half-wave) or a two-pulse (single-phase full-wave) circuit. Because of the lower ripple factor, the size of the filter is reduced. However, an increase in the number of phases in the input will decrease the *utility factor* (i.e., the ratio of the conducting period to the input period) of each SCR. With a six-phase input and using a twelve-pulse converter, the lowest order harmonic in the output will be 12 and the utility factor will be 1/6. The same output ripple factor can also be obtained by using two six-pulse converters as shown in Fig. 7.7a. The inputs to the two bridges are displaced by 30°, using star-star and star-delta transformer connections for the input, as shown in the figure. The corresponding phases in each bridge will conduct together and the output voltage will be the average of the two conducting phases. The load is connected between the common point of all the cathodes of the top group of SCRs and the midpoint of the interphase reactor. The induced voltage in this reactor will maintain the conduction of corresponding phases in the two bridges for a period of $2\pi/3$ radians. Figure 7.7b shows the output load voltage waveform with zero firing angle. The lowest order of harmonics in the output is 12 and the utility factor of each SCR is the same as that of a six-pulse bridge. The two bridges can also be connected in series directly without the interphase reactor to obtain a higher DC voltage at low ripple factor. The design of the interphase reactor is based on the same criterion as that used for designing the

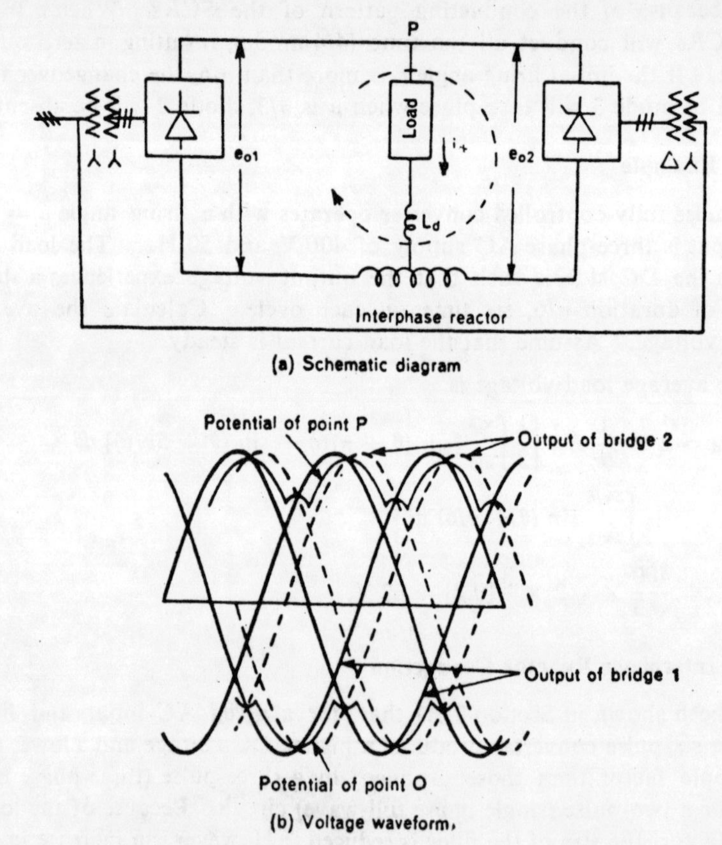

(a) Schematic diagram

(b) Voltage waveform.

Fig. 7.7 Interphase transformer connections.

reactor for limiting the circulating current in dual converters discussed in Chapter 6. The circulating current, shown by the dashed line in Fig. 7.7a, must be less than the load current supplied by either of the two converters. If the firing angle of the two bridges is the same, then from Eq. (7.18) it can be seen that the sixth harmonic component of the output voltages e_{o1} and e_{o2} will add up in the circulating-current path because of the phase difference in the input voltages of the converters. The inductance of the interphase reactor must be such that the sixth harmonic current ripple amplitude is sufficiently small.

7.2.9 Commutation with a Capacitive Source Impedance

In the normal operation of a converter, line commutation will take place only if the anode potential of the incoming SCR is more than that of the conducting SCR. Thus, referring to Fig 7.8a, if the firing angle is measured with respect to point O, then converter operation will not be possible if the firing angle is negative. However, if for each phase a capacitor is connected on the input side, then the capacitor voltage drop will aid in

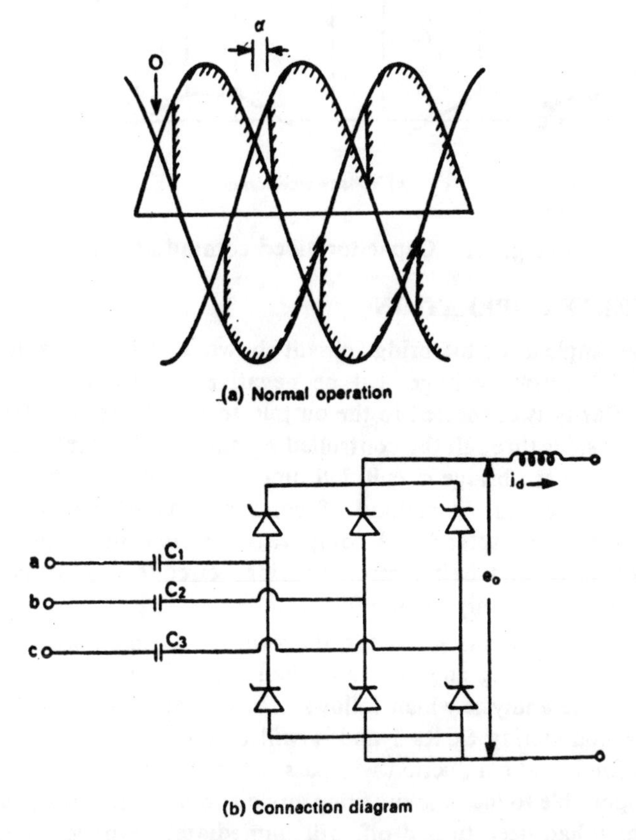

Fig. 7.8 Capacitor-aided commutation (*cont.*).

providing proper commutation even if the firing angle is negative. Figure 7.8b shows how the capacitors are connected; the voltage waveforms are shown in Fig. 7.8c. It is assumed that load current i_d is constant, and therefore the voltage across each capacitor will vary linearly when the corresponding SCRs are conducting. This capacitor-aided commutation is also referred to as *load commutation*. Details of various commutation circuits and their applications will be given in Chapter 8.

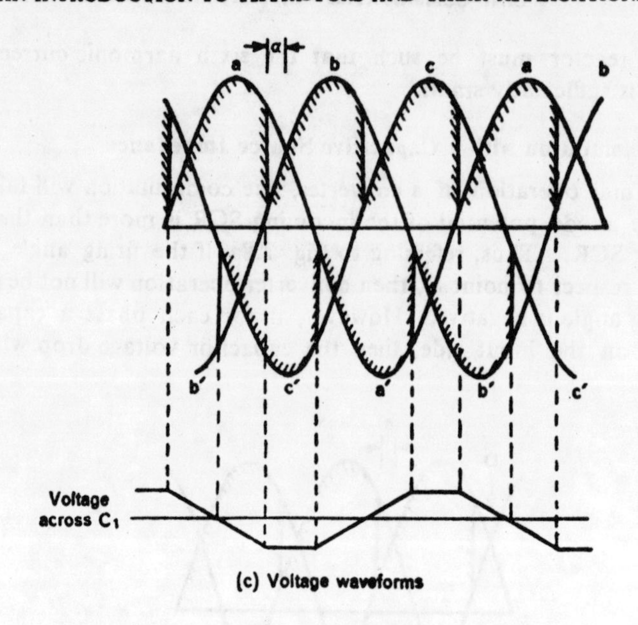

Fig. 7.8 Capacitor-aided commutation.

7.3 INVERTER OPERATION

If the firing angle α for the bridge circuit shown in Fig. 7.1 is more than $\pi/2$, the DC output voltage will be negative. Thus, if a DC source of negative polarity is connected to the output terminals, it will feed power to the AC system through the controlled circuit. This operation is known as *inversion* and the bridge circuit will function as an *inverter*. The principle of operation and the method of commutation of the inverter have been explained in Chapter 6. A comparison of the expressions for output voltage for fully- and half-controlled bridge circuits shows that inverter operation is possible only for the former.

Figure 7.9a shows the schematic diagram of a single-phase fully-controlled bridge for firing angles greater than $\pi/2$. Current i_d is assumed to be steady in the analysis which follows. In the positive half-cycle of the input, the conducting SCRs 3 and 4 will be reverse-biased when SCRs 1 and 2 are fired, and conduction will pass on to SCRs 1 and 2. Theoretically, it is possible to increase the firing angle α to π. If this is done, SCRs 3 and 4, which had been turned off, will immediately experience a forward voltage and start conducting again, producing a positive voltage at the output, and consequently disturbing the inverter action. As explained later in this section, the firing angle is limited to a value $(\pi - \gamma)$, where γ, called the *margin angle*, corresponds to the safe minimum time required for proper turn-off of the outgoing SCRs.

Figure 7.9b shows the output voltage and alternating line current waveforms. The output power factor is leading. The reactive power will increase as the firing angle α approaches $\pi/2$. The average value of the

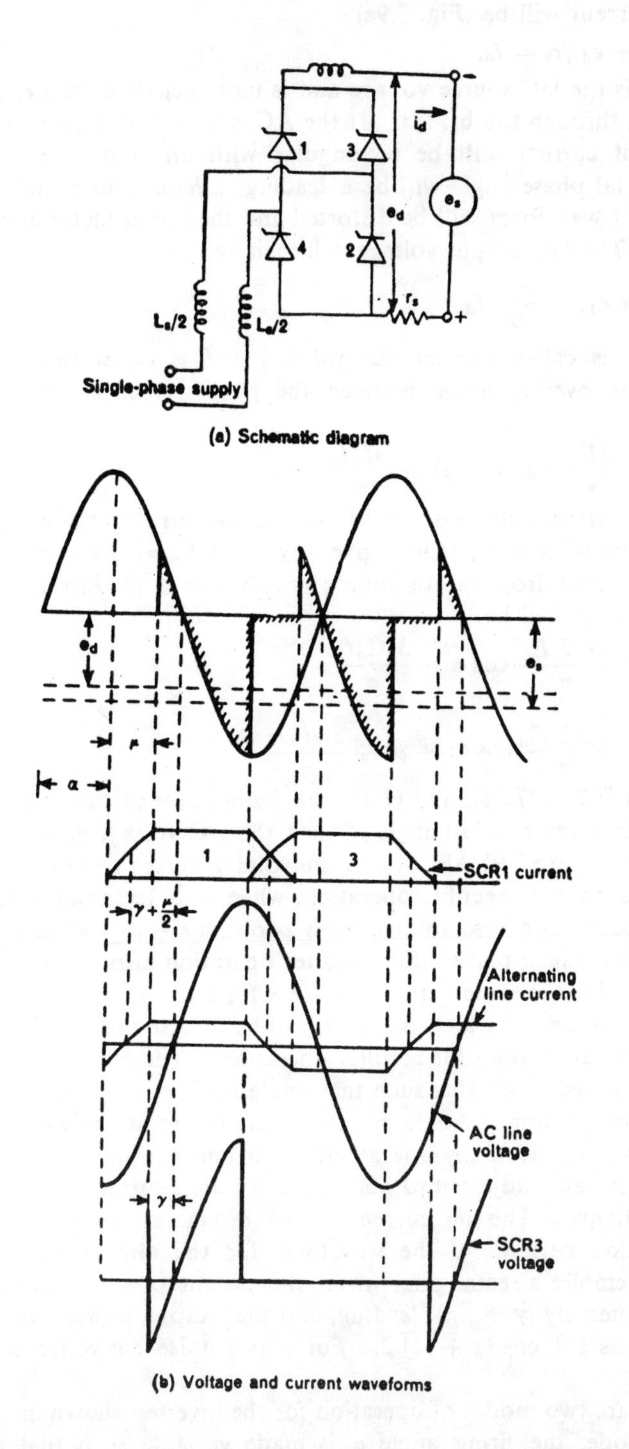

Fig. 7.9 Operation and characteristics of an inverter (*cont.*).

output current will be (Fig. 7.9a)

$$(e_s - e_d)/r_s = I_d, \tag{7.33}$$

where e_s is the DC source voltage and is more negative than e_d for forcing current I_d through the bridge. If the AC source inductance is neglected, the output current will be rectangular with an amplitude of I_d, and its fundamental phase angle will be α leading. With source inductance L_s, the current waveform will be distorted and the power factor angle will also change. The DC output voltage will then be

$$e_d = e_{do} - \frac{\omega L_s}{\pi} I_d; \tag{7.34}$$

where e_{do} is called the *interval voltage*, and is equal to $(2E_m/\pi) \cos \alpha$. If μ is the overlap angle between the phases, then the output voltage will be

$$e_d = \frac{2E_m}{\pi} \cos (\alpha + \mu) + \frac{\omega L_s I_d}{\pi}. \tag{7.35}$$

Similar equations can be obtained for a fully-controlled three-phase bridge. The internal no-load voltage is given by $(3\sqrt{3}\, E_m/\pi) \cos \alpha$ and the equivalent DC voltage drop due to source inductance is $(3\omega L_s/\pi)I_d$. Thus, the output voltage will be

$$e_d = \frac{3\sqrt{3}\, E_m}{\pi} \cos \alpha - \frac{3\omega L_s I_d}{\pi}$$

$$= \frac{3\sqrt{3}\, E_m}{\pi} \cos (\alpha + \mu) + \frac{3\omega L_s I_d}{\pi}. \tag{7.36}$$

Equations (7.34), (7.35), and (7.36) can be used for calculating the required firing angle α and overlap angle μ when the DC voltage e_d and current I_d are given. Figure 7.9b also shows the voltage across one of the SCRs. It can be seen that for rectifier operation, when α is less than $\pi/2$, the SCR will experience a reverse voltage for a period $(\pi - \mu - \alpha)$ after the commutation has taken place. For inverter operation also, the reverse voltage will appear for the same period. Since α for this operation is more than $\pi/2$, the duration for which the SCR will be reverse-biased will be less for inverter operation than for rectifier operation. Therefore, the firing angle α will be limited to a value such that angle $\gamma = (\pi - \mu - \alpha)$ is enough for proper commutation. Angle $(\pi - \gamma)$ is referred to as the *extinction angle*. The inverter will normally operate at a constant margin angle. The firing angle α is suitably adjusted to obtain the required current flow as the DC input e_S changes. The line current will go through zero at the middle of the commutation period. If the waveform for the alternating line current closely resembles a rectangular wave, the output power factor angle will be approximately $(\gamma + \mu/2)$ leading, and the output power factor will be close to $[\cos \alpha + \cos (\alpha + \mu)]/2$. For a two-pulse converter this will be $\pi \dot{e}_d/(4E_m)$.

There are two modes of operation for the inverter shown in Fig. 7.9a. In one mode, the firing angle α is made variable such that the direct current input I_d is maintained constant. As the source voltage e_S increases, the internal voltage e_{do} will also increase correspondingly so that the difference between the two will remain at a constant value, given by $[(\omega L_s/\pi + r_s)I_d]$. The characteristic of the inverter in this mode is shown by the vertical line (line 3) in Fig. 7.9c. To increase e_d, the firing angle α

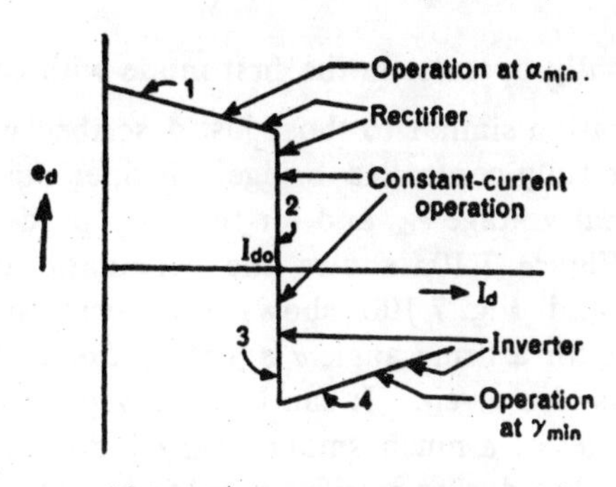

(c) Rectifier-inverter characteristics

Fig. 7.9 Operation and characteristics of an inverter.

has to be advanced. The duration for which the SCR is reverse-biased after commutation will reach the minimum value required at some level of the internal voltage. Any further increase in the firing angle, for maintaining current I_d constant, will result in commutation failure. This will be the limit of the constant-current mode. Thus, any further increase in I_d can be allowed only by reducing the firing angle α so that the margin angle γ remains constant. This is the second mode of operation for the inverter. Here, the inverter will operate with a constant margin angle or extinction angle. Line 4 in Fig. 7.9c represents the inverter characteristic in this mode. From Eq. (7.35) it can be seen that the DC voltage will decrease with an increase in the current, for a constant value of $(\alpha + \mu)$. Substituting $\gamma = \pi - (\alpha + \mu)$, Eq. (7.35) will get modified to

$$e_d = -\frac{2E_m}{\pi} \cos \gamma + \frac{\omega L_s}{\pi} I_d. \tag{7.37}$$

If the *V-I* characteristics are extended to the positive side, i.e., for rectifier operation with $\alpha < \pi/2$, the circuit will again exhibit two modes of operation. In one mode, the firing angle α is adjusted to maintain a constant current. Then, the internal voltage e_{do} is greater than the output voltage e_d by the drop $I_d(\omega L_s/\pi + r_s)$. This characteristic is shown by line 2 in Fig. 7.9c, in which current I_d (required to be maintained constant at I_{do}) is assumed to be the same for both the rectifier and inverter operations. At some value of the output voltage, the firing angle will be such that the forward voltage across the incoming SCR (before it is turned on) is just enough for it to be gated. If there is any further decrease in α, the incoming SCR will not turn on and commutation will not take place. Therefore, the firing angle is kept constant at this value ($=\alpha_{min}$), and if any increase in the DC voltage is required, it can be obtained by decreasing current I_d. This is the second mode of operation and its characteristic is given by line 1 in Fig. 7.9c. The output equation for this operation is

$$e_d = \frac{2E_m}{\pi} \cos \alpha_{min} - \frac{\omega L_s}{\pi} I_d. \tag{7.38}$$

The rectifier normally operates in the first mode with constant current.

Modes of operation similar to those just described will also take place with a three-phase fully-controlled bridge. Proper equations have to be used for the internal voltage e_{do} and for the drop produced by the source inductance L_s. Figure 7.10a shows the schematic diagram of a six-pulse converter, and Fig. 7.10b shows the corresponding voltage and current waveforms for a firing angle α equal to $5\pi/6$. The voltage across one of the SCRs is also given. It can be observed that the SCR experiences a reverse voltage for a much smaller period during inverter operation as compared with that during rectifier operation. Unless proper control, based on Eq. (7.36), is exercised to maintain a minimum margin angle, commutation failure may take place. Equation (7.36) can also be written as

$$\frac{\pi}{3\sqrt{3}} \left(e_d + \frac{3\omega L_s}{\pi} I_d\right) = E_m \cos \alpha, \tag{7.39}$$

where E_m is the peak value of the AC phase voltage. For given values of e_d and I_d, the firing angle α is fixed. To determine the instant of firing, replace α by ωt in Eq. (7.39). The signal $E_m \cos \omega t$ thus obtained can be easily generated in a three-phase system. For example, in the six-pulse converter under consideration, commutation from SCR1 to SCR2 is obtained by firing SCR2 at time t when the signal $E_m \cos \omega t$ (which is the negative value of phase-c-to-neutral voltage) becomes equal to the left-hand side of Eq. (7.39). The firing angle α will automatically get adjusted for different values of e_d and I_d until γ_{min} is reached, when the inverter will go into the second mode (constant margin angle). Combining the two forms of Eq. (7.36), and replacing α by ωt and $(\alpha + \mu)$ by $(\pi - \gamma_{min})$, we get

$$\frac{2\omega L_s}{\sqrt{3}} I_d - E_m \cos \gamma_{min} = E_m \cos \omega t, \tag{7.40}$$

where γ_{min} is constant. Therefore, the left-hand side of Eq. (7.40) will depend on current I_d. Again, the right-hand side of Eq. (7.40) is the negative value of phase c voltage for firing SCR2. This is also known as the *control voltage*. Whereas the right-hand side is the same for all SCRs, the control voltage will be different for each SCR. Therefore, separate firing circuits are required. Whenever the control voltage becomes equal to the right-hand side of Eq. (7.40), the corresponding SCR will be fired. The comparator shown in Fig. 6.9c can be used for this purpose. This will ensure inverter operation at a constant margin angle. The fundamental power factor of the AC output, assuming a rectangular waveform for the

current, will be cos $(\gamma + \mu/2)$ leading; it can also be approximated to $[\cos \gamma + \cos (\gamma + \mu)]/2$ leading.

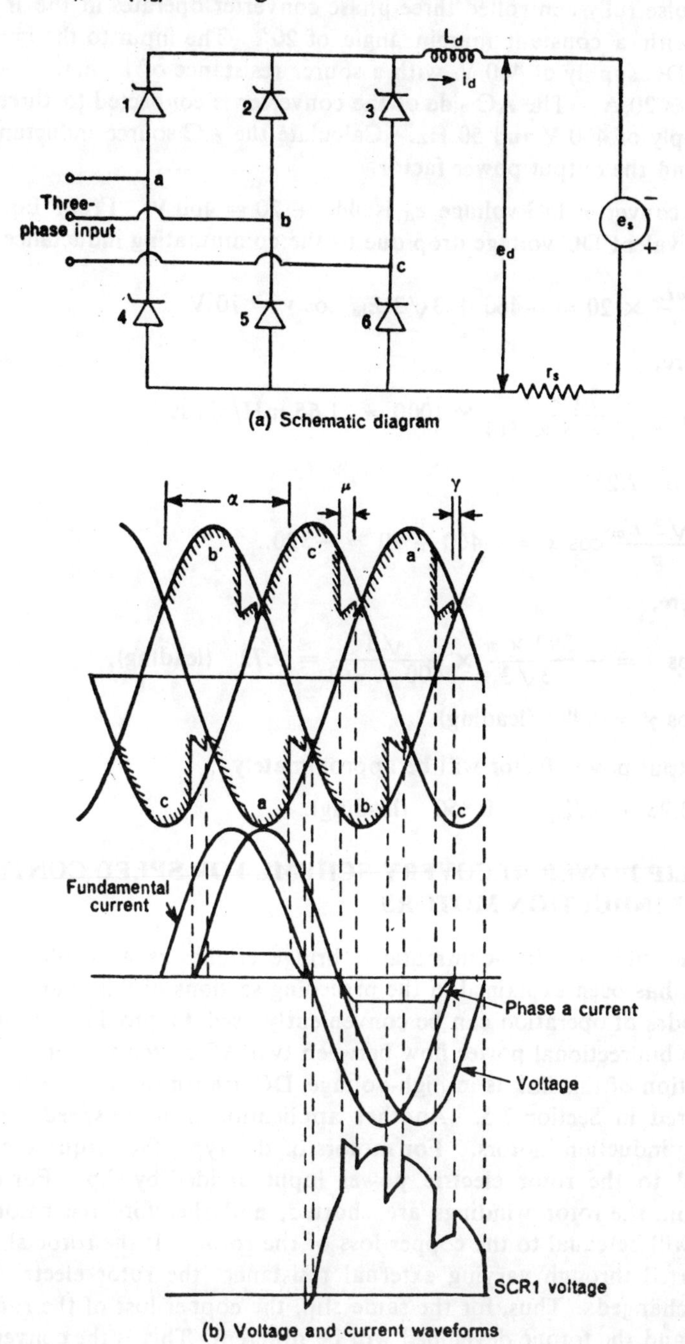

Fig. 7.10 Six-pulse inverter.

7.3.1 Example

A six-pulse fully-controlled three-phase converter operates in the inverting mode with a constant margin angle of 20°. The input to the inverter is from a DC supply of 440 V, with a source resistance of 1 ohm. The input current is 20 A. The AC side of the converter is connected to three-phase AC supply of 400 V and 50 Hz. Calculate the AC source inductance per phase and the output power factor.

The converter DC voltage e_d is 440 + 20 = 460 V. From Eq. (7.36), the equivalent DC voltage drop due to the commutating inductance is

$$\frac{3\omega L_s}{\pi} \times 20 = -460 + 3\sqrt{3}\, E_m \cos \gamma = 70 \text{ V}.$$

Therefore,

$$L_s = \frac{70 \times \pi}{20 \times 3 \times 314} \times 1000 = 11.65 \text{ mH/phase}.$$

From Eq. (7.23),

$$\frac{3\sqrt{3}\, E_m}{\pi} \cos \alpha = -460 + 70 = -390.$$

Therefore,

$$\cos \alpha = -\frac{390 \times \pi}{3\sqrt{3}} \times \frac{\sqrt{3}}{400 \times \sqrt{2}} = 0.72 \quad \text{(leading)},$$

$$\cos \gamma = 0.98 \quad \text{(leading)}.$$

The output power factor will be approximately

$$(0.98 + 0.72)/2 = 0.850 \quad \text{(leading)}.$$

7.4 SLIP POWER RECOVERY SCHEME FOR SPEED CONTROL OF INDUCTION MOTORS

The operation of a line-commutated bridge circuit as a rectifier and an inverter has been explained in the preceding sections of this chapter. These two modes of operation can be conveniently used to provide an electrical link for bidirectional power flow between two AC systems. One important application of this link is in high-voltage DC transmission, which will be considered in Section 7.5. Another application is in the speed control of slip-ring induction motors. For motors of this type, the torque developed is equal to the rotor electric power input divided by slip. For normal operation, the rotor windings are shorted, and therefore the rotor input power will be equal to the copper loss of the rotor. If the rotor slip-rings are shorted through varying external resistance, the rotor electric power can be changed. Thus, for the same slip, the copper loss of the rotor will change and the torque developed will be different. This is the conventional rotor-rheostat control. The effect of variable external resistance can be simulated by means of SCRs, using phase control as explained in Chapter 6, or on-off control as will be explained in Chapter 10. In all these methods, the total rotor electric input is dissipated as loss, and therefore the motor efficiency [which is approximately equal to $(1 - S)$, where S is the motor slip] will be poor at low speeds.

In the scheme shown in Fig. 7.11, the rotor terminals are connected to the AC input supply through two fully-controlled bridges. Bridge 1 operates as a rectifier and bridge 2 as an inverter to feed the power output of the rotor back into the AC mains. This is known as *slip-power recovery*. By varying the firing angles of both the bridges, the power output from the rotor can be changed. Therefore, the motor slip will also change for the

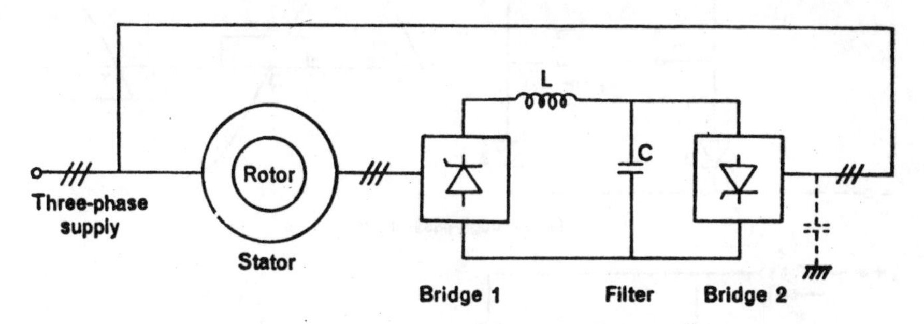

Fig. 7.11 Slip power recovery scheme.

same load torque. Neglecting losses in the bridge circuits, all the power output of the rotor will be fed back into the AC supply instead of being dissipated in the form of heat as would be the case if rotor-rheostat control were used, with or without SCRs. Thus, the efficiency of the slip power recovery control will be high. The only drawback of this scheme is that both rectifier and inverter take lagging reactive power from the mains (as the input power factor angle is equal to the firing angle). Thus, the total input power factor for the motor will be low. This problem is acute for operating the motor at speeds close to the synchronous speed when the firing angles for the bridge circuits are approximately 90°.

By reversing the operations of bridges 1 and 2, the power output of the rotor can be reversed. Therefore, for the same torque, the motor slip will be negative. This means that the motor will run at supersynchronous speeds. Such an operation is not possible with rotor-rheostat control. Thus, the conventional induction motor with slip power recovery control will produce the characteristics of a Schrage motor. In this scheme, the control of bridge 2 will not be difficult as it will operate from constant frequency mains, but the gating pulses for bridge 1 must be synchronised with the variable slip frequency supply of the rotor.

If only subsynchronous speeds are required, then bridge 1 may be uncontrolled (SCRs replaced by diodes) and the power output of the rotor can be controlled by varying the firing angle of the inverter (i.e., bridge 2). The leading power factor of the inverter can be compensated by providing a capacitor bank at its AC terminals as shown by the dashed line in Fig. 7.11. Another method of improving the inverter output power factor is by reducing the duration of conduction for each SCR. Such an operation can be obtained by a circuit known as a *through-pass inverter*. Figure 7.12a shows a single-phase through-pass inverter. Such a circuit can replace bridge 2 in Fig. 7.11 to improve the overall input power factor.

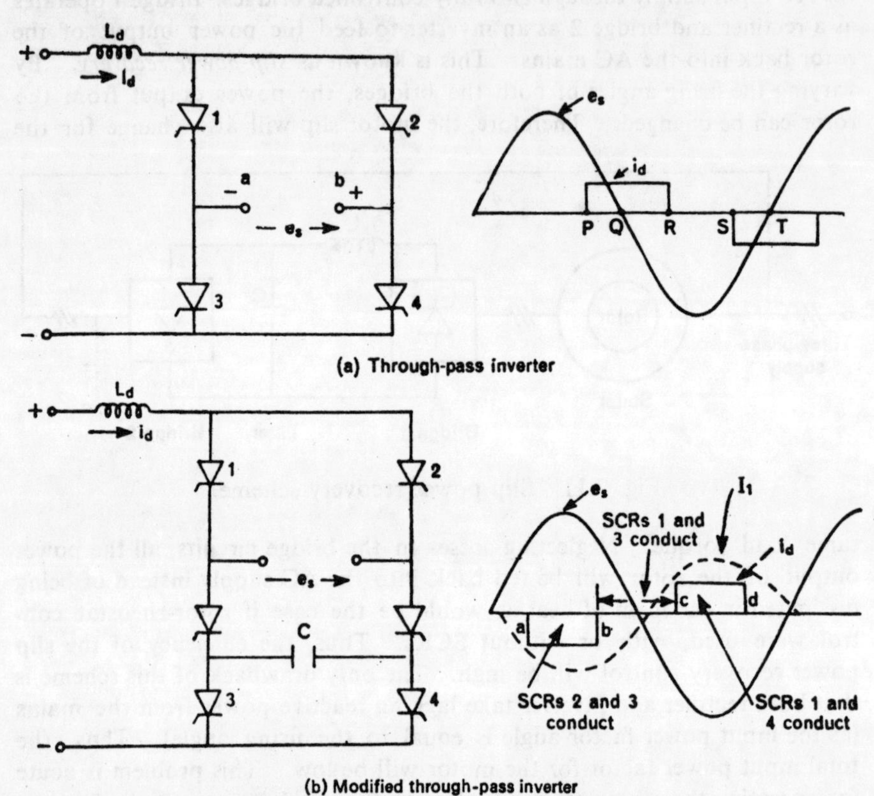

(a) Through-pass inverter

(b) Modified through-pass inverter

Fig. 7.12 Through-pass inverter.

7.4.1 Through-Pass Inverter

In Fig. 7.12a, SCRs 1 and 3 are gated together, producing a short-circuit on the DC input. The input reactor will keep the current ripple low. SCR4 is fired at point *P*, when the supply voltage is positive. This will turn off SCR3 by line commutation and current i_d will be forced through the output terminals *a* and *b*. When the AC voltage reverses its polarity at point *Q*, power will be fed from the DC side to the AC side. At *R*, SCR2 is fired, and SCR1 will be turned off by line commutation. The DC input will again be short-circuited through the reactor and SCRs 2 and 4. At *S*, when the supply voltage is still negative, SCR3 is fired, turning off SCR4, and power flow from the DC side to the AC side will take place from point T when the supply voltage again becomes positive. The output power factor can be improved by controlling the duration QR. The main difference between the operations of through-pass and conventional inverters is that the SCRs in the former conduct for a duration less than π. Line commutation is used in both types of inverters. Because of periodic short-circuits on the DC side, a large reactor L_d is required in through-pass operation to maintain a low current ripple. The output power factor will still be leading even though it will be better than that obtained with a conventional inverter. The output power factor can be further improved, and in fact made lagging, by using forced commutation (see Chapter 8) with a modified through-pass inverter as shown in Fig. 7.12b. The output current and voltage waveforms are also shown in the figure. It will be seen that the fundamental component I_1 of the current lags the voltage. Details of the operation of this circuit can be obtained from references given at the end of this chapter.

7.4.2 DC Transmission

High-voltage DC is used nowadays for bulk power transmission as it offers the following advantages over conventional AC transmission system: (a) only the thermal capacity of the line and equipment govern the stability limit, (b) the cost of transmission is less because of the fewer conductors used and the smaller towers required, (c) a smaller conductor can be used as there is no skin effect for the current, (d) two AC power systems of different operating frequencies can be interconnected because of the asynchronous nature of the DC line, (e) short-circuit detection and clearance are faster and the overall system stability can be very much improved since the power flow can be electronically controlled, and (f) it is ideal for cable transmission as there is no charging reactive power. The additional cost of converting and inverting the equipment makes DC transmission uneconomical at low-power levels and for short distances. Further, DC circuit breakers with high interrupting capacity are not yet available, and therefore, as of today, this system is used for point-to-point transmission of bulk power without any intermediate tap points. With the availability of high-power SCRs, solid-state converters are replacing mercury arc valves for DC transmission.

Figure 7.13a shows the schematic diagram of a DC bipolar transmission system. AC power systems 1 and 2 are interconnected by a DC link. Bridge 1 operates as a rectifier and bridge 2 as an inverter. The firing angles of the two bridges are suitably adjusted for such an operation. In every arm of each bridge a number of SCRs in series-parallel connection are used to obtain high ratings for current and voltage. Voltage and current equalising circuits and snubbers are used with the SCRs. To reduce the ripple factor in the output, and thereby the filter ratings, two six-pulse circuits, one with an input transformer having star-star connection and the other with an input transformer having star-delta connection, are used on each side of the DC link. This results in a twelve-pulse operation and therefore reduces the distortion in the input current.

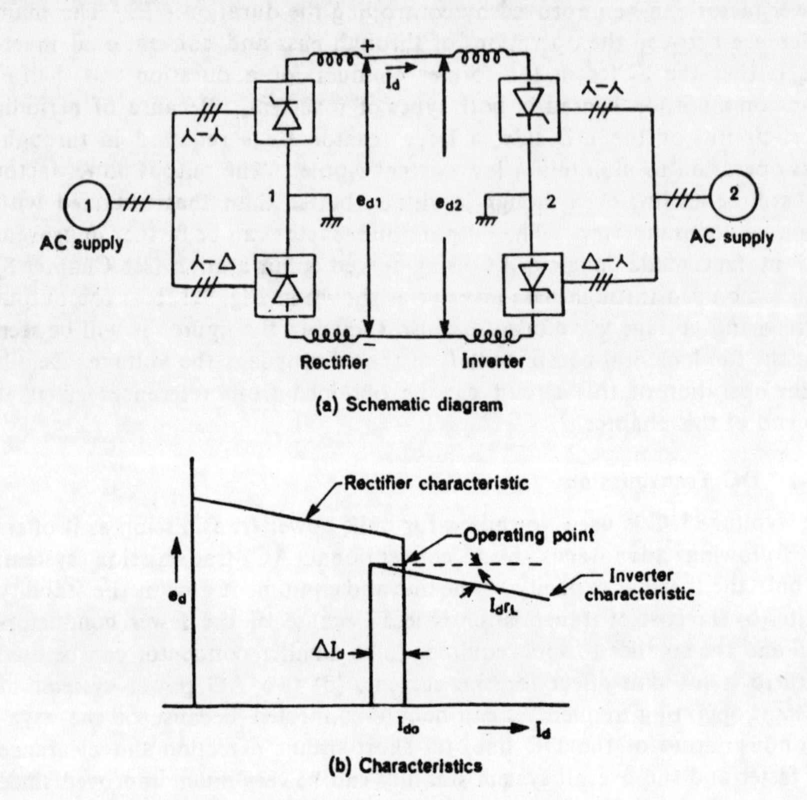

Fig. 7.13 DC transmission system (*cont.*).

The DC transmission system uses either unipolar or bipolar conductors. In unipolar transmission, the line conductor has either positive or negative polarity and the return conductor is grounded. In some cases, the return conductor can be omitted and the earth itself used for carrying the return current. This has problems of electrolytic action, higher conduction losses, and large potential gradients near the grounding point. In bipolar transmission, there are two conductors, one positive and the other negative with respect to the ground. The midpoints of the bridges at both ends of the DC line are grounded as shown in Fig. 7.13a. With these connections, the ground currents are normally small. If one of the lines is opened because of a fault, then unipolar transmission will be possible with the same set-up and power flow will be maintained. Obviously, the reliability of the bipolar system is better than that of the unipolar system.

When the power flow is from system 1 to system 2, then bridge 1 will operate in the rectifying mode and bridge 2 in the inverting mode. For given values of voltage e_{d1} and current I_d at the DC-sending end, the rectifier firing angle α can be computed if the source voltage and impedance are known [Eq. (7.23)]. The receiving-end DC voltage is obtained by subtracting the line drop from e_{d1}. Thus,

$$e_{d2} = e_{d1} - I_d r_L, \tag{7.41}$$

where r_L is the resistance of the DC line, including that of the reactor. The inverter normally operates at a specified margin angle γ or at a constant extinction angle $(\alpha + \mu)$ for all currents I_d to avoid any commutation failures. The method of controlling inverter operation has already been explained in Section 7.3. The required firing angle α for the inverter has to be computed from the DC input voltage e_{d2}, current I_d, margin angle $\gamma\,(= \pi - \mu - \alpha)$, source voltage, and source impedance [Eq. (7.36)]. The rectifier bridge operates in the constant-current mode and its firing angle α can be adjusted to produce the required current flow through the bridge. This is done by satisfying Eq. (7.23) at all operating points in the steady state. The control schemes discussed in Chapter 6 can be used for the automatic adjustment of the firing angle. In Fig. 7.13b, the inverter characteristic $(e_d - I_d)$ is plotted on the same side as the rectifier characteristic. The dashed line in the figure has been obtained by adding the voltage drop across the DC line resistance to the inverter DC voltage. The point of intersection of the rectifier characteristic and this dashed line gives the operating voltage and current of the rectifier. To obtain a stable operating point for the system, the constant-current mode for the inverter should take place at a current level $(I_{do} - \Delta I_d)$, where I_{do} is the current maintained constant by the rectifier and ΔI_d is the *current margin*. Since the current flow through the rectifier bridge and the inverter bridge must be the same, the inverter has to be operated with a constant margin angle at a current level I_{do}. It can be seen in Fig. 7.13b that for small perturbations in the voltage of the AC system, the operating point for the DC system is well defined and current I_{d1} is maintained constant at the specified value I_{do}. The internal voltage $[(3\sqrt{3}/\pi)E_{s2}\cos\gamma]$ of the inverter, for a six-pulse bridge, must be less than the internal voltage $[(3\sqrt{3}/\pi)E_{s1}\cos\alpha]$ of the rectifier; the difference ΔV will be

$$\Delta V = \frac{3\omega_1 L_{s1}}{\pi} I_{do} + I_{do} r_L + \frac{3\omega_2 L_{s2}}{\pi} I_{do}; \tag{7.42}$$

where E_{s1} and E_{s2} are the maximum phase-to-ground source voltages of AC systems 1 and 2, respectively, ω_1 and ω_2 are the respective system frequencies, L_{s1} and L_{s2} are the per phase inductances of the two sources, α is the firing angle for the rectifier bridge, and γ is the specified margin angle for the inverter bridge.

For a unipolar DC transmission system, separate bridge circuits 1′ and 2′ are used for reverse power flow. For these bridges, the SCRs are connected in a direction opposite to that shown for the SCRs in Fig. 7.13a so that the polarity of the DC voltage remains the same but the current flow is reversed. This eliminates the problem of electrolytic corrosion of the grounded conductor. The changeover from bridges 1 and 2 to bridges

1′ and 2′ is made through external switches. Then, bridge 1′ will operate as an inverter with a constant margin angle and bridge 2′ will function as a rectifier in the constant-current mode. For bipolar transmission, separate bridges are not required for power reversal.

It has already been mentioned that the input power factor of the rectifier bridge should be lagging and the output power factor of the line-commutated inverter must be leading. Therefore, proper equipment (e.g., shunt capacitors) for correcting the power factor must be connected to the AC terminals. In order to reduce the ripples on the DC side, a smoothing reactor is used. This results in a rectangular waveform for the current in the AC phases. The harmonic content of this current is reduced by using two six-pulse bridges, one with a transformer having star-star connection and the other with a transformer having star-delta connection, as shown in Fig. 7.13a. For such a system, the lowest order harmonic for the alternating line current is 11. To bypass this harmonic, shunt filters are used at the AC terminals so that the line currents may become fairly sinusoidal. In the case of commutation failures or of unsymmetrical firing angles, abnormal harmonics will also be generated in the AC lines and these may affect the operation of the system unless the faulty elements in the bridges are quickly isolated.

Figure 7.13c shows the block diagram of a rectifier controller which maintains constant current by adjusting the firing angles for the SCRs in the bridge. The firing frequency of these SCRs must be $6f$, where f is the input frequency. Since the frequency of the AC system may drift, the firing frequency also must track the system frequency. This is done by using the phase-locked loop (PLL) block in which the VCO output frequency automatically becomes six times the input frequency f. A detailed description of VCO control has been given in Chapter 6. The input signal for PLL is obtained from a comparator. The reference voltage for the comparator is the PI controller output. Whenever this voltage is less than the input voltage (a ramp derived from the AC system voltage), a pulse output will be obtained. The frequency of this pulse will be equal to that of the AC system. The PI controller will have a nonzero output voltage when the actual error in the current $(I_{do} - I_d)$ becomes zero, and it will thus produce zero static error for the control system. For inverter control, the margin angle error, in place of the current error signal, can be fed to the PI controller and the firing angle will automatically adjust itself to provide the required margin angle γ for all system voltages and the direct line current. The control scheme discussed here will produce *equal pulse spacing* since the interval between successive firing pulses from the PLL is $T/6$, where T is the period of the AC input. Controllers based on Eq. (7.40), where separate comparators are used for each SCR, will produce equal firing angles. A control system based on equal spacing of pulses will be better than that based on equal firing angles since in the former generation of abnormal harmonics in the alternating line current waveform is avoided because the duration of conduction of each

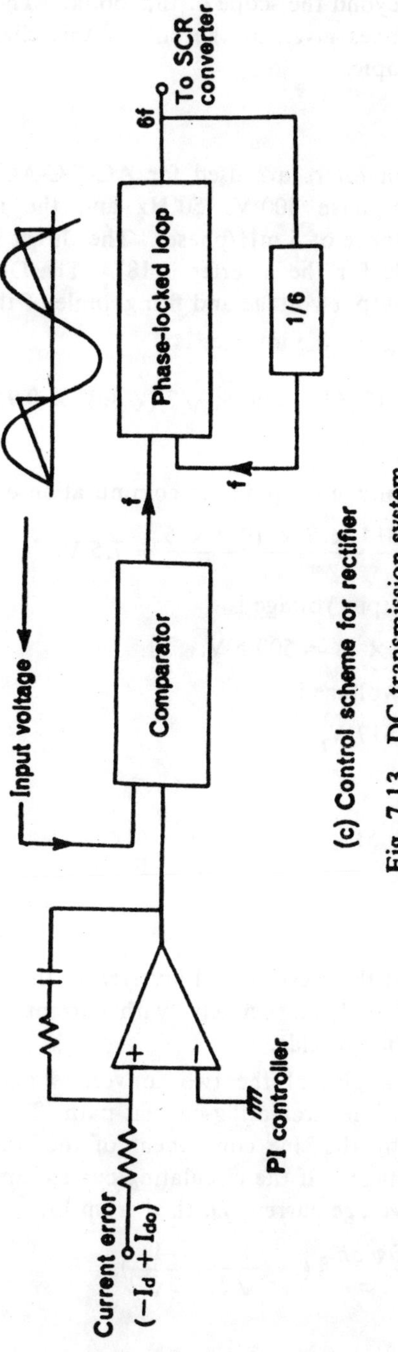

(c) Control scheme for rectifier

Fig. 7.13 DC transmission system.

SCR will be maintained constant even when the input voltages are distorted. Details of the dynamic performance of the DC transmission system and its controls are beyond the scope of this book. The interested reader should see the references given at the end of this chapter for additional information on this topic.

7.4.3 Example

(a) Two six-pulse converters are used for AC-DC-AC conversion. The AC system is three-phase 400 V, 50 Hz and the input transformers have a leakage inductance of 5 mH/phase. The direct line current is 5 A and the margin angle for the inverter is 18°. The DC line resistance is 1 Ω. Calculate the output voltage and firing angle of the rectifier.

The internal voltage of the inverter is

$$\frac{3\sqrt{3}E_m}{\pi}\cos\gamma = [3\sqrt{3}\times 400\times\sqrt{2}/(\sqrt{3}\pi)]\times 0.951$$
$$= 512\text{ V}.$$

The equivalent DC voltage drop due to commutation overlap is

$$\frac{3\omega L_s}{\pi}I_d = \frac{3\times 314\times 5\times 10^{-3}\times 5}{\pi} = 7.5\text{ V}.$$

Therefore, rectifier output voltage is

$$512 - 7.5 + (5\times 1) = 509.5\text{ V}.$$

The rectifier internal voltage is

$$509.5 + 7.5 = 517\text{ V}.$$

Therefore,

$$\frac{3\sqrt{3}E_m}{\pi}\cos\alpha = 517,$$

$$\cos\alpha = 0.98,$$

$$\alpha \approx 10°.$$

(b) Assuming that the rectifier and inverter in this example operate at firing angles $\pi/4$ and $3\pi/4$, respectively, with current I_d of 10 A, design a suitable reactor on the DC side.

Since the firing angles of the two converters add up to π, the DC voltages cancel out in the circulating-current path. The circulating current is produced mainly by the sine component of the sixth harmonic of the converter output voltage. If the circulating-current amplitude is taken to be one-fifth of the average current I_d, then from Eq. (7.18) we have

$$2 = \frac{2}{6\omega L_d}\times\frac{3\sqrt{3}E_m}{\pi}\left(-\frac{1}{7\sqrt{2}}+\frac{1}{5\sqrt{2}}\right).$$

Therefore,

$$L_d = \frac{3\times\sqrt{3}\times 400\times\sqrt{2}}{\sqrt{3}\times 6\times 314\times\pi}\times\frac{2}{35\sqrt{2}}\times 1000 = 11.6\text{ mH}.$$

(c) For the six-pulse converters connected in series as shown in Fig. 7.13a, obtain the alternating line current i_L, assuming that the load current I_d is steady and that the firing angle for both is α. What is the lowest order harmonic of i_L?

The input current through the converter fed from the transformer with the star-star connection is given by Eq. (7.11) to be

$$i_{L1} = \frac{2\sqrt{3}}{\pi} I_d[\sin(\omega t - \alpha) - \tfrac{1}{5}\sin 5(\omega t - \alpha) - \tfrac{1}{7}\sin 7(\omega t - \alpha) - \tfrac{1}{11}\sin 11(\omega t - \alpha) - \ldots].$$

The input voltages to the second converter are shifted by $\pi/6$ with respect to those to the first converter, and therefore, as the firing angle for both converters is the same, the input line current i_{L2} will be

$$i_{L2} = \frac{2\sqrt{3}}{\pi} I_d[\sin(\omega t - \alpha - \pi/6) - \tfrac{1}{5}\sin 5(\omega t - \alpha - \pi/6) - \tfrac{1}{7}\sin 7(\omega t - \alpha - \pi/6) - \tfrac{1}{11}\sin 11(\omega t - \alpha - \pi/6) - \ldots].$$

The primary line current i'_{L1} of the first converter will be equal to and in phase with the secondary line current i_{L1} since the transformer connection is star-star. However, for the second converter the primary line current i'_{L2} is phase-shifted and is given by (see Fig. 6.7c)

$$i'_{L2} = \frac{2\sqrt{3}}{\pi} I_d[\sin(\omega t - \alpha) + \tfrac{1}{5}\sin 5(\omega t - \alpha) + \tfrac{1}{7}\sin 7(\omega t - \alpha) - \tfrac{1}{11}\sin 11(\omega t - \alpha) - \ldots].$$

Therefore, the total current i_L $(= i'_{L1} + i'_{L2})$ will be

$$i_L = \frac{4\sqrt{3}}{\pi} I_d[\sin(\omega t - \alpha) - \tfrac{1}{11}\sin 11(\omega t - \alpha) - \ldots].$$

Thus, the lowest order harmonic in the input current is 11.

7.5 FREQUENCY CHANGERS

Frequency changers are referred to also as AC-to-AC converters. Figure 7.14a illustrates a very simple method of doubling the input frequency by

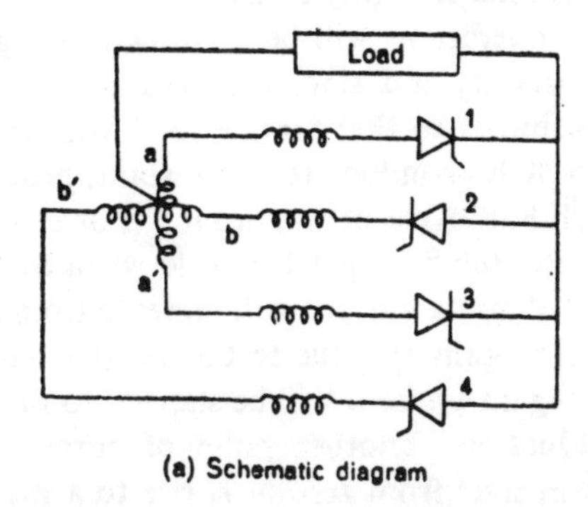

(a) Schematic diagram

Fig. 7.14 Frequency doubler (*cont.*).

means of line-commutated circuits. The input is a four-phase supply obtained by connecting the midpoints of the secondary windings of a two-phase transformer. The circuit shown produces a single-phase output. The configuration is similar to that for midpoint connection for rectification discussed in Chapter 6, except that half the SCRs here are connected in a direction opposite to that of the other half. A multiphase output can be obtained from *m* identical units supplying a common load. In our analysis, a single-phase circuit with a resistive load is considered. If the firing angle of each SCR is greater than $\pi/2$, then the SCRs will undergo natural commutation at the end of every half-cycle of input as shown in Fig. 7.14b. Therefore, the load current will be zero for a period $(\alpha - \pi/2)$. This is mode-1 operation where only one SCR conducts at a time. The sequence of firing is 1, 4, 3, and 2. The method of obtaining sequential firing will be explained in Chapter 9. The output voltage waveforms for different modes are shown in Fig. 7.14c. If the firing angle is $\pi/2$, then SCR4 will be fired immediately after SCR1 is turned off. Therefore, SCR1 will get forward-biased again since the potential of b' will be lower than that of a', and it will continue to conduct until SCR3 is fired. At this time, a reverse voltage will appear across SCR1 and it will be turned off, and SCRs 4 and 3 will conduct together. During the period SCRs 1 and 4 are conducting, the input phases a and b' will get shorted, a large circulating current will flow in the input, and the output load voltage will be the average of the phase voltages a and b'. The voltage waveform for this mode-2 operation is shown in Fig. 7.14b. Here, two SCRs conduct all the time. A current-limiting reactor is necessary to reduce the circulating current. Even if the load is open-circuited, there will be a continuous flow of line currents due to the periodic short-circuits in this mode of operation. If the firing angle is reduced still further, then SCR4 will get fired even before the current in SCR1 becomes zero. Therefore, for some period three SCRs will be conducting and the output voltage during this time will be zero. This is mode 3. Thus, the output voltage is maximum for mode 2, and in mode 3 it reduces fairly linearly with α. When $\alpha = \pi/4$, the output will be permanently short-circuited.

Assuming that the load is highly resistive, the input current waveforms for various modes of operations will be as shown in Fig. 7.15. In mode 1, there are no short-circuits, and since the load is resistive, the input phase current waveform is similar to that for the load voltage. In mode 2, for a firing angle $\pi/2$, two SCRs conduct at any instant, producing short-circuits on the phases in each half-cycle of the input. For example, phases a and b will be shorted during the first quarter-cycle when SCR1 is turned on (see Fig. 7.15) at point p. Phase a current will increase from zero to a maximum at q and return to zero again at r due to the reverse voltage applied. At r, SCR4 is fired and again phase a will be shorted to phase b'. Thus, SCR1 will continue to conduct and another pulse of current will flow through phase a; this pulse will start from zero at r, rise to a maximum value at s, and return to zero again at t. The other phase currents will also have the

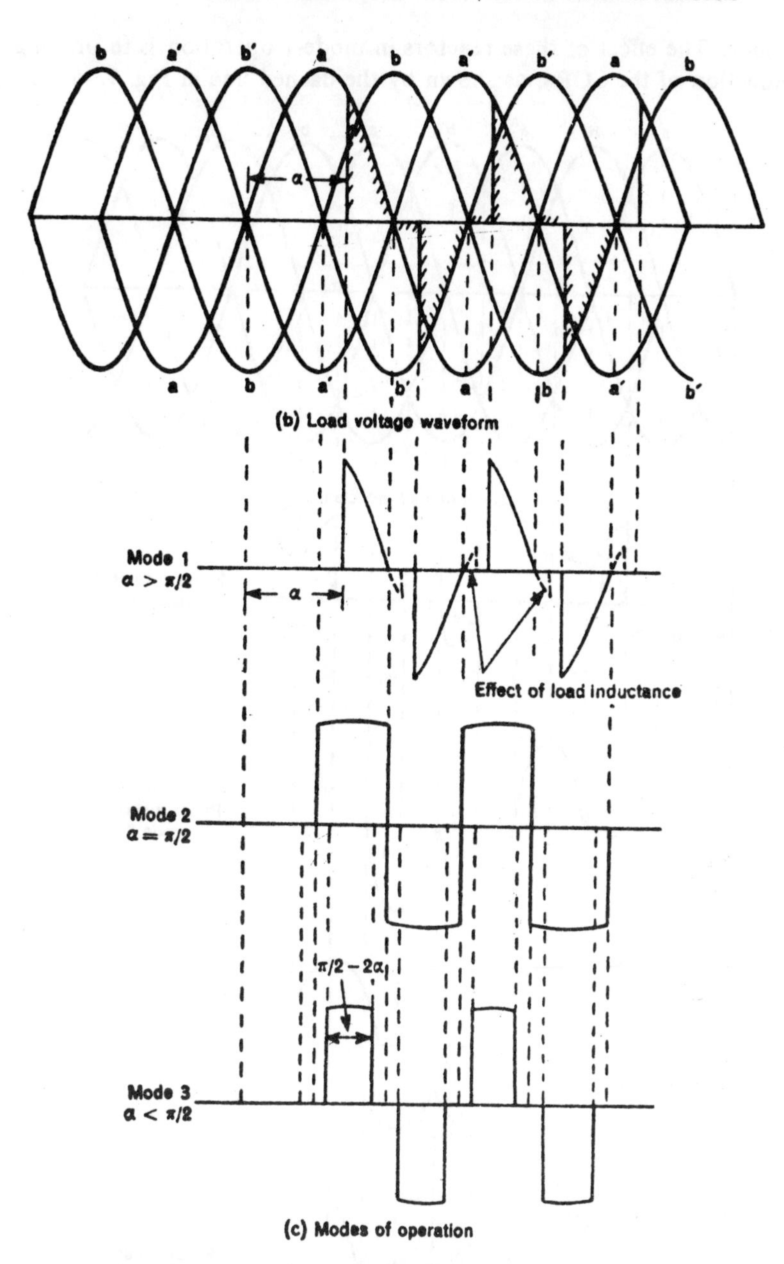

Fig. 7.14 Frequency doubler.

same waveform. In Fig 7.15, the current in phase a' is shown negative because both phases a and a' are derived from the same winding. In mode-2 operation, even if the load is open-circuited, there will be currents on the input side because of the short-circuits. Since reactors are used to reduce these currents, only reactive power will be consumed, and therefore the input power factor during mode-2 operation, with load connected, will be poor. The effect of these reactors in mode-1 operation is to prolong the conduction of the SCRs, as shown by the dashed line in Fig. 7.15. It can

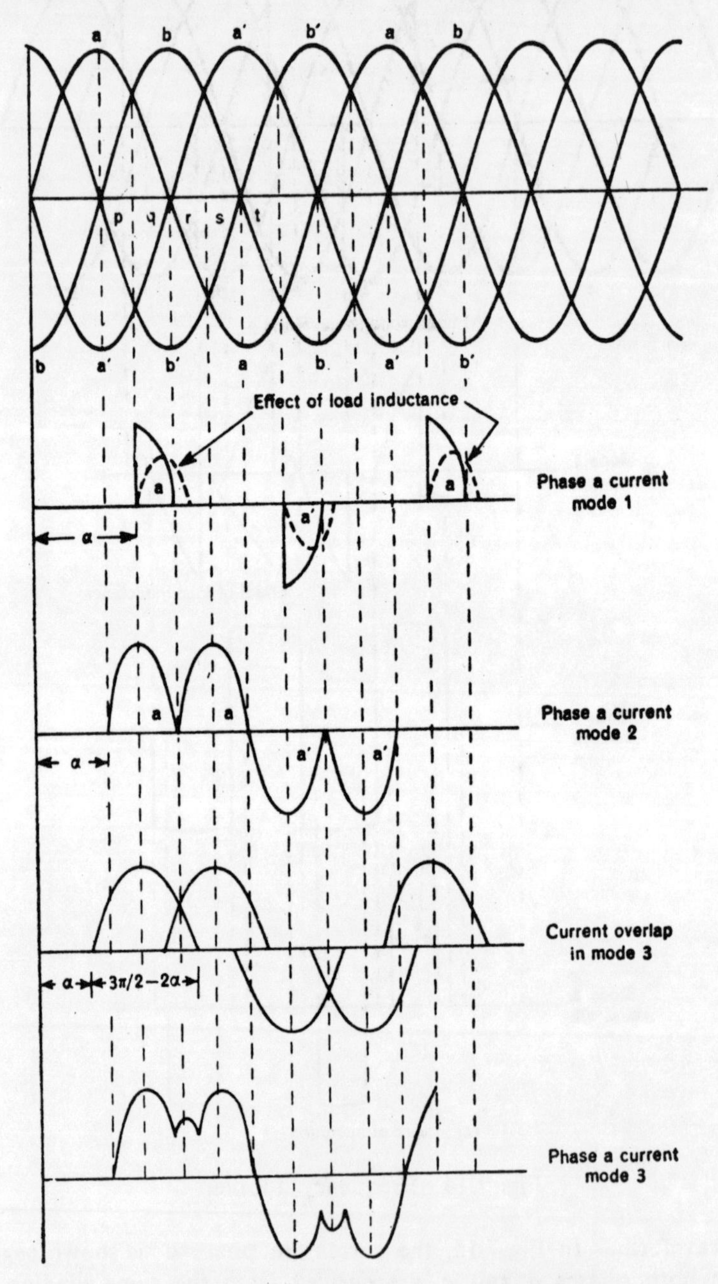

Fig. 7.15 Phase current waveforms.

be easily seen that, if the load power factor is lagging, the changeover from mode 1 to mode 2 will take place at firing angles larger than $\pi/2$. In mode-3 operation, the short-circuits overlap. For example, when SCRs 3 and 2 are conducting (phases b and a' shorted), the firing of SCR1 will increase phase b current and reduce the current in SCR3. Finally, SCR3 will get reverse-biased and only SCRs 1 and 2 will conduct, producing a short between b and a. Since the phase voltages of a and a' are equal and opposite, the change of current in SCR3 during the overlap period will be equal to that in SCR1 but opposite in direction because the load voltage is zero. Therefore, the total phase current will be obtained by drawing the individual current pulses, starting from the firing angle $\alpha < \pi/2$ and adding them as shown in Fig. 7.15 (for mode 3). The duration of conduction of each current pulse will be $(3\pi/2 - 2\alpha)$, and that of the voltage pulse will be $(\pi/2 - 2\alpha)$.

Neglecting the presence of the current-limiting reactors, the RMS value of the no-load voltage for different modes of operation (see Fig. 7.14) will be

$$
\begin{aligned}
e_{o\,\mathrm{RMS}} &= \frac{E_m}{\sqrt{\pi}}\sqrt{\pi - \alpha + \tfrac{1}{2}\sin 2\alpha}, && \text{mode 1, } \alpha > \pi/2, \\
e_{o\,\mathrm{RMS}} &= \frac{E_m}{\sqrt{(2\pi)}}\sqrt{\pi/2 + 1}, && \text{mode 2, } \alpha = \pi/2, \qquad (7.43) \\
e_{o\,\mathrm{RMS}} &= \frac{E_m}{\sqrt{(2\pi)}}\sqrt{2\alpha - \cos 2\alpha - \pi/2}, && \text{mode 3, } \alpha < \pi/2.
\end{aligned}
$$

In the foregoing expressions, α is the firing angle (measured as shown in Fig. 7.14b) and E_m is the maximum phase voltage.

Since the efficiency of the circuit (Fig. 7.14a) decreases in mode-2 and mode-3 operations due to the continuous flow of short-circuit currents, it is advisable to operate this circuit only in mode 1. The Fourier series expansion of the output voltage for this mode is

$$e_o = e_{ms}\sin m\omega t + e_{mc}\cos m\omega t, \qquad m = 2, 6, 10, \qquad (7.44)$$

where

$$e_{ms} = \frac{4E_m}{\pi} \times \frac{1}{m^2 - 1}(m \sin\alpha\cos m\alpha + \sin m\alpha\cos\alpha),$$

$$e_{mc} = \frac{-4E_m}{\pi} \times \frac{1}{m^2 - 1}(1 + \cos m\alpha\cos\alpha + m\sin m\alpha\sin\alpha),$$

and ω is the input frequency.

Figure 7.16a shows the schematic diagram of a frequency tripler. Here, the output frequency is three times the input frequency. This circuit requires a three-phase input and produces a single-phase output. The firing sequence of SCRs is 1, 3′, 2, 1′, 3, and 2′. All the modes of operation explained for a frequency doubler will also take place in this circuit on varying the firing angle α. Figure 7.16b shows the output voltage waveform for mode 1 when $\alpha > 2\pi/3$ (α is measured as shown in Fig. 7.16b). A resistive load is assumed. In this mode, only one SCR will conduct at a time. There will be no short-circuit on the input. Mode 2 will start when $\alpha = 2\pi/3$. Here, two SCRs will conduct simultaneously, producing

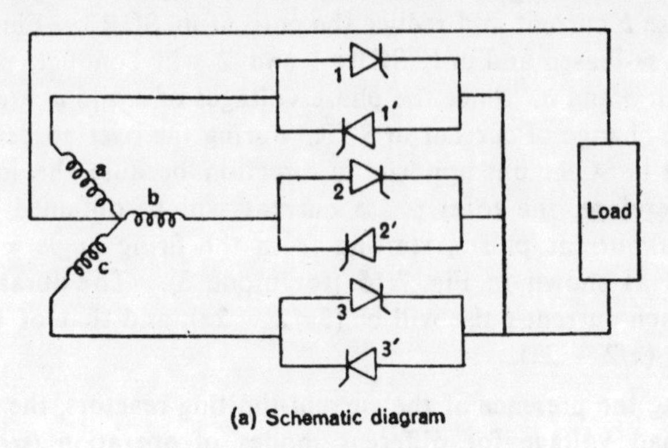

(a) Schematic diagram

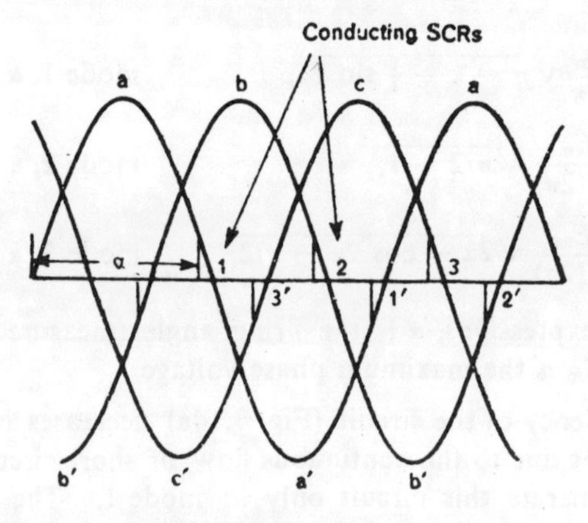

(b) Load voltage waveform (mode 1)

Fig. 7.16 Frequency tripler.

a short on two of the input phases. The load voltage will be the average voltage of the conducting phases and fairly rectangular. Mode 3 will start when $\alpha < 2\pi/3$, and the load voltage will become zero six times in every cycle, each time for a duration $(4\pi/3 - 2\alpha)$. When $\alpha = \pi/2$, the output will be permanently shorted and the load voltage will be zero.

It can now be generalised that, to obtain a single phase output n times the input frequency, a $2n$-phase input supply is needed when n is even, and an n-phase supply is required when n is odd. Therefore, this method of frequency conversion will not be suitable when output frequencies of more than six times the input frequency are desired. For such high frequencies, it is more convenient to rectify the input AC to DC, and use the forced-commutated circuits described in Chapters 8 and 9 for DC-to-AC inversion.

The main advantage of AC-to-AC conversion discussed here is that the SCRs are turned off either by natural commutation or by line voltage commutation, and no extra commutating components are needed as in forced-commutated inverters. Further, since there is no intermediate DC stage, a smaller number of components is required. These advantages must be weighed against the distortion in the output waveform, limited voltage variation, and the possibility of short-circuits on the input. Some applications of AC-to-AC converters are in high-speed drives and in electric heating and lighting.

7.5.1 Example

A four-phase supply is used for obtaining a single-phase output of twice the input frequency. The input phase voltage is 230 V at 50 Hz. The circuit operates in mode 1 with $\alpha = 2\pi/3$. Obtain the RMS value and fundamental frequency component of the output. Assume the load to be resistive.

In mode 1, only one SCR conducts at a time. Therefore, the load voltage will be the same as the conducting phase voltage. The RMS value of the load voltage V_L will be

$$V_L = \sqrt{\frac{2}{\pi}\int_{2\pi/3}^{\pi} (E_m \sin \theta)^2 \, d\theta}$$

$$= E_m \sqrt{\frac{2}{\pi}} \times \frac{1}{\sqrt{2}} \sqrt{\int_{2\pi/3}^{\pi} (1 - \cos 2\theta) \, d\theta}$$

$$= E_m \sqrt{\frac{2}{\pi}} \times \frac{1}{\sqrt{2}} \sqrt{\left[\frac{\theta - \sin 2\theta}{2}\right]_{2\pi/3}^{\pi}}$$

$$= \frac{E_m}{\sqrt{\pi}} \sqrt{\left(\pi - \frac{2\pi}{3}\right) + \tfrac{1}{2} \sin 4\pi/3}$$

$$= \frac{E_m}{\sqrt{\pi}} \sqrt{\frac{\pi}{3} - \frac{\sqrt{3}}{4}} = \frac{230\sqrt{2}}{\sqrt{\pi}} \times 0.78 = 215 \text{ V}.$$

The amplitude of the fundamental frequency component of the output is

$$V_{L1} = [V_{L1s}^2 + V_{L1c}^2]^{1/2},$$

where

$$V_{L1s} = \frac{4}{\pi}\int_0^{\pi-\alpha} E_m \sin(\theta + \alpha) \sin 2\theta \, d\theta = 119,$$

$$V_{L1c} = \frac{4}{\pi}\int_0^{\pi-\alpha} E_m \sin(\theta + \alpha) \cos 2\theta \, d\theta = 135.$$

Hence,

$$V_{L1} = \sqrt{119^2 + 135^2} = 182.5 \text{ V}.$$

Verify this result by using Eq. (7.44).

7.5.2 High-Frequency Conversion

The method of frequency conversion discussed in Section 7.5.1 is not suitable for high-frequency output due to the large number of input phases required and the poor utilisation of SCRs (since the duration of conduction for each SCR is considerably reduced). Figure 7.17 shows one method of frequency conversion using resonant turn-off or load commutation (see Chapter 8). Three-phase supply is used. Each SCR will turn off when the load current becomes zero. The SCRs are gated in a proper sequence

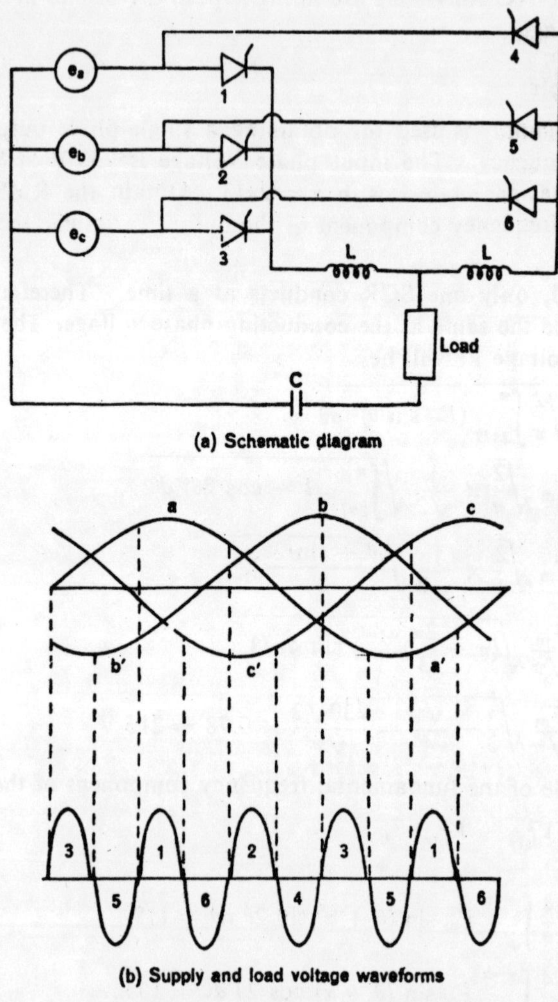

(a) Schematic diagram

(b) Supply and load voltage waveforms

Fig. 7.17 Cycloinverter.

to obtain symmetrical output at the required frequency. This principle is known as *cycloinversion* and is different from that of *cycloconverters* which are used for low-frequency generation. The capacitor and inductor make the output load circuit underdamped. The natural frequency of this circuit must be higher than the output frequency. These two criteria can be used for obtaining proper values of L and C. A similar circuit with DC input in place of three-phase supply can be used for high-frequency generation. Details of this are given in Chapter 8 in the section dealing with series inverters (Section 8.6).

7.6 CYCLOCONVERTERS

The principle of a cycloconverter can be easily understood with the help of Fig. 7.18a. The circuit shown is for obtaining a single-phase low-frequency output from a single-phase AC input. One group of SCRs produces positive polarity load voltage and other group produces negative half-cycle of the output. SCRs 1 and 1′ of the positive group are gated

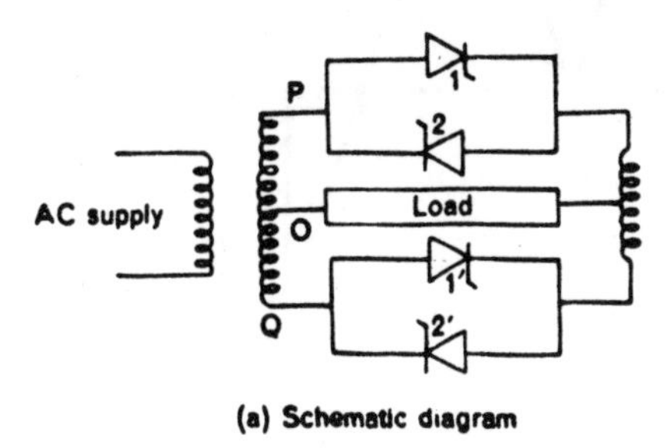

(a) Schematic diagram

Fig. 7.18 Cycloconverter (*cont.*).

together. Depending on the polarity of the input, only one of them will conduct. When P is positive with respect to O, SCR1 will conduct, and when P is negative, SCR1′ will conduct. Thus, in both half-cycles of input, the load voltage polarity will be positive. By changing the firing angle α, the duration of conduction of each SCR (and thereby the magnitude of the output voltage) can be varied. For the sake of simplicity, it is assumed that the load is resistive. Then, each SCR will have a conduction angle of $(\pi - \alpha)$ and turn off by natural commutation at the end of every half-cycle of the input. Figure 7.18b shows the output voltage waveform (hatched portion). At the end of each half-period of the output (at the frequency desired), the firing pulses to the SCRs of the positive group will be stopped and SCRs 2 and 2′ of the negative group will be fired. SCR2 will conduct when P is negative and SCR2′ will conduct when Q is negative. The last conducting SCR of the positive group will turn off when the voltage goes through zero. If the output frequency is not a submultiple of the input frequency, the firing of SCRs 2 and 2′ will start even before the last half-cycle of conduction of the SCRs in the positive group is completed. That is, SCRs 2 and 2′ will be fired when SCR1 or SCR1′ is conducting. This will produce a short-circuit on the input, and the load voltage will be reduced to zero. The duration of the short-circuit will be less than one-half the period of the input, and will take place once in every half-cycle of the output. Current-limiting reactors therefore must be introduced in the input lines to reduce the short-circuit current.

If the load is inductive, then the SCRs will not turn off at the end of every half-cycle of the input voltage. Instead, there will be a negative excursion of the load voltage as shown in Fig. 7.18b. If the firing angle is small, then the conduction will pass from one SCR to another in the same group by line commutation. For example, SCR1 which was fired in the positive half-cycle will continue to conduct during the negative half-cycle due to load inductance, making the load voltage negative, and will be turned off when SCR1′ is fired. The load voltage will again become positive. At the

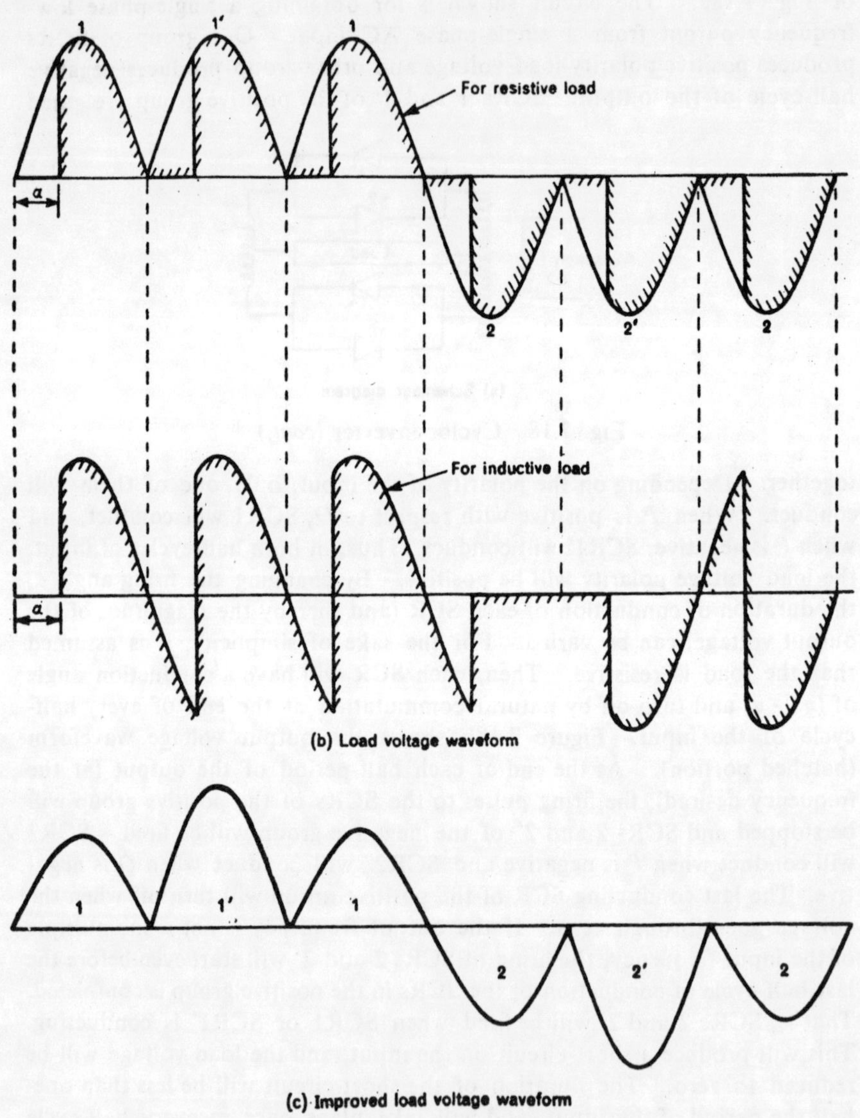

Fig. 7.18 Cycloconverter.

beginning of the next half-cycle of the output when SCRs 2 and 2′ are gated, SCR1 will still be conducting in the negative half-cycle and so, the SCRs in the negative group, being reverse-biased, will not turn on. It is assumed that all SCRs are gated by only a single pulse. Therefore, no voltage will appear across the load after SCR1 is turned off by natural commutation. The load will get open-circuited for a maximum duration of one-half the input period. Thus, the load voltage waveform will get distorted. It can be easily seen that, for the circuit shown in Fig. 7.18a, the distortion in the output voltage will increase as the load power factor decreases. (The operation of this circuit with large reactive loads will be discussed in Section 7.6.2.) In order to obtain symmetrical output voltage waveform, the output frequency must be an integral submultiple of the input frequency. Otherwise, in addition to the periodic short-circuit, waveform distortion will take place, which may result in the saturation of the input transformer due to asymmetrical currents and the generation of beat-frequency components. For these reasons, cycloconverters are used for producing very low-frequency output so that the duration of short- and open-circuits becomes a small portion of the total output period. Thus, the highest output frequency using cycloconverters is limited to one-third of the input frequency. Output power control is obtained by changing the firing angle of the SCRs.

A simple method of improving the output voltage waveform when the output frequency is one-third the input frequency is to connect SCRs 1′ and 2′ to a higher tap on the secondary winding of the input transformer. That is, the voltage across OQ is made twice that across OP (Fig. 7.18a). The resulting output voltage waveform of the cycloconverter with a resistive load and zero firing angle is shown in Fig. 7.18c.

7.6.1 Mathematical Analysis

Assume that the circuit shown in Fig. 7.18a has a resistive load. Let the output frequency be $1/m$ times the input frequency. Thus, there will be m half-cycles of input in each half-period of the output. Let the firing angle be α. Then, the RMS value of the output voltage will be

$$e_{o\,\mathrm{RMS}} = \sqrt{\tfrac{1}{2}\frac{E_m^2}{\pi}(\pi - \alpha + \tfrac{1}{2}\sin 2\alpha)}. \tag{7.45}$$

The fundamental amplitude is given by

$$\begin{aligned} e_{o_{1s}} &= \frac{2E_m}{m\pi}\Big\{\int_\alpha^\pi \sin\theta \sin\left(\frac{\theta}{m}\right) d\theta + \int_\alpha^\pi \sin\theta\sin\left(\frac{\theta+\pi}{m}\right) d\theta \\ &\quad + \dots + \int_\alpha^\pi \sin\theta \sin\left[\frac{\theta + (m-1)\pi}{m}\right] d\theta\Big\} \\ &= \frac{E_m}{m\pi}\sum_{n=1}^{m}\Big\{\frac{m}{m-1}\Big[\sin\frac{n\pi}{m} - \sin\Big(\alpha - \frac{\alpha+(n-1)\pi}{m}\Big)\Big] \\ &\quad + \frac{m}{m+1}\Big[\sin\frac{n\pi}{m} + \sin\Big(\alpha - \frac{\alpha+(n-1)\pi}{m}\Big)\Big]\Big\}. \end{aligned}$$

Similarly,

$$\begin{aligned} e_{o_{1c}} &= \frac{E_m}{m\pi}\sum_{n=1}^{m}\Big\{\frac{m}{m+1}\Big[\cos\frac{n\pi}{m} + \cos\Big(\alpha + \frac{\alpha+(n-1)\pi}{m}\Big)\Big] \\ &\quad + \frac{m}{m-1}\Big[\cos\frac{n\pi}{m} + \cos\Big(\alpha + \frac{\alpha+(n-1)\pi}{m}\Big)\Big]\Big\}; \end{aligned}$$

thus, we have

$$E_1^2 = e_{o_{1s}}^2 + e_{o_{1c}}^2, \tag{7.46}$$

where E_1 is the fundamental amplitude of the output voltage and E_m is the peak value of the input voltage. The foregoing expressions have been derived for a resistive load. Amplitudes of the harmonic components of the output can be obtained by dividing variable m appearing within the curly brackets of the expressions for $e_{o_{1s}}$ and $e_{o_{1c}}$ by the order of the harmonic under consideration.

7.6.2 Bridge Configuration

The input transformer used for the midpoint connection shown in Fig. 7.18a will not be required if the bridge configuration shown in Fig. 7.19a is used. Here, two single-phase fully-controlled bridges are connected in opposite directions, as in the dual converter schemes discussed in Chapter 6. Bridge 1 will produce positive load current, and when bridge 2 conducts, the load current will reverse. The two bridges should not conduct together as this will produce a short-circuit on the input. When the load current is positive, the firing pulses to the SCRs of bridge 2 will be inhibited and bridge 1 will be gated. Similarly, when the load current reverses, bridge 2 will be gated and the firing pulses will not be applied to the SCRs in bridge 1.

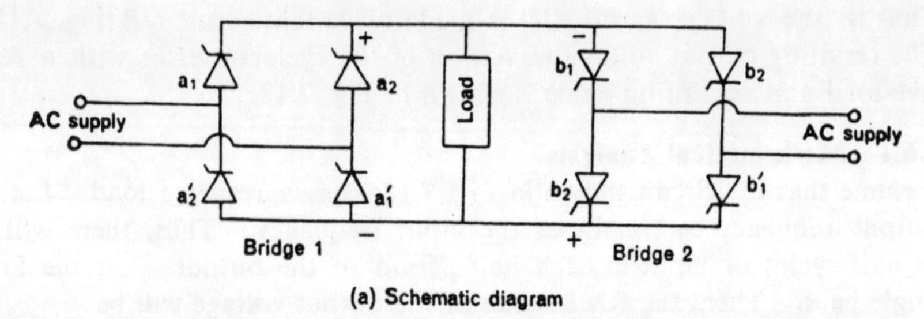

(a) Schematic diagram

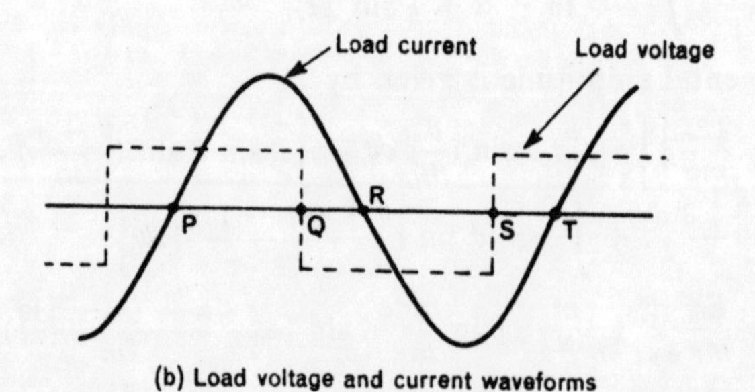

(b) Load voltage and current waveforms

Fig. 7.19 Bridge configuration for cycloconverter.

The firing angles of SCRs in the two bridges are kept the same so that the output load voltage is symmetrical. For resistive loads, the SCRs undergo natural commutation and produce discontinuous-current operation. For inductive loads, line commutation may take place and the load current may be continuous. Since for inductive loads the SCRs conduct for a duration more than $(\pi - \alpha)$ in each half-cycle of the input, the output voltage will become negative, as shown in Fig. 7.18b, during each positive half-cycle of the output. Free-wheeling diodes cannot be used across the load to cut off the negative excursions of the output voltage since this voltage is AC. Half-controlled bridges too will not permit the load voltage to become negative due to internal free-wheeling action. However, these bridges cannot be used as they will not operate in the inverting mode (see Chapter 6); but such a mode of operation is essential (as will be explained later in this section) for cycloconverters feeding an inductive load.

For the output waveforms shown in Fig. 7.19b, the load voltage is approximated by a rectangular wave, and an inductive load is assumed. Since the gating pulses are controlled by the load current, bridge 1 can conduct only during interval PR and bridge 2 during interval RT. From P to Q, both load voltage and current are positive, and therefore bridge 1 will operate as a rectifier with, say, a firing angle α. At Q, the load voltage will reverse. Since the load current will still be positive, bridge 2 will not be gated. Therefore, to generate negative load voltage with positive current, bridge 1 must operate in the inverting mode with a firing angle $(\pi - \alpha)$ so that the negative load voltage is equal in magnitude to the positive load voltage. The operation of the SCR bridge as an inverter, using line commutation, has been explained in Section 7.3. At R, the load current will reverse because of the negative voltage applied to the load and bridge 2 will receive the gating pulses. Since the SCRs are fired at discrete intervals (once in each half-cycle of the input), it is quite likely that by the time the load current has reversed its polarity at R, the firing of the SCRs in bridge 2 for the first half-cycle may not take place. Thus, the load will be open-circuited until the next firing pulse is applied to bridge 2. If the output frequency is low, the duration of the open-circuit condition for the load will be a very small fraction of the output period. This condition is not represented in Fig. 7.19b and the load current is assumed to reverse immediately and flow through bridge 2. Between R and S, bridge 2 will operate as a rectifier with a firing angle α. At S, the voltage will reverse and the current will be negative. So, bridge 1 will not be gated and bridge 2 will operate in the inverting mode with a firing angle $(\pi - \alpha)$.

The following logic is used to control the firing of the SCRs:

	Bridge 1	*Bridge 2*
Load voltage positive, load current positive or zero	Firing angle α	No firing
Load voltage negative, load current positive	Firing angle $(\pi - \alpha)$	No firing
Load voltage negative, load current negative or zero	No firing	Firing angle α
Load voltage positive, load current negative	No firing	Firing angle $(\pi - \alpha)$

With this logic, the periodic short-circuits on the input (which take place when the output frequency is not a submultiple of the input frequency) are avoided because the gating of the bridges is controlled by the current. All SCRs belonging to any group or bridge are fired together but only two will conduct at a time. As long as bridge 1 is conducting, the SCRs in bridge 2 will not receive any firing pulses. This type of control is known as the *noncirculating-current scheme* because only one bridge conducts at a time and there is no circulating current between the bridges. The only disadvantage of this scheme is the dead-time during the changeover of current, when the second bridge does not conduct immediately, and the load gets open-circuited for a maximum duration of one-half the input cycle.

7.6.3 Control Circuit

Figure 7.20a shows the schematic diagram of a control circuit for gating cycloconverter bridges. UJT1 is a relaxation oscillator whose output frequency is synchronised with the AC supply (see Fig. 6.9). It produces one

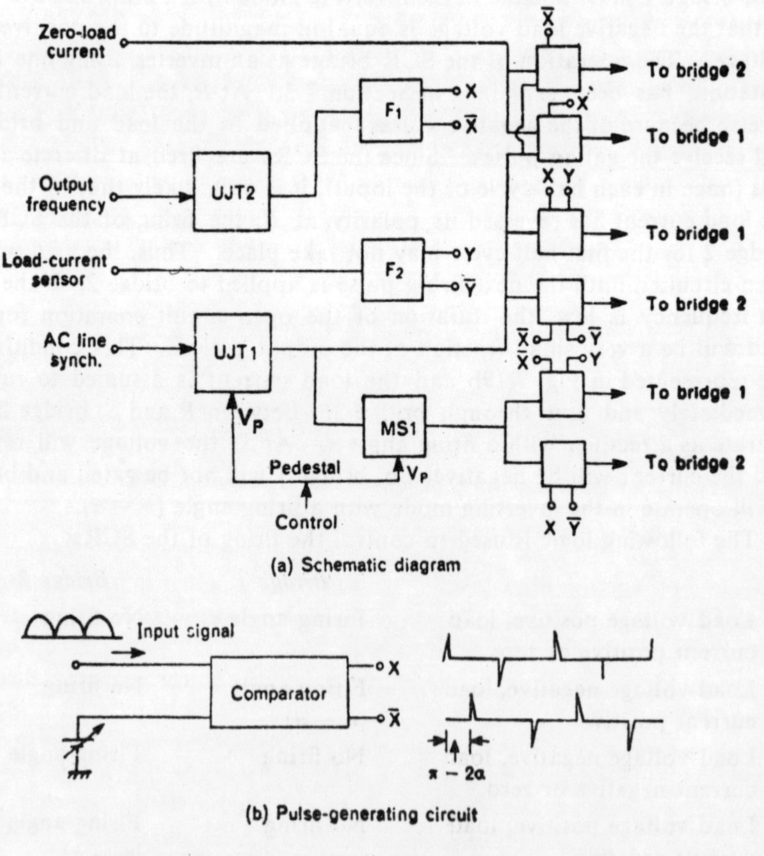

Fig. 7.20 Control circuit for cycloconverter.

pulse in every half-cycle of the input. The firing angle α of each pulse can be varied by changing pedestal control voltage V_P. The maximum value of α is limited to $\pi/2$. MS1 is a variable-width monostable circuit which is driven by the pulses from the UJT and produces output pulses with a delay of $(\pi - 2\alpha)$. Pedestal control voltage V_P is used also for controlling the delay produced by MS1 so that the phase angle between the pulses from UJT1 and MS1 is $(\pi - 2\alpha)$.

The circuit shown in Fig. 7.20b can be used also for producing two pulses differing by an angle $(\pi - 2\alpha)$ if the firing angle α is less than $\pi/2$. Here, a rectified sinusoidal signal, which is in phase with the input voltage of the cycloconverter, is compared with a DC voltage. The comparator output is differentiated to produce two pulse trains with a phase difference of $(\pi - 2\alpha)$. The firing angle α is changed by varying the DC voltage.

The two pulse trains produced by the aforementioned circuits are properly applied to the SCRs in each bridge by the six logic gates (Fig. 7.20a). Each gate allows the pulses to pass through if the two inputs to it are positive. UJT2 is another relaxation oscillator whose frequency is twice the desired output frequency. It drives a flip-flop F_1 whose complementary outputs X and $\bar{X}$ are used as inputs to the logic gates. During the period the load voltage is positive, X will be positive, and in the negative half-cycle of the load voltage, $\bar{X}$ will be positive. Similarly, F_2 is another flip-flop which is driven by a load-current sensor. The outputs of F_2 are Y and $\bar{Y}$, and these also are fed to the logic gates. If the load current is positive, Y will be positive; if the load current becomes negative, $\bar{Y}$ will be positive. The condition that the load may be open-circuited is also included in the control circuit.

With this control scheme, the cycloconverter can deliver load at all power factors. Figure 7.21a shows the actual output voltage waveform for an inductive load if the firing angle is maintained constant. The output frequency is one-fourth the input frequency. At the end of the positive half-cycle (point Q), the current (shown by the dashed line) will still be positive, and so the firing pulses to bridge 2 will be blocked and bridge 1 will operate in the inverting mode with a firing angle $(\pi - \alpha)$. Thus, SCRs a_1 and a_1' in bridge 1 (Fig. 7.19a), which were fired in the previous positive half-cycle, will continue to conduct in the negative half-cycle. When SCRs a_2 and a_2' are fired at an angle $(\pi - \alpha)$ (point R), SCRs a_1 and a_1' will get reverse-biased and turn off. Because of the negative voltage appearing across the load, the load current will decrease and go through zero at point S. Thereafter, the gating pulses will be applied to bridge 2. If point S occurs between S' and T, then SCRs b_2 and b_2' will get fired at T and the load current will reverse as shown in Fig. 7.21a. If point S occurs after T, then SCRs b_2 and b_2' will not be gated during that half-cycle and the current will continue to be zero till the next pulse fires SCRs b_1 and b_1' at point U.

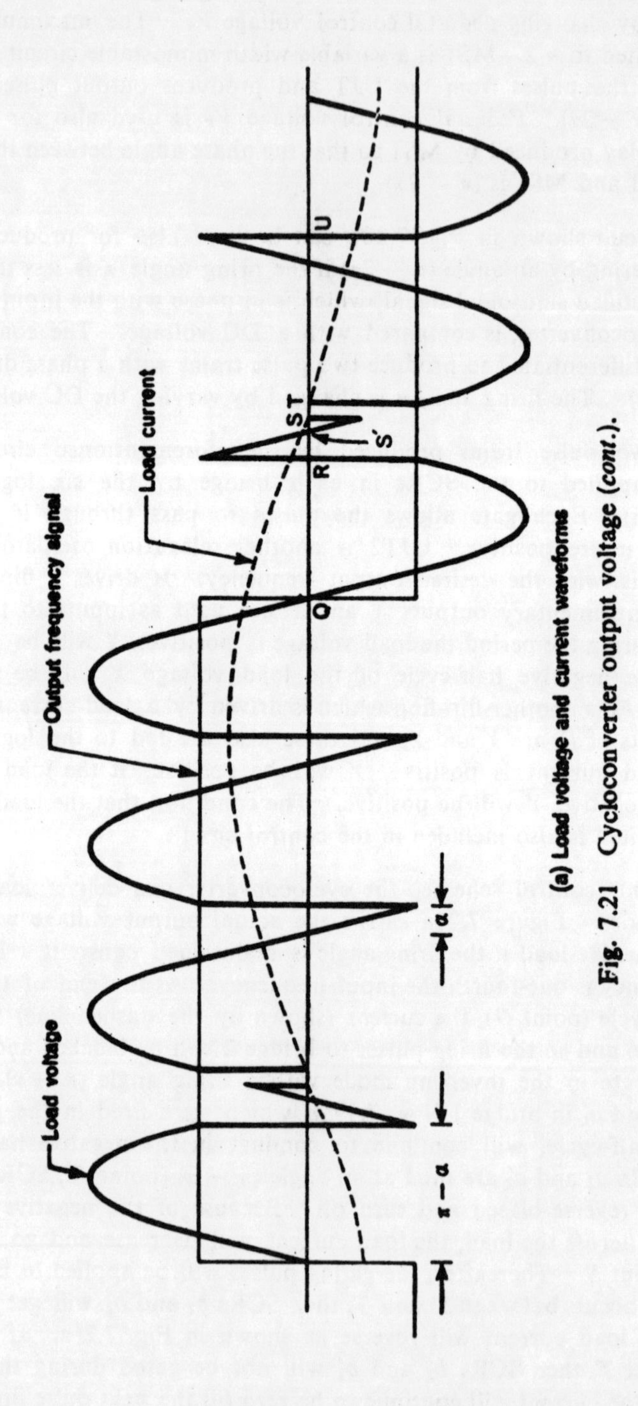

(a) Load voltage and current waveforms

Fig. 7.21 Cycloconverter output voltage (*cont.*).

7.6.4 Improved Cycloconverter Circuits

Figure 7.21b shows a method of controlling the firing angle α for the SCRs in each bridge to reduce harmonic distortion in the output load

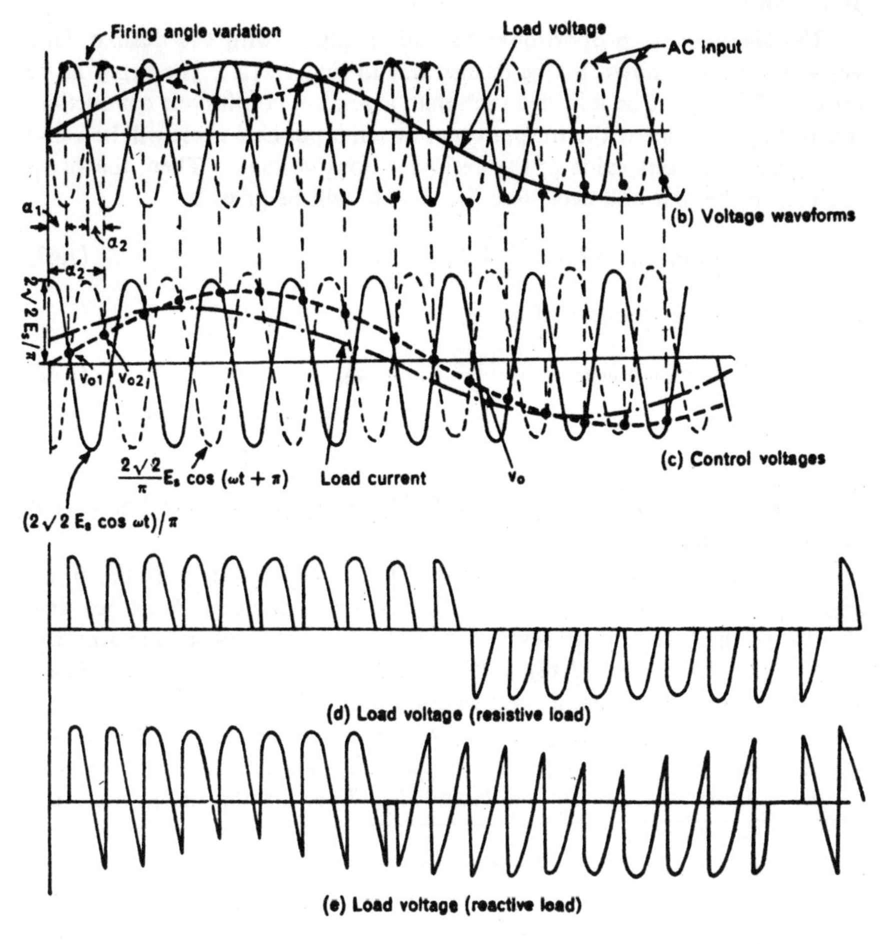

Fig. 7.21 Cycloconverter output voltage.

voltage. The variation of α in each half-cycle approximates to a cosine curve of a frequency twice the output frequency, as shown by the dashed line. The controller required to achieve this type of variation in α uses a comparator in place of a relaxation oscillator (UJT1 in Fig. 7.20a). This comparator compares the instantaneous magnitudes of the phase-shifted AC voltage e_1 of peak value $2\sqrt{2}E_s/\pi$ (where E_s is the RMS value of the input voltage to the bridge) and the sinusoidal signal v_o of peak value $\sqrt{2}V_o$ and output frequency ω_o. An output pulse is generated by the comparator when v_o is greater than e_1. This pulse is used to fire SCRs a_1 and a_1' in bridge 1 (Fig. 7.19a). Another comparator with signal e_1 reversed in polarity is used for triggering SCRs a_2 and a_2'. A similar control scheme is used for triggering the SCRs in bridge 2. However, for this controller, signal v_o is inverted before applying it to the comparators. The output pulses from the comparators are used to fire SCRs b_1 and b_1', and b_2 and b_2'.

The signal v_o is proportional to and in phase with the desired load voltage. This is shown by the dashed line in Fig. 7.21c. The firing angle for each SCR must be adjusted such that, assuming continuous conduction, the average voltage at the output due to each input half-cycle will be equal to the instantaneous value of the desired output voltage. Then, the firing angle α_1 in the first half-cycle (see Fig. 7.21b) will be given by

$$v_{o1} = \sqrt{2}V_o \sin \omega_o t_1 = \frac{2\sqrt{2}}{\pi} E_s \cos \alpha_1. \tag{7.47}$$

Similarly, for the second half-cycle,

$$v_{o2} = \sqrt{2}V_o \sin \omega_o t_2 = -\frac{2\sqrt{2}}{\pi} E_s \cos \alpha_2', \tag{7.48}$$

where

$$\alpha_2' = (\alpha_2 + \pi).$$

In these equations, V_o and E_s are the RMS values of the control and input voltages, respectively.

For the bridge circuit shown in Fig. 7.19a, Eq. (7.47) gives the firing angle for SCRs a_1 and a_1', and Eq. (7.48) the firing angle for SCRs a_2 and a_2'. When the load voltage reverses and the current polarity remains the same, Eqs. (7.47) and (7.48) will automatically give the required firing angles which result in the inverter operation of bridge 1. When the load current reverses, the gating pulses will be removed from bridge 1 and applied to the SCRs in bridge 2. Equation (7.47) is used for obtaining the firing angle for SCRs b_1 and b_1' and Eq. (7.48) for the firing angle for SCRs b_2 and b_2'.

The method of obtaining angles α_1 and α_2 from Eqs. (7.47) and (7.48) is shown in Fig. 7.21c. When the instantaneous value of v_o becomes equal to $(2\sqrt{2}E_s \cos \omega t)/\pi$, a firing pulse will be generated. (The cross-over points and the firing angles are indicated in the figure.) Evidently, this pulse will be generated in alternate half-cycles and will fire SCRs a_1 and a_1'. The comparators discussed earlier (Fig. 7.20b or Fig. 6.9c) can be used for generating the pulses. The firing angles will be

$$\begin{aligned} \alpha_1 &= \cos^{-1}\left(\frac{v_{o1}\pi}{2\sqrt{2}E_s}\right), \\ \alpha_3' &= \cos^{-1}\left(\frac{v_{o3}\pi}{2\sqrt{2}E_s}\right) + 2\pi, \\ \alpha_5' &= \cos^{-1}\left(\frac{v_{o5}\pi}{2\sqrt{2}E_s}\right) + 4\pi, \end{aligned} \tag{7.49}$$

where v_{o1}, v_{o3}, and v_{o5} are the instantaneous values of the control voltage at the instants of firing. Similarly, the firing angles α_2' and α_4' for SCRs a_2 and a_2' can be obtained by comparing the instantaneous value of the control signal with $(-2\sqrt{2}/\pi)E_s \cos \omega t$, where ω is the input frequency. It can be seen that the output load voltage amplitude may be controlled by varying V_o, and is maximum when V_o is equal to $2E_s/\pi$. Figures 7.21d and 7.21e show the output voltage waveform for two load conditions. In Fig. 7.21d, the load is resistive and each SCR turns off by natural commutation. In Fig. 7.21e, the load is reactive and produces continuous current so that the SCRs turn off by line commutation and the average output voltage is fairly sinusoidal.

Another convenient method of improving the output voltage waveform is to use a three-phase input. Figure 7.22a shows the schematic diagram

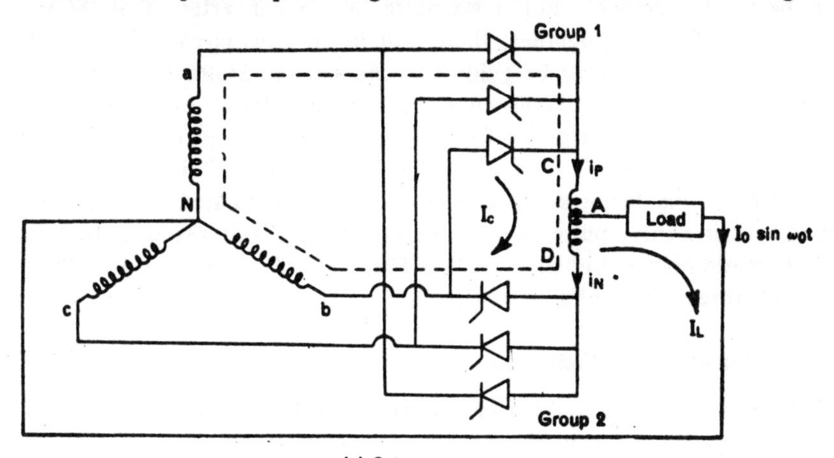

(a) Schematic diagram

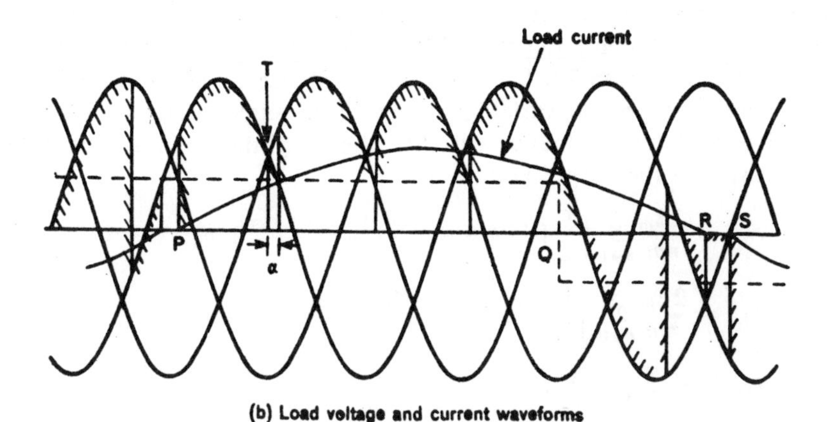

(b) Load voltage and current waveforms

Fig. 7.22 Three-pulse cycloconverter.

for a three-phase half-wave circuit using a midpoint connection. SCR groups 1 and 2 operate in both the rectifying and inverting modes depending on the polarity of the load current. Figure 7.22b shows the output voltage waveform for a fixed firing angle α. As explained in Section 7.6.3, there are three modes of operation for this circuit. In mode 1, the SCRs of group 1 with a firing angle α will operate as rectifiers in region *PQ*. From *Q* to *R*, the SCRs in group 1 will be fired at $(\pi - \alpha)$. This is mode 2. During mode 3, from *R* to *S*, none of the SCRs in either group will conduct, and the load will get open-circuited. In Fig. 7.22b, angle α is assumed to be less than $\pi/6$ (measured from point *T*). So, there will be no negative excursion for the load voltage during the positive half-cycle of the output. The output waveform can be improved still further with six-pulse converters. Since the firing angle α is restricted to a value less than $\pi/2$, the same control circuit as that shown in Fig. 7.20a can be used also for the three-pulse converter, and all the SCRs in each group can be fired together. For a six-pulse bridge converter, if all SCRs in each group are fired together, there will be a short-circuit on the input. So, the firing pulses must be guided to the proper SCRs through a logic circuit consisting of six-stage counters and decoders. Logic circuits are similar to the circuits used for forced-commutated bridge inverters and are described in Chapter 9. As explained in the case of single-phase input (Fig. 7.21b), the firing angle can be suitably modulated for three-phase input by using three-pulse or six-pulse converters, and the output voltage can be made reasonably sinusoidal. Three-phase output can be obtained by using three single-phase circuits.

7.6.5 Harmonic Analysis

The output voltage waveform of a cycloconverter is highly distorted due to the presence of a number of chopped sinusoids in each half-cycle. Assuming a purely resistive load and a constant firing angle, the harmonic components of the output voltage are as given by Eq. (7.46). To reduce the harmonic content, the firing angles should be suitably modulated as explained in Section 7.6.4. Equation (7.46) has to be modified to take into consideration the variation in the firing angles. Again assuming a resistive load (discontinuous current), the amplitude of the fundamental component of the output voltage is

$$e_{o_{1s}} = \frac{2E_m}{\pi m} \sum_{n=1}^{m} \int_{\alpha_n}^{\pi} \left\{\sin\theta \times \sin\frac{[\theta + (n-1)\pi]}{m}\right\} d\theta,$$

$$e_{o_{1c}} = \frac{2E_m}{\pi m} \sum_{n=1}^{m} \int_{\alpha_n}^{\pi} \left\{\sin\theta \times \cos\frac{[\theta + (n-1)\pi]}{m}\right\} d\theta, \tag{7.50}$$

$$e_{o_1} = \sqrt{e_{o_{1s}}^2 + e_{o_{1c}}^2},$$

where *m*, an integer, is the number of half-cycles of input in each half-cycle of the output (the output frequency being $1/m$ times the input frequency), and α_n is the firing angle for the *n*-th half-cycle [Eq. (7.49)].

It has already been mentioned that the distortion in the output is minimised by using bridges having a high pulse number. Basically, there are two types of distortion terms in the cycloconverter output. These are: (a) necessary distortion terms, and (b) unnecessary distortion terms. The former are produced by the basic operating mechanism of the converter, where the output voltage waveform is pieced together from segments of the input voltage waves. These terms include (in addition to the multiples of output frequency) multiples of the components of beat frequency $(\omega - \omega_o)$. The presence of beat-frequency components will not be perceptible if the output frequency is sufficiently low. Output distortion also results when consecutive half-cycles are not identical (see Figs. 7.21d and 7.21e). Such distortion will occur if the output frequency is not an integral submultiple of the input frequency. Therefore, Eqs. (7.46) and (7.50) for harmonic components have to be applied with caution. The unnecessary distortion terms are generated as a result of modulating the firing angle by a process that does not control the timing of the firing pulses in a theoretically ideal manner.

7.6.6 Circulating-Current Scheme

The cycloconverter-control scheme described in Section 7.6.5 does not permit simultaneous conduction of SCRs in groups 1 and 2 of the circuit shown in Fig. 7.22a. That is, as long as any SCR in group 1 is conducting, the firing pulses to SCRs in group 2 will be blocked, thereby preventing a short-circuit on the input. The drawback in this scheme is that the load gets open-circuited for a small duration in each half-cycle. As in the dual converter schemes explained in Chapter 6, the load current polarity can be changed from positive to negative without any discontinuity, or without producing open-circuits, by operating the two groups of SCRs in the *circulating-current mode*. In this operation, both groups will conduct continuously. Group 1 will work as a rectifier with a firing angle α, and group 2 as an inverter with a firing angle $(\pi - \alpha)$, so that the terminal voltage, i.e., the load voltage or potential of point A (Fig. 7.22a), of one group is approximately the same as that of the other. The angle can be modulated, as already explained. Under these conditions, the current through group 1 will be $(I_L + I_c)$, where I_L is the load current, and I_c is the circulating current which flows through the SCRs in group 2, and both the load voltage and current will be positive (region PQ in Fig. 7.22b). The circulating current I_c can be controlled by correctly adjusting the firing angles. When the voltage reverses and the load current is still positive, the firing angles for the two groups get interchanged so that SCRs of group 1 will go into the inverting mode and SCRs of group 2 will function as rectifiers, with the current distribution (region QR in Fig. 7.22b) remaining the same as before. When the current reverses, because the voltage polarity is still in a reversed state, no change in the firing angles takes place. The current in each group of SCRs will be automatically adjusted in a manner such that group 1 will carry current I_c and group 2 will supply current $(I_L + I_c)$. Thus, there will be no break in the load current. Since this mode of operation requires a continuous flow of circulating current through the two groups of SCRs, the internal losses will be more than those in the noncirculating-current mode of operation.

The main difference between the dual converter explained in Chapter 6 and the cycloconverter discussed here is that whereas the load current for the former is fairly steady that for the cycloconverter is alternating. The alternating load current produces additional problems when the circulating-current mode of operation is used. If i_P and i_N are the currents from the positive and negative groups of converters connected to a common load and operating in the circulating-current mode, then the following equations will be valid in the steady state:

$$i_P + i_N = I_o,$$

$$i_P - i_N = I_o \sin \omega_o t,$$

where I_o is the peak value of the load current. Therefore,

$$\begin{aligned} i_P &= \frac{I_o}{2} + \frac{I_o}{2} \sin \omega_o t, \\ i_N &= \frac{I_o}{2} - \frac{I_o}{2} \sin \omega_o t. \end{aligned} \tag{7.51}$$

It can be seen that the alternating component of the load current sets up a circulating current $I_o/2$ through the two converters. [The derivation of Eq. (7.51) can be obtained from the references given at the end of this chapter.] This circulating current is in addition to the current flowing through the converters because of the instantaneous difference in the potentials of points C and D (Fig. 7.22a), even though the sum of the firing angles of the converters is π. Thus, the total circulating current for cycloconverters will be more than that for dual converters. Further, the component of circulating current $I_o/2$, which is proportional to the load current, cannot be reduced by the current-limiting reactor. Even in the case of dual converters used for reversible DC drives, whenever the load current changes during transient conditions, there will be an alternating component of load current which induces circulating currents. This results in trapped energy in the reactor and increases the response time of the controller. Thus, the circulating-current scheme is recommended only where large circulating currents can be tolerated.

7.6.7 Input Characteristics of a Cycloconverter

An infinite pulse number is assumed for the cycloconverter so that the output waveform may be purely sinusoidal and without any distortion terms. If this ideal converter system is operated as a dual converter to produce DC output voltage, then the input current will be sinusoidal. When it is operated to produce a DC output voltage of magnitude $V_{d\,max}$ and a load current I_d with zero firing angle, then the input line current will be sinusoidal with an RMS value I_1 and in phase with the respective phase voltage. When it is operated as a cycloconverter feeding a resistive load, it will produce a sinusoidal output voltage of amplitude $V_{d\,max}$ and a sinusoidal load current with an RMS value I_d. Then, the output power will be

$$P_D = \frac{V_{d\,max}\,I_d}{\sqrt{2}} - \frac{V_{d\,max}\,I_d}{\sqrt{2}} \cos 2\omega_o t, \tag{7.52}$$

where ω_o is the output frequency. For a rectifier with zero firing angle, the output power is $V_{d\,max}\,I_d$. Since the RMS values of the load current are the same for both the cycloconverter and the rectifier, and since the real output power for cycloconverter operation is 0.707 times that for rectifier operation, the maximum possible input power factor for an ideal cycloconverter is only 0.707.

7.6.8 Cycloconverter Drives

Cycloconverters can be used for controlling the speed of AC motors where variable frequency supply is required. Since line commutation is used, additional commutating components, as required for other types of inverters, are not necessary. The conversion is directly from AC to AC, without any intermediate DC stage. Thus, the overall efficiency is high. Only low frequencies are possible with cycloconverters, and therefore they are ideally suited for AC motor speed control in the subsynchronous range.

In the circulating-current scheme, regenerative braking can be very easily incorporated since one of the SCR groups operates as an inverter at any instant of time. Thus, the load power can be fed back into the input. This is one of the important advantages of using a cycloconverter drive. In view of the distortion, the maximum output frequency is limited to one-third the input frequency. At low frequencies, the output voltage waveform will be better than that obtained directly from a forced-commutated bridge inverter (see Chapter 9). Therefore, cycloconverter drives are being recommended for electric traction. The control circuit required for incorporating regenerative braking when forced-commutated inverters are used is involved. The proper choice of commutating components will depend on the operating frequency, and therefore there is a possibility of commutation failure or higher commutation losses when using variable frequency forced-commutated circuits. A detailed discussion of these inverters and the characteristics of AC motors supplied from variable frequency sources will be given in Chapters 8 and 9. The method of speed control for motors with a cycloconverter is the same as that for motors with any other form of inverter; this aspect therefore will also be considered in these chapters.

7.7 EXAMPLE.

(a) A cycloconverter is made from two single-phase fully-controlled bridges. The input to the bridges is 230 V, 50 Hz, single-phase AC. If the output frequency is one-fifth the input frequency, and if the firing angle α is $\pi/4$, calculate the fundamental amplitude of the output. Assume a resistive load.

Using Eq. (7.46), we have

$$e_{o_{1s}} = \frac{230\sqrt{2}}{5\pi} \sum_{n=1}^{5} \{\frac{5}{4} [\sin \frac{n\pi}{5} - \sin (\frac{\pi}{4} - \frac{\pi/4 + (n-1)\pi}{5})]$$

$$+ \frac{5}{6} [\sin \frac{n\pi}{5} + \sin (\frac{\pi}{4} + \frac{\pi/4 + (n-1)\pi}{5})]\}$$

$$= \frac{230\sqrt{2}}{5\pi} (6.24 + 4.75) = 228 \text{ V},$$

$$e_{o_{1c}} = \frac{230\sqrt{2}}{5\pi} \sum_{n=1}^{5} \{\frac{5}{6} [\cos \frac{n\pi}{5} + \cos (\frac{\pi}{4} + \frac{\pi/4 + (n-1)\pi}{5})]$$

$$+ \frac{5}{4} [\cos \frac{n\pi}{5} + \cos (\frac{\pi}{4} - \frac{\pi/4 + (n-1)\pi}{5})]\}$$

$$= \frac{230\sqrt{2}}{5\pi} (-2.36 + 2.03) = -8.3 \text{ V}.$$

Therefore, the fundamental amplitude will be

$$E_1 = \sqrt{e^2_{o_{1s}} + e^2_{o_{1c}}}$$

$$= \sqrt{228^2 + 8.3^2} \approx 228 \text{ V}.$$

(b) An ideal cycloconverter delivers a sinusoidal current $i_L = I_o \sin \omega_o$ to a resistive load. The output voltage is given by $v_L = V_{max} \sin \omega_o t$, where V_{max} is the maximum DC voltage produced when the converter operates in the rectifying mode with zero phase angle. Calculate the input fundamental power factor.

The output power is

$$P_o = v_L i_L$$

$$= \frac{V_{max} I_o}{2} - \frac{V_{max} I_o}{2} \cos 2\omega_o t.$$

Since the converter is an ideal one, the total RMS value of the input current I_1 assuming three-phase input can be expressed in terms of the output current as

$$\frac{V_{max} I_o}{\sqrt{2}} = \frac{3 V_n I_1}{\sqrt{2}},$$

where V_n is the peak phase-to-neutral voltage. For cycloconverter operation, let the input current be

$$i_L = I_P \sin \omega t + I_Q \cos \omega t + I_H \sin (\omega t \pm 2\omega_o t).$$

In this expression, the first term (in-phase component) on the right-hand side produces the real power $(V_{max} I_o/2)$; the second term gives the fundamental reactive power; and the last term (beat-frequency components) produces oscillatory power corresponding to $(V_{max} I_o/2) \cos 2\omega_o t$. Equating the respective power components, we get

$$I_P = I_1,$$

$$I_H = I_1/2.$$

Since the total RMS value of the input must be equal to I_1, we have

$$I_P^2 + 2I_H^2 + I_Q^2 = 2I_1^2.$$

Therefore,

$$I_Q = I_1/\sqrt{2}.$$

Thus, the fundamental input power factor will be

$$PF = \frac{I_P}{\sqrt{I_P^2 + I_Q^2}} = \frac{1}{\sqrt{1 + 1/2}} = 0.817.$$

It can be seen that the overall input power factor given by

$$\frac{\text{active power input}}{\text{total volt-amperes}}$$

will be 0.707.

A rigorous analysis of cycloconverter circuits is beyond the scope of our discussion. For details on their operation and control, see McMurray (1972) under General References.

REFERENCES

Ainsworth, J. D., The phase locked oscillator: A new controlled system for static converters, *IEEE Trans.* (PAS), 1968, p. 859.

Basu, P. R., A variable speed induction motor using thyristor in the secondary circuit, *IEEE Trans.* (PAS), 1971, p. 509.

Bedford, R. D., Nene, V. D., Voltage control of the three-phase induction motor by thyristor switching: A time domain analysis using the $\alpha\beta o$ transformation, *IEEE Trans.* (IGA), 1970, p. 553.

Dewan, S. B., Kankam, M. D., A method of harmonic analysis of cyclo-converters, *IEEE Trans.* (IGA), 1970, p. 455.

Erlicki, M. S., Ben Uri, J., Wallach, Y., Switching drive of induction motors, *Proc. IEE*, 1963, p. 1441.

Hamblin, T. M., Barton, T. H., Cycloconverter control circuits, IEEE Conference Record (IGA), 1970, p. 559.

Lavi, A., Polze, R., Induction motor speed control with static inverter in rotor, *IEEE Trans.* (PAS), 1966, p. 76.

Lipo, T. A., Analog computer simulation of a three-phase full wave controlled rectifier bridge, *Proc. IEEE*, 1969, p. 2137.

Machida, T., Kaminosono, K., Umezu, T., Sugimoto, M., Yokoyama, K., Control and protective system of HV DC transmission system, *IEEE Trans.* (PAS), 1971, p. 2778.

Mellgren, G., Thyristor converters for motor drives: Some experience in design and operation, IEE Conference Publication, No. 17, 1965, p. 230.

Miljanic, P. N., The through-pass inverter and its applications to the speed control of wound rotor induction machines, *IEEE Trans.* (PAS), 1968, p. 234.

Paice, D. A., Induction motor speed control by static voltage control, *IEEE Trans.* (PAS), 1968, p. 585.

Ramamoorty, M., Ilango, B., Application of state space techniques to the steady state analysis of thyristor controlled single-phase induction motors, *International Journal of Control* (UK), 1972, p. 353.

Ramamoorty, M., Parihar, J. S., Digital model of an induction motor for the study of switching transients, *JIE* (India), 1972, p. 302.

Ramamoorty, M., Samek, M. F., Steady state analysis of phase controlled induction motor with isolated neutral, *IEEE Trans.* (IECI), 1969, p. 178.

Robinson, C. E., Redesign of DC motors for applications with thyristor power supplies, *IEEE Trans.* (IGA), 1968, p. 508.

Schofield, J. R. G., Smith, G. A., Whitmore, M. G., The application of thyristors to the control of DC machines, IEE Conference Publication, No. 17, 1965, p. 219.

Shepherd, W., Stanway, J., Slip power recovery in an induction motor by the use of thyristor inverter, *IEEE Trans.* (IGA), 1969, p. 74.

Slabiak, W., Lawson, L. J., Precise control of a three-phase squirrel cage induction motor using a practical cycloconverter, *IEEE Trans.* (IGA), 1966, p. 274.

Turnbull, F. G., Carrier frequency gating circuit for static inverters, converters and cycloconverters, *IEEE Trans.* (Magnetics), 1966, p. 14.

8

Parallel-Series Inverters

8.1 FORCED-COMMUTATED INVERTERS

As explained in Chapter 2, the SCR is turned off when its forward current is reduced below the level of the holding current. Forward voltage can be reapplied after allowing a certain time for the excess carriers in the outer and inner layers of the SCR to decay. The decay and recombination of these excess carriers may be accelerated by applying a reverse voltage across the SCR.

In AC circuits, when the current in the SCR goes through a natural zero, a reverse voltage automatically appears across the SCR. This is known as *natural commutation*. No external circuits are required for turning off the SCR. The phase-controlled circuits and the line-commutated rectifiers and inverters discussed in Chapters 6 and 7 make use of the available AC line voltage for commutating the conducting SCR.

In DC circuits, the forward current has to be forced to zero by an external circuit to turn off the SCR. This is known as forced commutation. DC input is required for SCR controlled circuits used for DC-to-DC converters (choppers) and for DC-to-AC inverters. The performance of choppers and details of commutation circuits will be discussed in Chapter 10. The line-commutated inverters described in Chapter 7 require at the output terminals an existing AC supply which is used for commutation. This means that such inverters cannot function as isolated AC voltage sources operating from a DC supply, or as variable frequency generators. On the other hand, forced-commutated inverters independently provide an AC output of variable frequency, and so have much wider applications. Because of their requirement for separate commutation circuits, forced-commutated inverters need more control components.

There are broadly three types of inverters that use forced commutation: (a) the parallel inverter, (b) the series inverter, and (c) the bridge inverter. This classification is based on the configuration of the SCRs and the arrangement of commutating capacitors. The first two will be discussed in this chapter and the bridge inverter in Chapter 9.

In parallel inverters, the commutating capacitor is connected in parallel with the load. When this capacitor applies a reverse potential across the conducting SCR, commutation is achieved. This is referred to also as *impulse* or *voltage commutation.* In series inverters, the commutating components (inductors and capacitors) are connected in series with the load, thus forming an underdamped circuit. When this circuit is excited by firing the SCR, the current in the circuit will, after reaching the maximum, go through a zero value. When the forward current of the SCR touches zero, the device will go into the blocking state. This method of turn-off is known as *resonant turn-off* (since the forward current is made zero by a resonant circuit) and also as *current commutation.* When the current is zero, the capacitor will get charged to a voltage higher than the supply voltage, and a reverse voltage will appear across the SCR after it is turned off. Therefore, when the SCR is fired, it will automatically get turned off after approximately one-half period of the resonant circuit. No other SCR need be fired to commutate the conducting SCR as is done in parallel or bridge inverters. This method of turn-off is also referred to as *self commutation.* Series inverters are therefore classified in our discussion as *self-commutated inverters.* For self commutation, a resonant circuit is essential, and the capacitor required for underdamping can be connected in series or in parallel with the load.

8.2 CLASSIFICATION OF CIRCUITS FOR FORCED COMMUTATION

When commutation is desired in DC circuits, there are many ways by which the forward current in the SCR can be forced to zero. Shunting the conducting SCR by a low resistance element (mechanical switch or transistor), and thereby reducing the forward current to a level below I_h in order to turn off the SCR, requires large off-times since no reverse voltage is applied across the SCR. Therefore, this method is not convenient when periodic switching off of the SCR is required. For efficient forced commutation, reverse voltage must appear across the SCR. This reverse voltage can be obtained from a charging circuit consisting of an inductor and a capacitor which are called the *commutating components.* The voltage across the capacitor is used for obtaining forced turn-off of the SCR.

The classification of the methods of forced commutation is based, as we shall now see, on the arrangement of the commutating components and the manner in which zero current is obtained in the SCR.

Class A: Resonant Commutation Here, the commutating components L and C are connected to the load as shown in Figs. 8.1a and 8.1b, so that the overall circuit becomes underdamped and zero current is obtained. When such underdamped circuits are excited by applying a DC voltage, the resulting current waveforms will be as shown in the figures. If an SCR is used in series with the circuit, it will be turned off when its forward current is zero at point *a*. Figure 8.1a gives a typical configuration of a series inverter in which the load is in series with the capacitor. The performance and design details of different forms of series inverters will be discussed in Section 8.6. In Fig. 8.1b, the capacitor is in parallel with the load. The commutation of the SCR is however due to the resonant behaviour of the overall circuit. A high-frequency inverter using this commutation arrangement is described in Section 8.8. Such an inverter is referred to also as a *modified series inverter* or a *self-commutated inverter.*

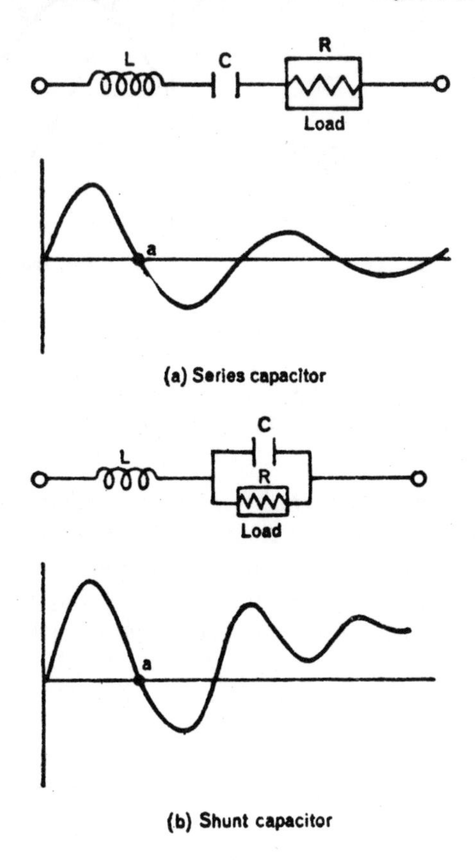

Fig. 8.1 Resonant turn-off methods.

Class B: Self Commutation In this method, the commutating components need not form a resonant circuit with the load. A commutating circuit appears as shown in Fig. 8.2a. Initially, capacitor C is charged to voltage e_C. When the SCR is fired, the capacitor will discharge through it, and at the end of the discharge will have a reverse voltage. Since the SCR is conducting, the negative voltage in the capacitor will produce a negative current i. When this current is equal to the load current I_L (point a in Fig. 8.2b), the SCR will be turned off. This is a typical example of current commutation. It is clear that the SCR, once it is turned on, will keep conducting for a specified period before automatically turning off.

The main difference between class A and class B methods of commutation is that in the latter the commutating components do not carry the load current. This method of commutation is often used for DC-to-DC choppers (discussed in detail in Chapter 10). In circuits of both classes A

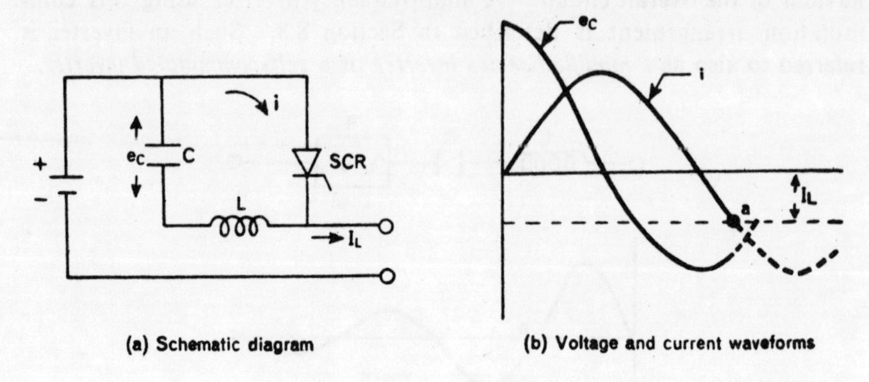

Fig. 8.2 Self-commutated circuit.

and B, the SCR turns off by itself shortly after it is turned on. Hence, inverters with this type of commutation belong to the general category of self-commutated inverters.

Class C: Auxiliary Commutation Here, an auxiliary SCR is turned on to commutate the main SCR. The commutating circuit is shown in Fig. 8.3a. It is assumed that the capacitor is initially charged to voltage e_C. When SCR1 is turned on, the capacitor will discharge through it and also through inductor L. At the end of the discharge, the capacitor voltage will be reversed. The reverse discharge through SCR1 will be prevented by diode D. When SCR1a is fired, the capacitor will discharge through SCR1 and turn it off. Since a reverse voltage is applied to SCR1 immediately after turning on SCR1a, this is known as *voltage commutation.* The design procedure for obtaining the values of the commutating components L and C of this circuit is the same as that for the circuit used for class B commutation. This circuit is used also for choppers whose on-time can be varied.

Auxiliary commutation is used in certain types of bridge inverters which are explained in detail in Chapter 9.

Class D: Complementary Commutation Figure 8.3b shows an arrangement using more than one load-carrying SCR for complementary commutation. The firing of one such SCR commutates the other. When SCR1 is conducting, load 1 will get energised and capacitor C will get charged to the supply voltage. The voltage polarity is as shown in the figure. When SCR2 is fired, the capacitor will apply a reverse potential across SCR1 and turn it off. Then, load 2 will get energised and capacitor C will be charged in the opposite direction.

Complementary commutation is used for parallel inverters and certain forms of bridge inverters.

Class E: External-Pulse Commutation Here, a pulse of current obtained from an external voltage turns off the conducting SCR. The peak amplitude of the current pulse must be greater than that of the load current through the SCR, and the duration of the reverse voltage applied following the turn-off of the SCR must be longer than the SCR turn-off time. Alternating current choppers (discussed in Chapter 10) make use of this type of commutation.

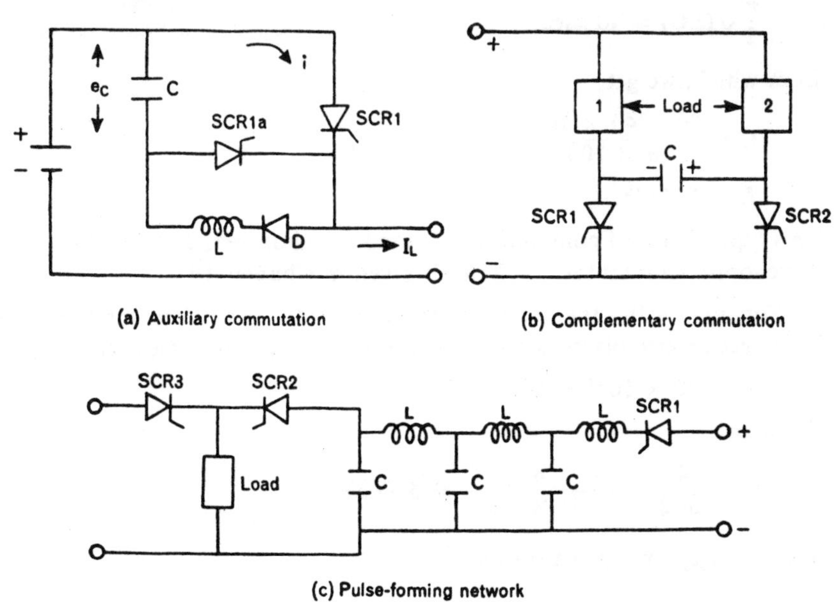

Fig. 8.3 Methods of forced commutation.

A pulse-forming network is as shown in Fig. 8.3c. When SCR1 is fired, a current pulse of peak value $E\sqrt{(C/L)}$ will flow through the SCR and charge all the capacitors to a voltage level $2E$. The duration of the current pulse will be $n\sqrt{(LC)}$, where n is the number of LC sections. When SCR2 is fired, the charged network will discharge through SCR3 in the opposite direction and after SCR3 is turned off, apply a reverse potential across it for a period of about $n\sqrt{(LC)}$.

Class F: Line Commutation In this method, the available AC line voltage is used for turning off the conducting SCR. This method of commutation is applied in line-commutated rectifiers and inverters (discussed in Chapter 7).

8.3 EXAMPLE

Obtain the proper values of the commutating components for the circuits shown in Figs. 8.2a and 8.3b, given that the load current to be commutated is 10 A, the turn-off time required is 40 μsec, and the supply voltage is 100 V.

For Fig. 8.2a, the peak discharge current of the capacitor is assumed to be twice the load current I_L and the time for which the SCR is reverse-biased is approximately equal to one-quarter period of the resonant circuit. Therefore,

$$E\sqrt{\frac{C}{L}} = 20 \text{ A},$$

$$\frac{\pi}{2}\sqrt{(LC)} = 40 \ \mu\text{sec},$$

from which, we get

$$C = \frac{20 \times 40 \times 10^{-6}}{\pi \times 100} \times 2 = 5.1 \ \mu\text{F},$$

$$L = 127 \ \mu\text{H}.$$

This type of forced commutation is used for choppers; a detailed design of the chopper-commutation circuit is given in Chapter 10.

For Fig. 8.3b, the load resistance is 10 Ω. The time for which the SCRs get reverse-biased after being turned off is given by (see Section 5.2.1)

$$t_q = 40 \times 10^{-6} = 10C \ln 2.$$

Therefore,

$$C = \frac{4}{\ln 2} = \frac{4}{2.303 \times 0.301} = 5.77 \ \mu\text{F}.$$

8.4 PARALLEL INVERTER

Figure 8.4a shows the schematic diagram of a single-phase parallel inverter. SCRs 1 and 2 are the main load-carrying SCRs. The commutating components are L and C. Diodes 1′ and 2′ permit the load reactive power to be fed back to the DC supply. These are called the *feedback diodes.* When SCR1 is conducting, neglecting the small voltage drop across L, the supply voltage E_{DC} will appear across the left-half of the transformer primary winding OA. Terminal O is positive with respect to A. By transformer action, terminal B will be at a potential of $2E_{DC}$ with respect to A. Thus, capacitor C will get charged to twice the supply voltage. The load voltage will be positive and of magnitude E_{DC} if the ratio of turns PQ and OA is unity. The load current will also be positive and will have a magnitude I_L. At the end of a half-period, SCR2 is fired. Capacitor C will immediately apply a reverse voltage of $2E_{DC}$ across SCR1 and turn it off.

When SCR1 is turned off, the capacitor will discharge through SCR2, inductor L, diode 1′, and a portion of the transformer winding LA. Thus, the energy stored in the capacitor will be fed back to the load through the transformer coupling of windings LA and PQ. During this period, the potential of point L will be fixed by the DC input supply and the load voltage will still be positive but more than E_{DC}. The load current, which earlier was flowing through SCR1, will now flow through OL and diode 1′ to the negative input terminal. This can happen only if diode 1′ is forward-biased and the capacitor discharge current is more than the load current. As the potential of point L increases sufficiently to reverse-bias diode 1′, the capacitor will no longer discharge through 2 and point L will not get connected to the negative supply terminal. The current through inductor L will now flow through 2′, MB, and 2, and the trapped energy in L will be fed back to the load. The load current I_L, which earlier was flowing through OL, will now flow from M to O through diode 2′, and the load reactive energy will be returned to the DC supply. Since point M is now connected to the negative supply terminal, the load voltage polarity will be reversed and more than E_{DC}. Also, the capacitor will be charged in the opposite direction to slightly more than twice the supply voltage. SCR2 will stop conducting after all the energy in the commutating inductor L has been completely dissipated in the load. Immediately following the commutation of SCR1, energy is transferred from the capacitor and inductor to the load; during this period, high-frequency oscillations will be superimposed on the normal rectangular waveform of the load voltage. After this transient period, only diode 2′ will continue to conduct. This will cause application of a reverse voltage across SCR2, and thereby help in turning it off. When the load current becomes zero, diode 2′ will be blocked and SCR2 will have to be triggered again to reverse the direction of the load current. When SCR2 starts conducting, the load voltage will again become equal to E_{DC}. The load voltage and current waveforms are as shown in Fig. 8.4b; the transient waveforms during the commutation period are indicated by the dashed lines.

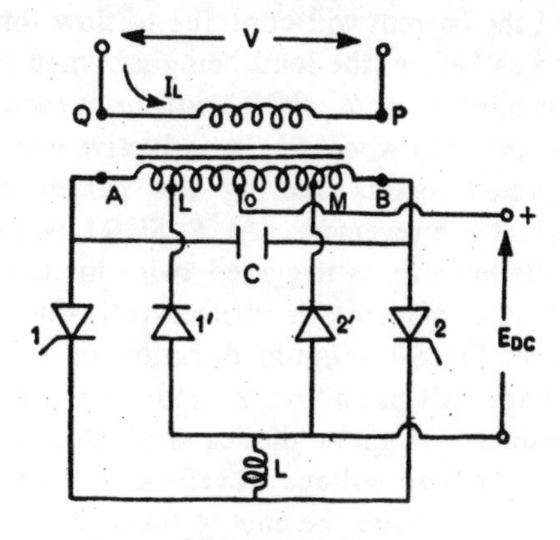

(a) Schematic diagram

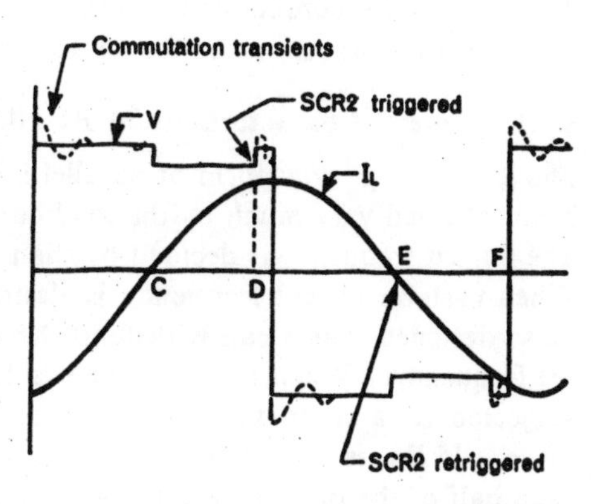

(b) Load voltage and current waveforms

Fig. 8.4 Parallel inverter.

In Fig. 8.4b, SCR1 will conduct during the period CD, when both load voltage and load current are positive. At D, SCR2 is fired to commutate

SCR1. After the commutation transient, the load voltage will be reversed and the current will continue to flow through diode 2′ in the same direction as before (the load being assumed inductive). Because of the reverse bias applied by 2′, SCR2 will be turned off. At E, the load current will become zero when all the inductive energy is dissipated and SCR2 will be triggered again. During the period EF, both load voltage and load current are reversed. At F, SCR1 is fired to turn off SCR2. Since the SCRs have to be triggered twice in each half-cycle, and since interval DE is load-dependent, it is necessary that the SCRs be gated by a train of pulses for a minimum duration of a quarter-cycle. Note that the load voltage will rise above E_{DC} during period DE when the feedback diodes conduct. If these diodes are connected to points A and B (Fig. 8.4a), then the load voltage waveform will be rectangular. But, such a connection will require the energy trapped in the commutating components to be dissipated as heat in the SCRs and diodes, thereby necessitating the derating of the components. Thus, the efficiency of the circuit can be increased by connecting the diodes to the tap points.

8.4.1 General Characteristics of Parallel Inverter

The load voltage waveform of parallel inverters is nearly rectangular and is not affected very much by the load current or by the nature of the load. The output frequency is decided by the triggering frequency of the SCRs. When variable output frequency is desired, the output transformer must be so designed that it can withstand the rated voltage at the lowest possible frequency. When the frequency is low, the transformer core will be subjected to a large flux excursion and magnetic saturation may take place. If E_{DC} is the DC input voltage and N is the number of turns in each half of the primary winding, the maximum flux level reached in the core will be

$$\phi_{max} = \frac{E_{DC} \times T/2}{2N}, \tag{8.1}$$

where T is the period of the output at the lowest frequency. If transformer saturation takes place, the output voltage will become zero and the voltage across the commutating capacitor will collapse, leading to commutation failure.

Another important consideration in the operation of the inverter is that, while turning off the inverter, the DC input must be switched off before disconnecting the gate supply to the inverter. If the gate supply is stopped first, the conducting SCR will not be turned off until the DC input is removed. This will result in transformer saturation followed by a large line current. Similarly, at the time of starting the inverter, if the DC supply is connected after the gating pulses are applied, there may not be sufficient time for the capacitor to be fully charged before the first commutation occurs and there may be commutation failure again. To avoid this, the DC supply is given before the gate control circuit is energised. The application of DC voltage will produce a large *dv/dt* across the SCRs. Therefore, snubber circuits (discussed in Chapter 11) must be used to reduce the *dv/dt*.

As mentioned in Section 8.4, the SCRs must be fired for a minimum duration of a quarter-period by either a train of pulses (carrier frequency gating) or a continuous gate signal of proper amplitude. Figure 8.5a shows the schematic block diagram of such a control scheme. UJT2 is a relaxation oscillator. Its output is a train of high-frequency pulses which

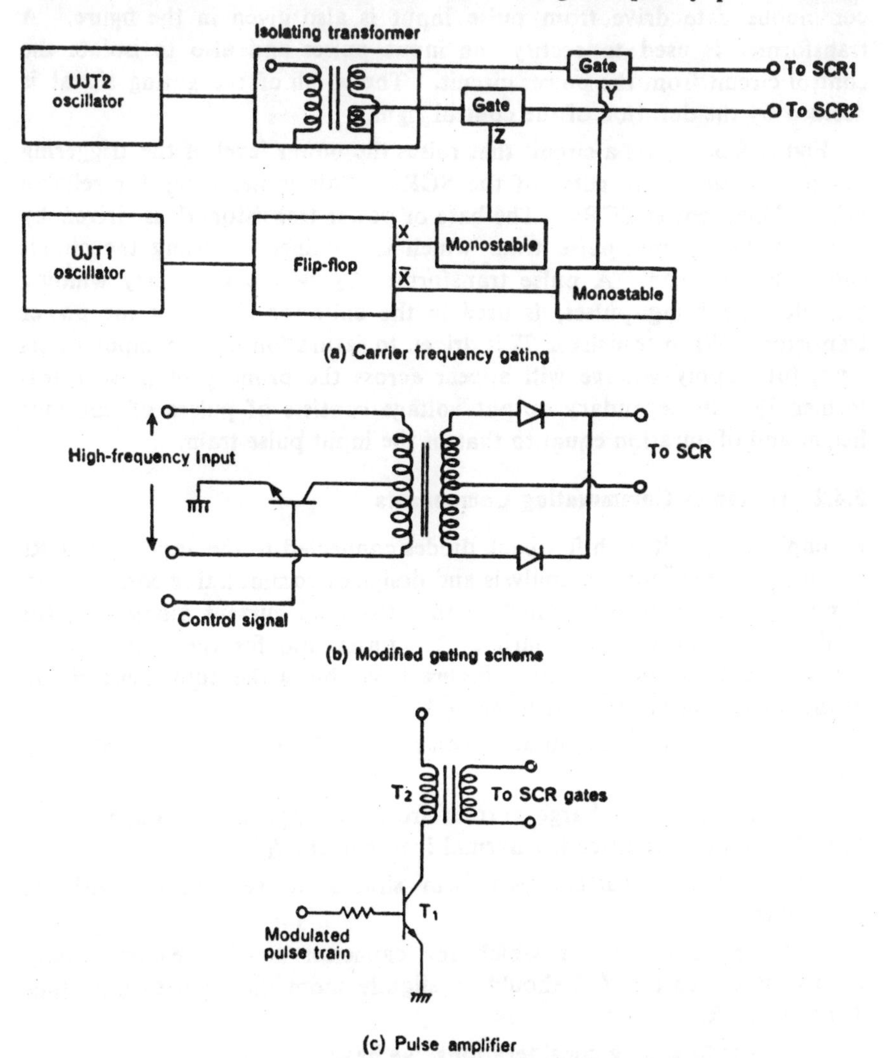

(a) Carrier frequency gating

(b) Modified gating scheme

(c) Pulse amplifier

Fig. 8.5 Inverter control scheme.

are gated to the proper SCR by the outputs of two variable-width monostable circuits. The duration of the outputs at *Y* and *Z* is so chosen that the SCRs receive gating pulses for a minimum period of $\pi/2$ at the desired output frequency. Oscillator UJT1 decides the triggering frequency. The output of this oscillator drives a flip-flop. The two complementary outputs at X and $\bar{X}$ are connected to the monostable circuits to provide the proper pulse width.

Figure 8.5b shows a modified gating scheme suitable for inverters supplying highly-inductive loads. Here, continuous gate drive should be used for the SCRs (see Chapter 2). The pulse width must be equal to $T/4$, where T is the output period. The circuitry required for obtaining continuous gate drive from pulse input is also given in the figure. A transformer is used to rectify the input pulses and also to isolate the control circuit from the power circuit. The width of the gating signal is decided by the duration of the control signal.

Figure 8.5c shows a circuit that raises the power level of the triggering pulses applied to the gates of the SCRs. This is necessary for reliable firing of high-power SCRs. The base of power transistor T_1 is driven by a modulated carrier pulse train which is obtained by using the circuit shown in Fig. 8.5a. A pulse transformer T_2, whose secondary winding provides the firing pulses, is used in the collector circuit of the power transistor. When transistor T_1 is driven to saturation by the input to its base, full supply voltage will appear across the primary of pulse transformer T_2. Its secondary output voltage consists of pulses of constant height and of duration equal to that of the input pulse train.

8.4.2 Design of Commutating Components

A simplified circuit with feedback diodes connected to the anodes of SCRs is considered here for the analysis and design of commutating components. A pure resistive load is assumed so that the load current along with the load voltage also reverses polarity. The turns ratio for the output transformer is assumed to be unity. Figure 8.6a shows the equivalent circuit of the inverter after SCR1 is turned off.

The design of the commutating components is based on the following criteria:

(a) The peak discharge current from the capacitor through SCR2, L, and 1′ should be twice the normal load current I_L.

(b) The load current waveform should be rectangular with an amplitude I_L.

(c) The duration for which the capacitor applies reverse voltage across the turned-off SCR should be slightly more than the turn-off time t_q of this SCR.

From the foregoing considerations, we have

$$2E_{DC}\sqrt{\frac{C}{L}} = 2I_L,$$

$$\frac{\pi}{3}\sqrt{(LC)} = t_q. \tag{8.2}$$

These two equations can be used for obtaining the appropriate values for L and C for given values of E_{DC} and I_L.

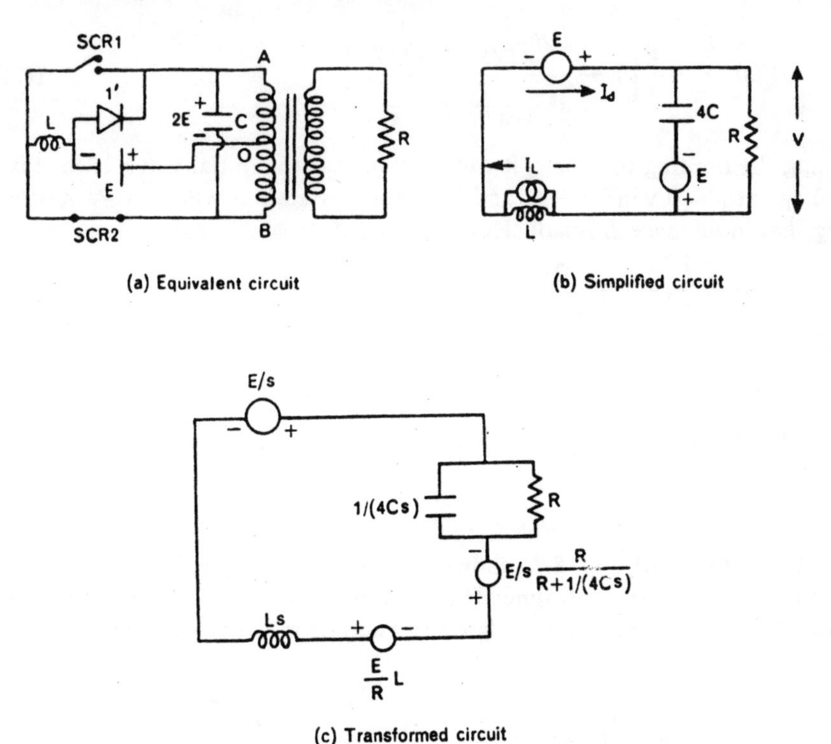

(c) Transformed circuit

Fig. 8.6 Representation of inverter circuit.

The mathematical analysis of inverter operation is based on the equivalent circuit shown in Fig. 8.6a. The commutation interval is neglected in this analysis. Therefore, SCR1 will be turned off when SCR2 is fired. The voltage on the capacitor will be $2E$ and the current in the inductor immediately after SCR1 is turned off will be $I_L = E/R$. These are the initial conditions required for the analysis.

Figure 8.6b shows the simplified circuit where all the components and variables are referred to winding OB in Fig. 8.6a. In Fig. 8.6c, this circuit is modified using Thevenin equivalents and transformed variables. The Laplace transforms, $I_d(s)$ of the input current and $V(s)$ of the load voltage, are given by

$$I_d(s) = \frac{\dfrac{E}{s} + \dfrac{E}{s}\left[\dfrac{R}{R + 1/(4Cs)}\right] + (E/R)L}{Ls + \dfrac{R[1/(4Cs)]}{R + 1/(4Cs)}},$$

$$V(s) = \frac{[R/(4Cs)]I_d(s)}{R + 1/(4Cs)} - \frac{E}{s}\left[\frac{R}{R + 1/(4Cs)}\right], \tag{8.3}$$

where s is the Laplace variable. Simplification of the expression for $V(s)$ gives

$$V(s) = -\frac{E}{s}\left[1 - \frac{2\left(\frac{1}{4RC}s + \frac{1}{4LC}\right)}{s^2 + \frac{1}{4RC}s + \frac{1}{4LC}}\right]. \tag{8.4}$$

Using the limiting theorems, it can be seen that the initial value of the load voltage at $t = 0$ will be $-E$, and in the steady state it will be $+E$. Assuming that inductance L is sufficiently large, Eq. (8.4) will reduce to

$$V(s) = \frac{E}{s}\left[-1 + \frac{2\tau}{s+\tau}\right], \tag{8.5}$$

where

$$\tau = 1/(4RC).$$

The inverse transform of Eq. (8.5) gives

$$v(t) = E[1 - 2e^{-\tau t}], \tag{8.6}$$

where $v(t)$ is the load voltage at time t.

Using Eq. (8.6), Fig. 8.7a shows, for two values of τ, the variation of the terminal voltage with time. Figure 8.7b gives the corresponding voltage waveforms obtained from a rigorous analysis in which the initial voltage

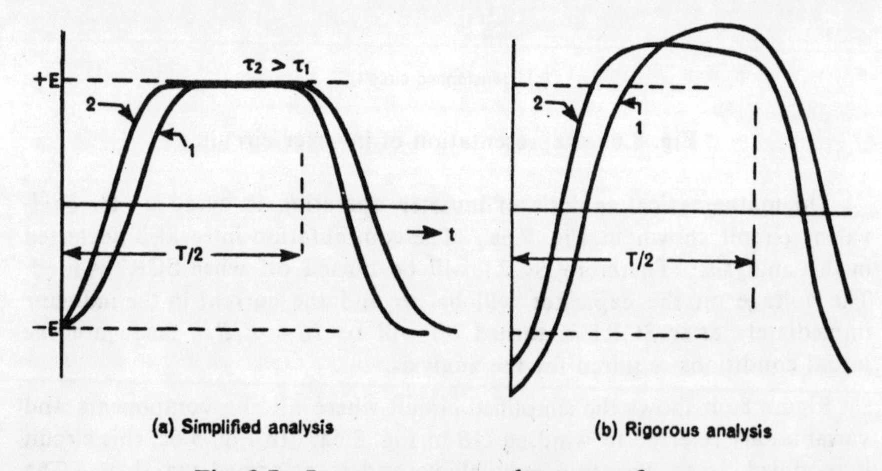

Fig. 8.7 Inverter output voltage waveforms.

across capacitor e_{Co} and the current through inductor i_{Lo} are not made equal to $2E$ and I_L, respectively, but are treated as functions of the circuit parameters, input voltage, and load. The expression we get for the load voltage is

$$v(t) = \frac{E(1 + e^{-\tau T/2} - 2e^{-\tau t})}{1 + e^{-\tau T/2} - \frac{4}{\tau T}(1 - e^{-\tau T/2})}, \tag{8.7}$$

where T is the period of the inverter output. Using Eq. (8.6), the duration t_0 for which the SCR will be reverse-biased is given by

$$t_0 = \frac{1}{\tau} \ln 2. \tag{8.8}$$

For proper commutation, this duration must be longer than the turn-off time of the SCR.

8.4.3 Output Voltage and Waveform Control

The RMS value of the AC output of a parallel inverter can be controlled from the DC side or from the AC side. In the first case, the DC voltage is varied. This changes the amplitude as also the RMS value of the AC output voltage. The variable DC input voltage can be very conveniently obtained from the controlled rectification and filtering of the alternating current. In the second case, a tap-changing transformer is used at the output. Neither of these methods affects the load voltage waveform. The output voltage control can also be obtained by having multiple commutation in each half-cycle or by using a chopper in the input. When the latter arrangement is employed, the DC voltage will become zero whenever the chopper is off. During the on-time, the input voltage to the inverter will be the supply voltage E. Therefore, the output voltage will be zero during the off-time of the chopper. (Chopper circuits are discussed in Chapter 10.) If the inverter is subjected to multiple commutation in each half-cycle, the load voltage will swing both to positive and to negative. Figures 8.8a and 8.8b show the output voltage waveform for these cases. By varying the time interval t_1 for a fixed value of T, the RMS value of the output voltage can be changed. However, it will be observed that this method of voltage control produces a lot of distortion in the output. Therefore, some form of harmonic filter is required at the output to obtain a sinusoidal load voltage.

Figure 8.8c shows a common LC filter used for obtaining the fundamental frequency. This is known as the *Ott filter*. The values of capacitances and inductances used in the filter are as follows:

$$\begin{aligned} L_2 &= \frac{Z_d}{\omega_d}, \qquad C_1 = \frac{1}{6Z_d}\,\omega_d, \\ L_1 &= \frac{9Z_d}{2\omega_d}, \qquad C_2 = \frac{1}{3Z_d}\,\omega_d, \end{aligned} \tag{8.9}$$

where ω_d is the desired output frequency, and Z_d is the base impedance (taken to be one-half the load impedance) of the filter. The filter components required for the configuration in Fig. 8.8c are smaller than those for the conventional series-parallel filter shown in Fig. 8.8d.

When multiple commutation is used, the time interval t_1 (Fig. 8.8a) can be modulated in a sinusoidal manner. The firing circuits for cycloconverter control (discussed in Chapter 7) can be employed for this purpose. The variable pulse width further reduces the harmonic content.

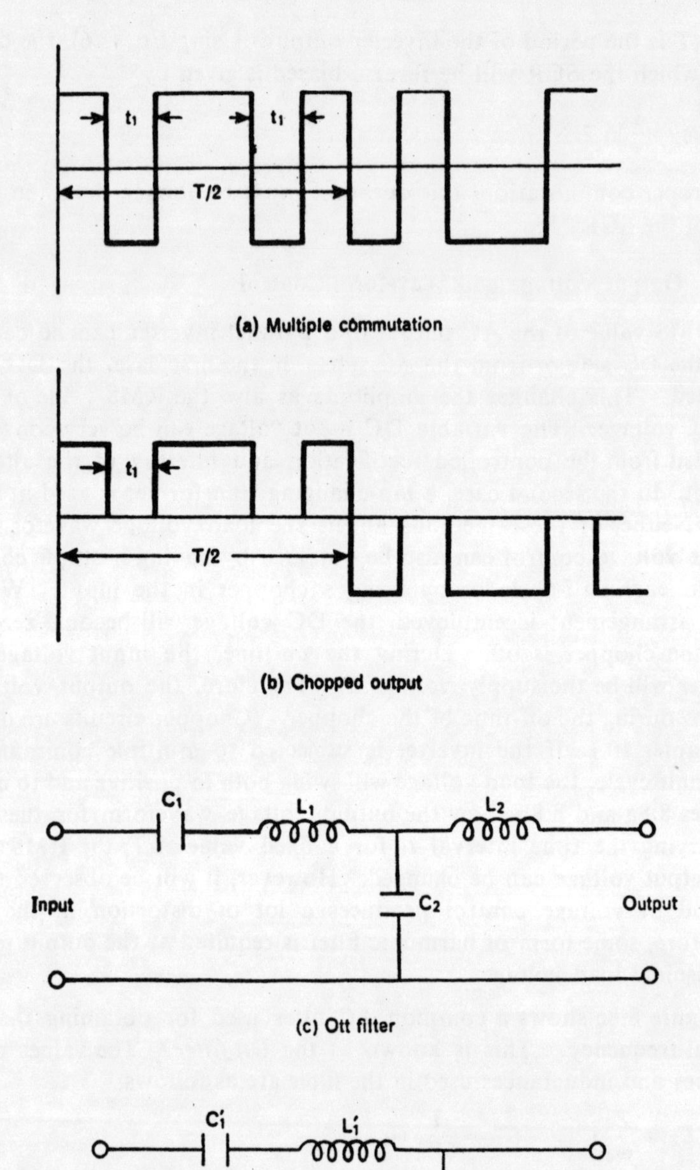

Fig. 8.8 Output voltage control (*cont.*).

Figure 8.8e shows one method of obtaining triggering pulses for the SCRs to produce multiple commutation. The frequency of the triangular waveform decides the number of output pulses in each half-cycle; the reference signal to the comparator is sinusoidal; and the off-time in the output pulses is equal to the time during which the instantaneous value of the reference signal is lower than that of the triangular wave. If the reference signal is flat, the duration of each output pulse will be the same. By changing the amplitude of the reference signal, the RMS value of the output voltage can be altered.

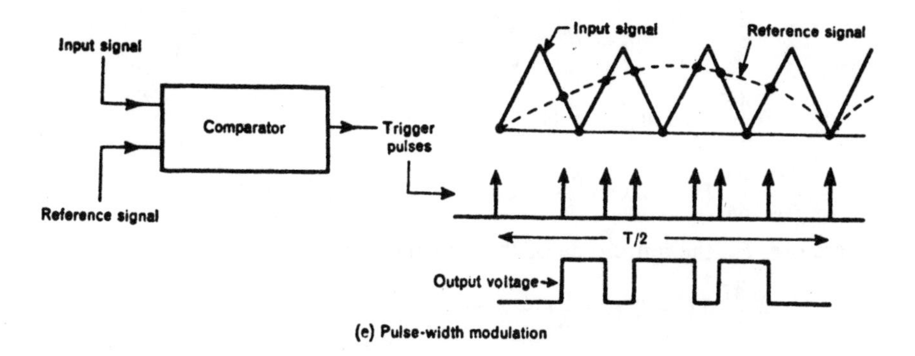

(e) Pulse-width modulation

Fig. 8.8 Output voltage control.

Another method for smooth control of output voltage uses two identical inverter outputs in series. The firing of SCRs in one of the inverters is delayed with respect to the firing of those in the other. Thus, there is a phase-shift in the two inverter outputs and thereby their phasor sum will result in a rectangular waveform of variable width. By changing the firing angle delay between the two inverters, the RMS value of the output can be varied. This method also produces distortion in the output. However, by controlling the pulse width, certain harmonics in the output can be eliminated. For example, if the pulse width is $2\pi/3$, all triplen harmonics will be absent in the load voltage waveform. Similarly, if the pulse width is $2\pi/n$, the n-th harmonic will be filtered out. This type of control for output voltage is known as *pulse-width modulation* (PWM). By automatically controlling the width of the output pulse, the RMS value of the load voltage can be made constant against variations of input voltage or changes in load.

8.4.4 Example

Obtain the harmonic components for the voltage waveforms shown in Figs. 8.8a and 8.8b. The output frequency is 50 Hz and each half-cycle has three pulses, placed symmetrically with respect to the centre of the half-cycle. Compare these harmonics with those of a rectangular waveform of the same amplitude.

The Fourier series for a rectangular voltage waveform $e(t)$ is given by

$$e(t) = \frac{4E}{\pi}[\sin \omega t + \tfrac{1}{3} \sin 3\omega t + \tfrac{1}{5} \sin 5\omega t + \ldots + \frac{1}{2n-1} \sin (2n-1)\omega t + \ldots].$$

For the waveforms shown in Figs. 8.8a and 8.8b, there are only odd harmonics of sine terms because of waveform symmetry.

The Fourier analysis of the waveform shown in Fig. 8.8a gives the n-th harmonic amplitude as

$$E_n = \frac{4E}{\pi}\left[\int_0^{\pi/6} \sin n\theta\, d\theta - \int_{\pi/6}^{\pi/3} \sin n\theta\, d\theta + \int_{\pi/3}^{\pi/2} \sin n\theta\, d\theta\right]$$

$$= \frac{4E}{n\pi}\{[\cos n\theta]_0^{\pi/6} + [\cos n\theta]_{-\pi/6}^{\pi/3} + [-\cos n\theta]_{\pi/3}^{\pi/2}\}$$

$$= \frac{4E}{n\pi}[1 - 2 \cos \frac{n\pi}{6} + 2 \cos \frac{n\pi}{3} - \cos \frac{n\pi}{2}].$$

Therefore, E_n, the amplitude of the n-th harmonic, is

$$E_n = \frac{4E}{n\pi}[1 - 2 \cos \frac{n\pi}{6} + 2 \cos \frac{n\pi}{3}].$$

Hence,

$$e(t) = \frac{4E}{\pi}[0.268 \sin \omega t - 0.333 \sin 3\omega t + 0.746 \sin 5\omega t + 0.532 \sin 7\omega t + \ldots].$$

It can be observed that this waveform has higher harmonic content than a rectangular waveform.

The Fourier analysis of the waveform shown in Fig. 8.8b gives

$$E_n = \frac{4E}{\pi}\left[\int_0^{\pi/6} \sin n\theta\, d\theta + \int_{\pi/3}^{\pi/2} \sin n\theta\, d\theta\right]$$

$$= \frac{4E}{n\pi}\{[-\cos n\theta]_0^{\pi/6} + [-\cos n\theta]_{\pi/3}^{\pi/2}\}$$

$$= \frac{4E}{n\pi}[1 - \cos \frac{n\pi}{6} + \cos \frac{n\pi}{3}].$$

Therefore,

$$e(t) = \frac{4E}{\pi}[0.734 \sin \omega t + 0.473 \sin 5\omega t + 0.338 \sin 7\omega t + \ldots].$$

Hence, the harmonic content of this waveform is much less than that of the waveform in Fig. 8.8a, and comparing it with the harmonic content of a rectangular waveform, it is seen that, since the third harmonic is eliminated, the output waveform will improve after filtering. In general, it can be stated that with multiple pulses, or with pulse-width modulation, the lower harmonics will be attenuated and the higher harmonic amplitudes will be increased.

8.5 POLYPHASE INVERTER

The single-phase parallel inverter (discussed in the preceding sections) can be used as a building block for generating polyphase output. Corresponding SCRs in each unit must be sequentially gated at intervals of $2\pi/(m\omega)$, where m is the number of phases and ω is the output frequency. A more convenient configuration for the polyphase output is the *ring inverter* shown in Fig. 8.9. In this, parallel capacitors provide complementary commutation of the SCRs; and a diametrical connection with the required number of

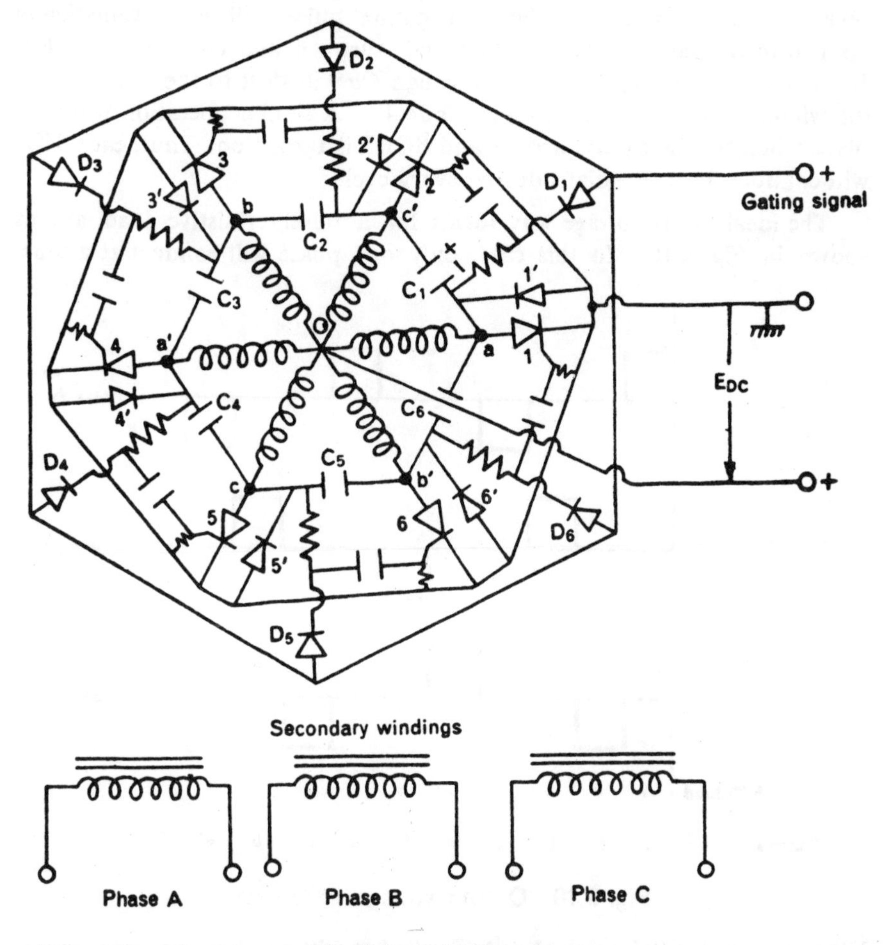

Fig. 8.9 Ring inverter.

phases is used for the output transformer. A three-phase inverter has been used in this figure to facilitate description. The midpoints of the primary windings aa', bb', and cc' of the three-phase transformer are connected to the

DC positive terminal; their ends are connected to the anodes of the SCRs. For three-phase output, six SCRs are needed, the cathodes of all the SCRs being connected to the negative DC terminal. Diodes 1′ to 6′ serve as feedback diodes. The secondary windings of the transformer are connected to the three output phases *A*, *B*, and *C*.

The control circuit consists of an oscillator producing pulses at the rate of six times the input frequency. A starting circuit directs the first pulse to the gate of SCR1, turning it on. Then, winding *Oa* will get energised, and induced voltage will appear in phase *A* of the secondary winding. Capacitor C_1 will get charged to voltage $+E_{DC}$ with the polarity as shown. The potential of each of the terminals *c*′, *b*, *a*′, *c*, and *b*′ will be E_{DC}, whereas that of terminal *a* will be zero. So, diodes D_2 to D_6, but not D_1, will be reverse-biased. Therefore, the next gating pulse will be automatically applied to the gate of SCR2, which will turn on and commutate SCR1. If the load is reactive, the current through *Oa* will shift to the other half of the winding *a*′*O*, and flow through diode 4′. A similar operation will take place when the next pulse arrives and SCR3 is turned on. Thus, each SCR will conduct for $\pi/3$ radians during every cycle.

The ideal phase voltage waveforms for a purely resistive load are as shown in Fig. 8.10. In this case, only one phase will conduct at a time.

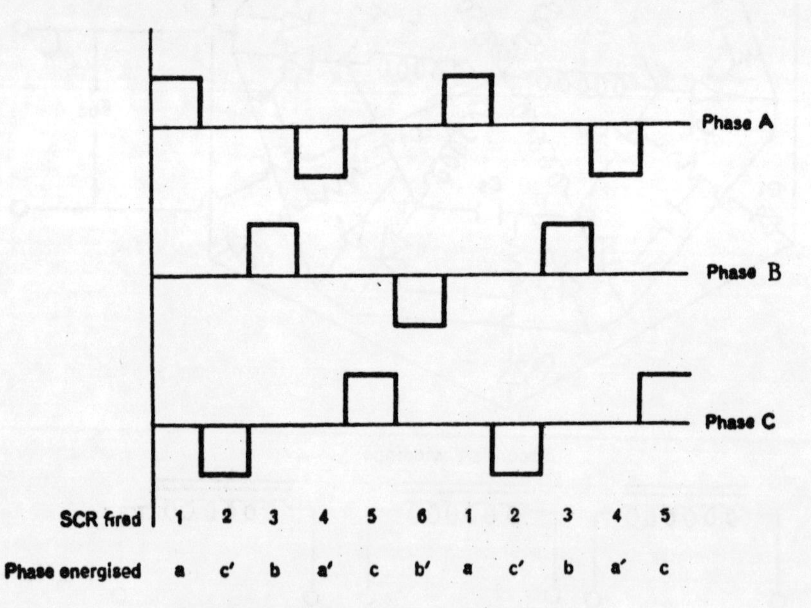

Fig. 8.10 Output voltage waveform.

When the load is inductive, the phase current will not become zero immediately after the SCR is turned off. The free-wheeling diodes carry the load current and apply a negative voltage across the load, thereby decreasing the harmonic content in the load voltage waveform. The conducting duration of phases *A*, *B*, and *C* will be more than $\pi/3$ in each half-cycle.

If the voltages at the anodes of the six SCRs are observed, the following pattern will emerge as the pulses are applied to the control circuit. Here,

Pulse number	*SCR anode voltage*					
	1	*2*	*3*	*4*	*5*	*6*
1	0	1	1	1	1	1
2	1	0	1	1	1	1
3	1	1	0	1	1	1
4	1	1	1	0	1	1
5	1	1	1	1	0	1
6	1	1	1	1	1	0
1	0	1	1	1	1	1

0 and 1 correspond to respectively low voltage and high voltage across the SCR. This pattern will repeat itself after every sixth pulse. Thus, the circuit in Fig. 8.9 can also be used as a six-state sequential circuit where each state is independent of the others. As it is possible to move from one state to another through a fixed number of steps, we can, by decoding the states, use this circuit to count the number of pulses in each state. When used in this manner, the circuit is called a *ring counter*. Chapter 9 describes the application of this circuit in triggering SCR bridge converters and inverters.

8.5.1 Example

(a) Design a parallel inverter for an output voltage of 230 V and 50 Hz, and a peak load current 1 A. The DC input voltage is 30 V. Specify the ratings of the SCRs, and derive the equations used for obtaining the values of the commutating components.

When SCR2 is turned on (see Fig. 8.4a), the capacitor discharge current will be

$$i(t) = A \sin \omega t + B \cos \omega t,$$

where $\omega = 1/\sqrt{LC}$, and L and C are the commutating components. The constants A and B in the foregoing equation are obtained from the initial conditions, namely,

$$i(0) = I'_{\mathrm{L}} \quad \text{(the reflected load current)},$$

and

$$\left.\frac{di}{dt}\right|_{\text{at } t=0} = \frac{2E_{\mathrm{DC}}}{L}.$$

Using these initial conditions, current $i(t)$ is obtained as

$$i(t) = 2E_{\mathrm{DC}} \sqrt{\frac{C}{L}} \sin \omega t + I'_{\mathrm{L}} \cos \omega t.$$

Diode 1′ conducts till current $i(t)$ reaches the peak value, and SCR1 will be reverse-biased as long as diode 1′ conducts. This duration must be at least equal to the turn-off time t_q of SCR1. With these considerations, and assuming that

$$E_{DC}\sqrt{\frac{C}{L}} = I'_L,$$

t_q is approximately given by

$$t_q \approx \frac{\pi}{3}\sqrt{LC}.$$

Using these two equations, commutating components L and C can be calculated.

For the given problem,

$$E_{DC} = 30\text{ V},\qquad I_L = 1\text{ A}.$$

Therefore,

$$\sqrt{\frac{C}{L}} = \frac{230}{30}\times\frac{1}{30} = 0.255.$$

Assuming the turn-off time t_q for SCRs to be 40 μsec, we have

$$\sqrt{LC} = 40\times\frac{3}{\pi}\times 10^{-6} = 38.0\times 10^{-6}.$$

Therefore,

$$C = 9.75\ \mu\text{F},$$
$$L = 149\ \mu\text{H}.$$

The minimum load resistance, referred to the primary side, is 3.9 Ω. From Eq. (8.5), the value of $1/\tau$ is $4\times 3.9\times 9.75\times 10^{-6} = 153\times 10^{-6}$. The turn-off time as calculated from Eq. (8.8) is 143 μsec which is more than 40 μsec assumed earlier. Therefore, the values obtained for L and C are satisfactory. The SCR ratings are 100 V, 10 A, and $t_q = 40$ μsec.

(b) Design an output transformer for the inverter resulting from (a).

The output transformer has a centre-tapped primary winding with a voltage rating of 60 V for each side and a secondary winding with a voltage rating of about 500 V. The turns ratio of the secondary winding to one-half of the primary winding is 8.

The design of the transformer is based on the fundamental frequency terms. A square wave of amplitude 230 V has a fundamental frequency amplitude given by 415 V. Assuming a core flux density $B_{max} = 1.0$ Wb/m² and a core cross section of 25 sq cm, the number of turns on the secondary side will be

$$N = \frac{415}{\sqrt{2}\times 4.44\times 50\times 25\times 10^{-4}} = 530.$$

From this, the primary turns can be calculated. The reader can verify that this design also satisfies Eq. (8.1).

(c) Obtain the values of L and C for an output filter (which has a configuration as shown in Fig. 8.8c) for the inverter specified in (a).

Since the output frequency ω_d is 314 rad/sec, and the base impedance Z_d is 115 Ω, using Eq. (8.9), we have

$$L_2 = \tfrac{115}{314} = 365 \text{ mH},$$

$$L_1 = 4.5L_2 = 1.643 \text{ H},$$

$$C_1 = \frac{1}{6 \times 115 \times 314} = 4.6\ \mu\text{F},$$

$$C_2 = 2C_1 = 9.2\ \mu\text{F}.$$

8.6 SERIES INVERTER

The series inverter uses class A type of commutation. Commutating components L and C are applied in series with the load to form an underdamped circuit. Since the SCRs turn off by themselves when the current becomes zero, this inverter is classified as a self-commutated inverter.

Figure 8.11a shows the schematic arrangement of a simple series inverter. Let the initial voltage on capacitor C be E_C with the polarity as shown in this figure. When SCR1 is turned on, the waveform for current i will be as in Fig. 8.11b. The necessary condition to obtain this load current is that the series circuit consisting of commutating components C and L, and load R (assumed resistive), must be underdamped. Therefore, $R^2 < 4L/C$, and the time period of oscillation will be

$$\frac{T}{2} = \frac{\pi}{\sqrt{\dfrac{1}{LC} - \dfrac{R^2}{4L^2}}}. \tag{8.10}$$

At a (Fig. 8.11b), load current i is zero and SCR1 will be turned off; also, capacitor C will be charged to voltage V_C in the reverse direction. Duration ab is the off-period when the load is open-circuited. So, the capacitor will retain voltage V_C. At b, SCR2 is fired. As SCR1 had already been turned off, duration ab (T_{off}) should be more than the turn-off time it requires. Capacitor C will now discharge through SCR2 and the underdamped series circuit. Load current i will be in the opposite direction and again becomes zero at point c. SCR2 will then be turned off. A similar operation will occur when SCR1 is turned on. The output frequency is given by $1/(T/2 + T_{off})$. Thus, with the same LC components, variable-frequency output can be obtained by changing the off-time.

The main limitations of the series inverter discussed here are as follows:

(a) Its maximum possible output frequency is limited to the ringing frequency $\sqrt{1/(LC) - R^2/(4L^2)}$ of the resonant circuit. This is because SCR2 can be triggered only after SCR1 is turned off; otherwise there will be a short-circuit on the DC supply and commutation of SCRs will not take place, with the result that the circuit will not operate as an inverter.

(b) For output frequencies much lower than the ringing frequency, the distortion in the load voltage waveform is high. This is because the off-time is large in comparison with the duration of conduction of the SCRs.

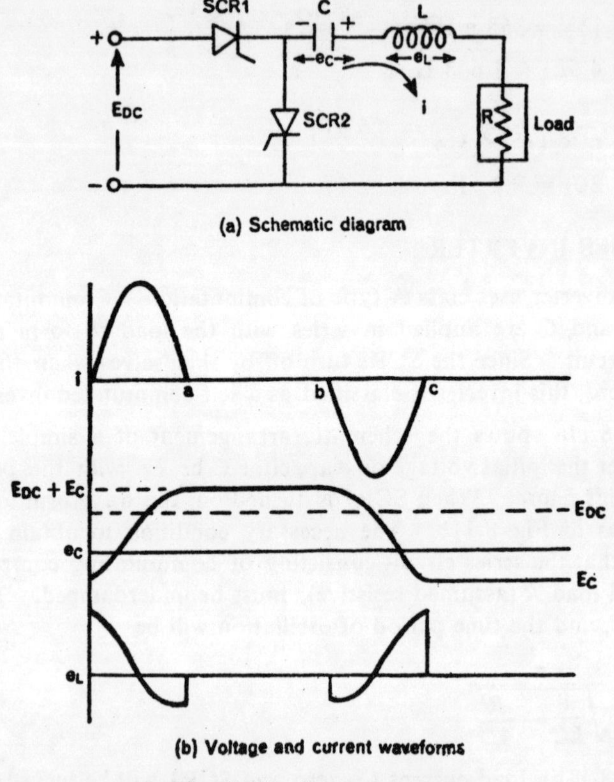

Fig. 8.11 Series inverter.

(c) High rating for the commutating components is required because these components carry the load current continuously and the capacitor supplies the load current in every alternate half-cycle.

(d) The power flow from the DC source is intermittent. This is because current is supplied to the load only when SCR1 is fired. Therefore, the DC supply must have a large peak current rating, and the input current has a high harmonic content.

(e) The peak amplitude and duration of the load current in each half-cycle depend on the load parameters, resulting in poor output regulation for the inverter.

Of these limitations, (b), (c), and (e) are inherent in all types of series inverters and cannot be overcome. However, limitations (a) and (d) can be relaxed by the modified series inverters discussed in Section 8.6.3.

For series inverters, if the load is inductive, the load inductance can be considered part of the commutating inductance L since both are connected in series. No separate feedback diodes are required for inductive loads as in parallel inverters.

8.6.1 Circuit Operation

The differential equation describing the operation of the circuit when SCR1 is fired is

$$E_{DC} + E_C = iR + L\frac{di}{dt} + \frac{1}{C}\int i\,dt, \tag{8.11}$$

where E_C is the initial voltage on the capacitor and the initial value of current $i(0) = 0$. Since the circuit is underdamped, the solution for i will be

$$i(t) = Ae^{-Rt/(2L)}\sin \omega t, \tag{8.12}$$

where

$$\omega = \sqrt{\frac{1}{LC} - \frac{R^2}{4L^2}},$$

and A is obtained by equating di/dt (at $t = 0$) to $(E_C + E_{DC})/L$ as

$$A = \frac{E_C + E_{DC}}{\omega L}.$$

When current $i(t)$ again becomes zero at $t = T/2$, SCR1 will be turned off. Capacitor C will be charged to voltage V_C. When SCR2 is next fired, the circuit operation will be described by

$$V_C = iR + L\frac{di}{dt} + \frac{1}{C}\int i\,dt, \tag{8.13}$$

with i equal to zero initially. Therefore, in the steady state, when the positive and negative load current waveforms must be identical, Eqs. (8.11) and (8.13) have to be the same. Thus, the necessary condition for the steady state is

$$V_C = E_{DC} + E_C. \tag{8.14}$$

The steady-state voltage waveforms across L and C are as shown in Fig. 8.11b. Equation (8.14) can be used for obtaining E_C in the steady state. The capacitor voltage at the end of a half-cycle is given by

$$V_C = -E_C + \frac{1}{C}\int_0^{T/2} i\,dt. \tag{8.15}$$

Thus, from Eq. (8.14) we have

$$\frac{1}{C}\int_0^{T/2} i\,dt = E_{DC} + 2E_C.$$

Therefore,

$$E_{DC} = \frac{E_{DC} + E_C}{C(R^2 + 4\omega^2 L^2)} \times 4L\{1 + e^{[-R/(2L)](\pi/\omega)}\} - 2E_C, \tag{8.16}$$

$$E_C = \frac{E_{DC}\{1 - [\frac{4L}{C(R^2 + 4\omega^2 L^2)}][1 + e^{[-R/(2L)](\pi/\omega)}]\}}{-2 + \{\frac{4L}{C(R^2 + 4\omega^2 L^2)}\}\{1 + e^{[-R/(2L)](\pi/\omega)}\}}.$$

Substituting for ω in this equation, we get

$$E_C = E_{DC} \frac{e^{[-R/(2L)](\pi/\omega)}}{1 - e^{[-R/(2L)](\pi/\omega)}}. \tag{8.17}$$

8.6.2 Design Considerations

From Eq. (8.17), we can calculate voltage ($E_C + E_{DC}$), which will appear as the forward voltage across the SCRs during the off-period. This voltage should not be more than the blocking voltage rating of the SCRs. Voltage E_C also must be of sufficient magnitude to turn off the SCRs within time T_{off}. The ringing frequency ω must be so chosen as to approximate the value of the desired output frequency ω_o, such that the off-time $[=\pi(1/\omega_o - 1/\omega)]$ of the SCRs is greater than their turn-off time. The peak current rating of the SCRs is given by A in Eq. (8.12), and can be evaluated once E_C is known. If the load is variable, the maximum value of R must be used in Eq. (8.10) so that the circuit remains underdamped for all load conditions. To obtain the peak current rating A from Eq. (8.12) and the peak reverse voltage E_C from Eq. (8.17), the minimum value of R should be used.

8.6.3 Improved Series Inverters

Figure 8.12 shows two modifications for the series inverter configuration described earlier. In Fig. 8.12a, inductors L_1 and L_2 have the same inductance and are closely coupled. Therefore, when SCR1 is fired and current i_1 begins to rise during the first quarter of the cycle, the potential across L_1 will be positive with polarity as shown in the figure. The induced voltage in L_2 will now add to the capacitor voltage in reverse-biasing SCR2. Since L_1 and L_2 have the same inductance, the equivalent circuit of the inverter and the differential equations describing the circuit operation will be identical to those given for Fig. 8.11a if the SCRs are triggered after the load current has become zero. For this mode of operation, the circuit has no special advantage over that in Fig. 8.11a except that each SCR in the former will experience a reverse voltage for a longer period. However, the important feature of this circuit is that SCR2 can be triggered even before the load current has touched zero or before SCR1 has been turned off. That is, the output frequency can be made higher than the ringing frequency, thereby increasing the frequency range of the inverter. This mode of operation is possible because of the induced voltage in the commutating inductors L_1 and L_2, whereas in the series inverter configuration described earlier, the same operation results in a short-circuit on the DC supply.

Let us suppose that SCR2 is triggered shortly before SCR1 is turned off. At the instant of firing, the voltage across the capacitor will be slightly less than $(E_{DC} + E_C)$ and the load voltage and current will be close to zero. Therefore, a voltage equal to the voltage across the capacitor minus the load voltage will appear across L_2. Since L_1 is closely coupled to L_2, the same voltage will appear across L_1. The cathode potential of SCR1 will be raised to a level higher than its anode potential, and therefore SCR1 will be reverse-biased and turn off. A similar operation will take place if SCR1 is triggered before SCR2 is turned off. Thus, there will be no danger of a short-circuit on the DC supply.

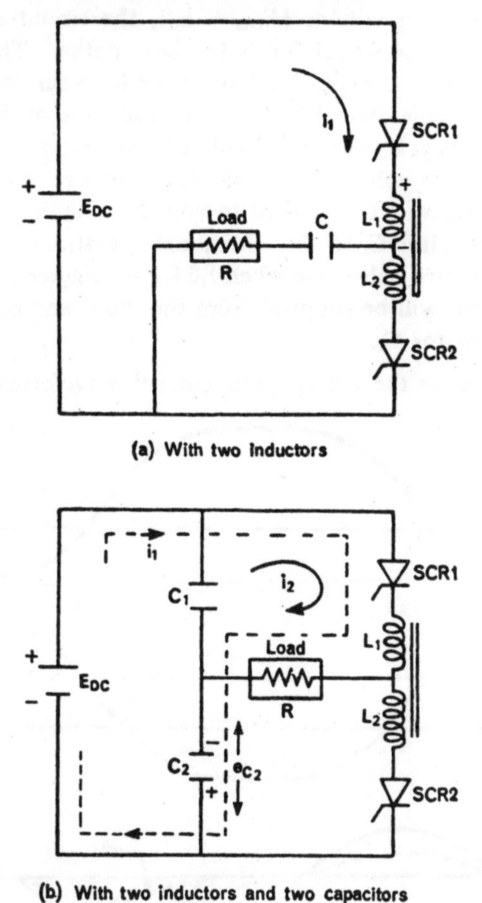

Fig. 8.12 Improved series inverter.

In this circuit, as in the one shown in Fig. 8.11a, the power flow from the DC source is intermittent. This drawback is overcome in the circuit shown in Fig. 8.12b where, during both half-cycles of the output, power is drawn from the input. One-half of the load current is supplied by capacitor C_1 or C_2, and the other half flows from the DC supply. Inductors L_1 and L_2 are identical, and so also are capacitors C_1 and C_2. If the two inductors are closely coupled, then the output frequency for this circuit, as for the circuit shown in Fig. 8.12a, can be made higher than its ringing frequency.

In Fig. 8.12b, let the initial voltage across capacitor C_2 be E_C, with the polarity as shown. Then, capacitor C_1 will be charged to voltage $(E_{DC}+E_C)$ in the opposite direction. When SCR1 is fired, there will be two parallel paths for load current $i_L (=i_1 + i_2)$. Current i_1 will flow from the positive DC terminal, through SCR1, L_1, load, and capacitor C_2, to the negative supply terminal. Current i_2 will flow from C_1, through SCR1 and L_1, to the load. The driving voltage $(E_{DC} + E_C)$, the circuit elements, and the initial conditions are identical for both these paths. Therefore, the two currents will be equal. Hence, one-half of the load current will come from the DC supply and the other half from the discharge of the capacitor. At the end of the half-cycle, when the load current becomes zero, SCR1 will be turned off and the voltage across the capacitors reversed. In the steady state, capacitor C_2 will be charged to voltage $(E_{DC} + E_C)$ in the opposite direction and capacitor C_1 to E_C. Identical operations will take place in the following negative half-cycle when SCR2 is triggered. Then, one-half of the load current will be supplied from the input and other half from the discharge of capacitor C_2.

Figure 8.13 shows the voltage and current waveforms across different

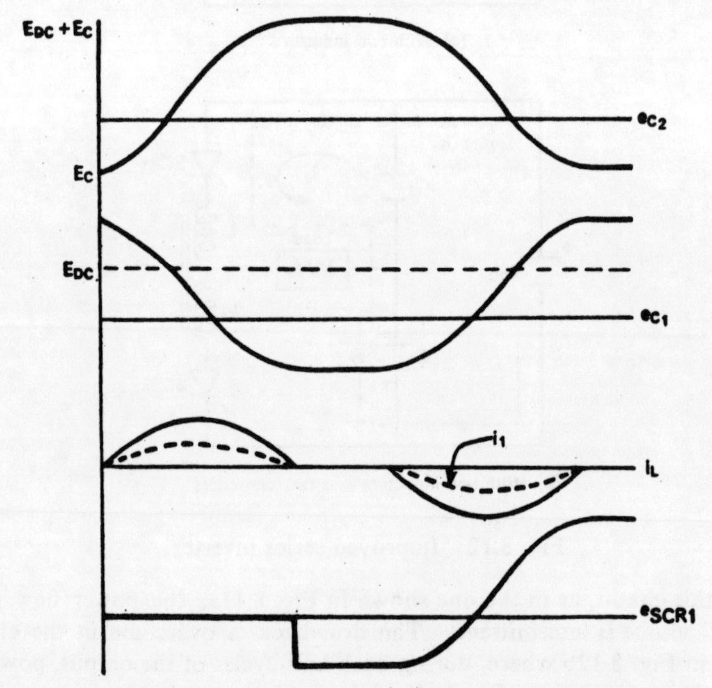

Fig. 8.13 Voltage and current waveforms.

elements in the circuit shown in Fig. 8.12b. Voltage E_C in the steady state is given by Eq. (8.17). The design criteria for the commutating components L_1 and C_1 are the same as those discussed in Section 8.6.2. The peak forward off-state voltage for the SCRs is $(E_{DC} + E_C)$ and the peak reverse voltage is E_C. Figure 8.13 also shows the voltage waveform across SCR1.

8.6.4 Three-Phase Series Inverter

Three single-phase series inverters of the type shown in Fig. 8.11a can, when used with proper connections, give a three-phase output as shown in Fig. 8.14. This circuit may be very easily analysed if the capacitors across

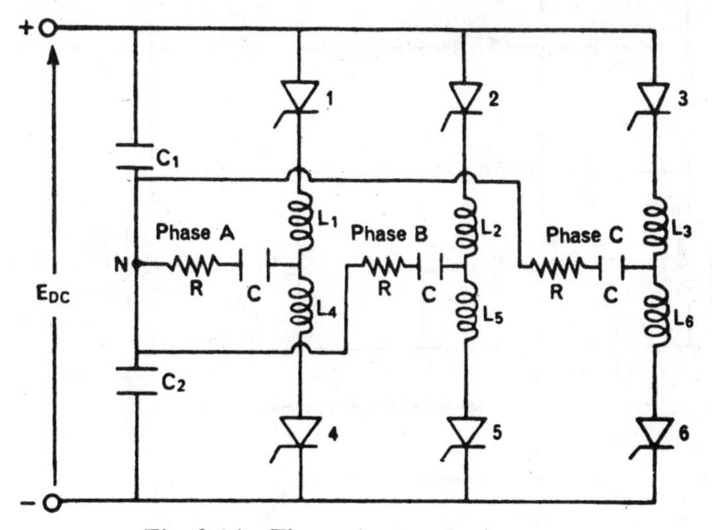

Fig. 8.14 Three-phase series inverter.

the DC supply are large enough to establish a constant neutral voltage. Then, each phase will function as an independent series inverter. Also, the capacitors in series with the load will resonate with the centre-tapped reactors thereby providing series-capacitor-type commutation.

The SCRs are fired in the sequence 1, 6, 2, 4, 3, and 5, the firing frequency being six times the output frequency. Therefore, the interval between successive firings will be $T/6$, where T is the period of the output. SCR1 should be turned off before SCR4 is turned on. The design considerations are the same as those for a single-phase series inverter. The drawbacks and merits listed for single-phase series inverters in the preceding sections apply also for three-phase series inverters. (For a discussion on control circuits for the sequential firing of SCRs, see Chapter 9.)

8.6.5 High-Frequency Series Inverter

The output frequency of the series inverter (discussed in the preceding sections) is limited to its resonant frequency. Any attempt to increase this frequency by decreasing the value of commutating components will be faced with the problem of satisfying the minimum turn-off time requirement of the SCRs. With inverter-grade SCRs having an average turn-off time of 15 μsec, the output frequency can go upto 30 kHz. One method of obtaining high-frequency output, with normal SCRs which have a large turn-off time, is to use a number of series inverters in parallel as shown in Fig. 8.15a, and to operate them one at a time in a specified sequence.

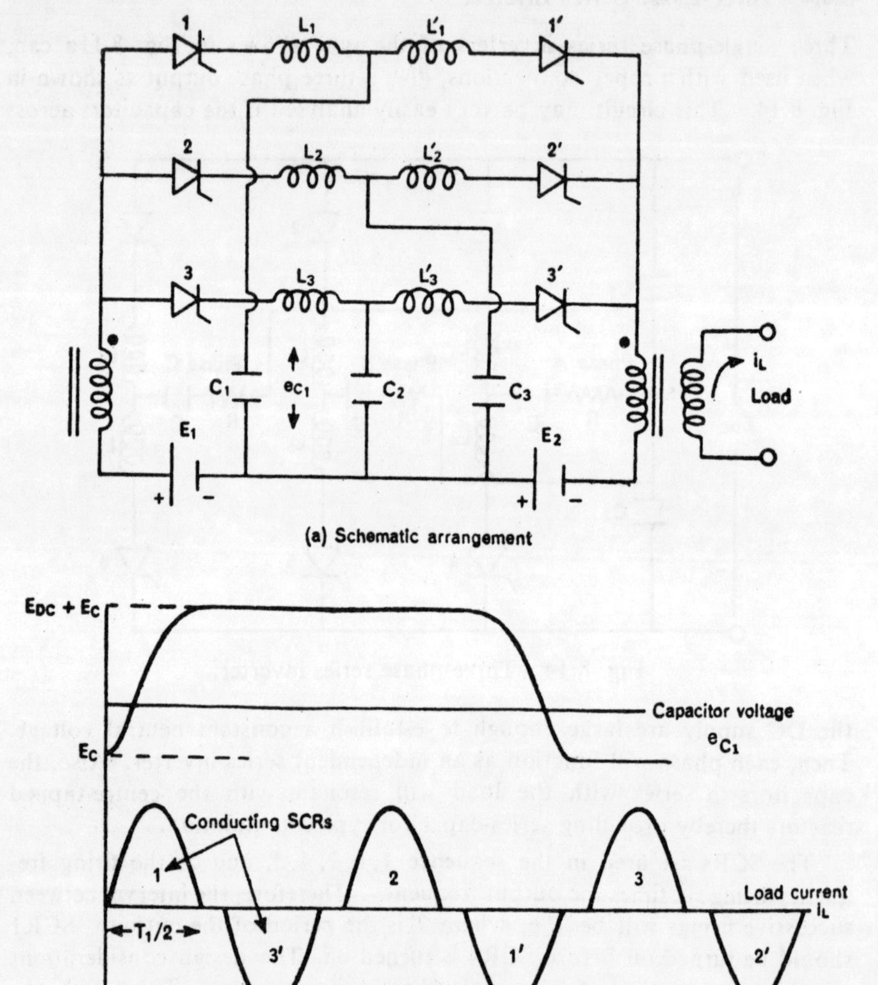

Fig. 8.15 High-frequency series inverter.

This operation is similar to that for the ring inverter (discussed in Section 8.5). In an *n*-stage series inverter, each inverter will be active and will supply the load current for (1/*n*)-th of the whole period. This is also known as the *time sharing operation*. The outputs of all the inverters are coupled to the load through a transformer. For example, when SCR1 is fired, the current will flow from the DC source E_1 to the load through SCR1, L_1, and C_1, charging the capacitor to voltage E_C, which will be more than E_1. After SCR1 is turned off, SCR3′ is fired. As can be seen from Fig. 8.15a, the load current will now flow in the opposite direction. After SCR3′ is turned off, SCR2 is turned on. Thus, the SCRs are fired in the order 1, 3′, 2, 1′, 3, and 2′, producing alternate positive and negative half-cycles of current in the load. In the steady state, voltage E_C across every capacitor, at the end of the conduction period of the corresponding SCR, will be the same and therefore both the positive and negative half-cycle load current waveforms will be identical, as shown in Fig. 8.15b.

A specific advantage of this circuit is that after each SCR is turned off a reverse voltage appears across the SCR until its complementary SCR is fired, since the voltage E_C across the capacitor will be more than the supply voltage. For example, SCR1 will have a reverse voltage after it is turned off until SCR1′ is fired. Assuming that the output frequency is equal to the ringing frequency, the reverse bias on SCR1 will exist during the time SCRs 3′ and 2 conduct. In other words, if each SCR has a firing frequency $1/T$ and the output frequency is $1/T_1$ (for optimum design, $T = nT_1$ for an n-stage inverter, where T_1 is the period of each resonant circuit), then the SCRs will be reverse-biased for approximately a duration of $(T - T_1)/2$. This specifies the turn-off time requirement of each SCR. By choosing a high value for T, T_1 can be made as small as possible for a given turn-off time of the SCR, and thus high-frequency output can be produced.

8.6.6 Example

Design a series inverter which has a configuration as shown in Fig. 8.11a. The maximum output frequency required is 3 kHz. The load resistance may vary from 300 ohms to 100 ohms, and the supply voltage is 100 V.

Assume the ringing frequency of the resonant circuit to be 5 kHz. A suitable value of L is chosen on the basis of the attenuation factor $e^{[-R/(2L)]t}$. For example, when $\omega t = \pi/2$, the peak value of A will be reduced by the factor $e^{-R\pi/(4\omega L)}$. For the maximum value of load resistance, inductance L is chosen to be 10 mH, so that the attenuation factor becomes 0.475. With $L = 10$ mH, capacitance C is calculated from the expression for ω, using $R = 300$ ohms. Since

$$\omega^2 = \frac{1}{LC} - \frac{R^2}{4L^2},$$

we get

$$C = 0.082\ \mu\text{F}.$$

From Eq. (8.17), the value of E_C in the steady state is given by

$$E_C = 100\left[\frac{e^{[-R/(2L)](\pi/\omega)}}{1 - e^{[-R/(2L)](\pi/\omega)}}\right].$$

Using the minimum value of R in this equation, we get $E_C = 150$ V. Therefore, the maximum forward blocking voltage for the SCR is 250 V and its rating must be about 300 V. The peak current in the load is given by

$$I_{peak} = \frac{E_{DC} + E_C}{\omega L} e^{[-R/(2L)][\pi/(2\omega)]}.$$

Using the minimum value of resistance, we get $I_{peak} = 0.62$ A. Therefore, the current rating of the SCR is 1 A.

8.7 SELF-COMMUTATED INVERTERS

The inverters discussed in the preceding sections also fall into the general category of self-commutated inverters since they do not use separate commutation circuits. Figure 8.1 shows two methods for connecting the commutating components to form an underdamped circuit. Inverters that use the connection shown in Fig. 8.1a are termed *series inverters* because the commutating components are in series with the load. These have already been discussed in detail. For the configuration in Fig. 8.1b, the capacitor is connected in parallel with the load and the inductor is used in series to form a resonant circuit. Inverters using this commutation arrangement also do not require any separate commutating mechanism. The SCRs will turn off when their forward current goes to zero (class A type). A detailed discussion of the performance characteristics of this type of self-commutated inverter follows.

8.7.1 Inverter Connections

Figure 8.16a shows the schematic diagram of a self-commutated inverter which uses class A type of commutation. Capacitor C and inductor L_1 form an underdamped resonant circuit with the load. SCRs 1 and 2 are fired together. The capacitor will get charged with the polarity as shown, and when the SCR current becomes zero the SCRs will be turned off. The necessary conditions for this operation will be derived in Section 8.7.2. At the instant of zero current, the voltage across the capacitor will be more than the supply voltage. So, the capacitor will begin discharging into the supply through feedback diodes 1′ and 2′. These diodes apply a reverse voltage across SCRs 1 and 2. At the end of the half-period of the output, SCRs 3 and 4 will be fired. Diodes 1′ and 2′ will be reverse-biased and the current will shift to SCRs 3 and 4. The capacitor will then begin charging in the opposite direction. It is necessary, as in other series inverters, that the output frequency f_o be lower than the resonant frequency $f_r\,[=1/(2\pi\sqrt{2L_1C})]$ of the commutating circuit. In other words, the effective load power factor must be leading so that the SCR current may become zero before the load voltage polarity reverses.

Figure 8.16b shows the load current waveforms for different output-to-resonant frequency ratios. It will be observed that, since the capacitor is connected in parallel with the load, the load current waveform is continuous over a wide range of output frequency. This is an advantage over the series inverters (discussed in Section 8.6).

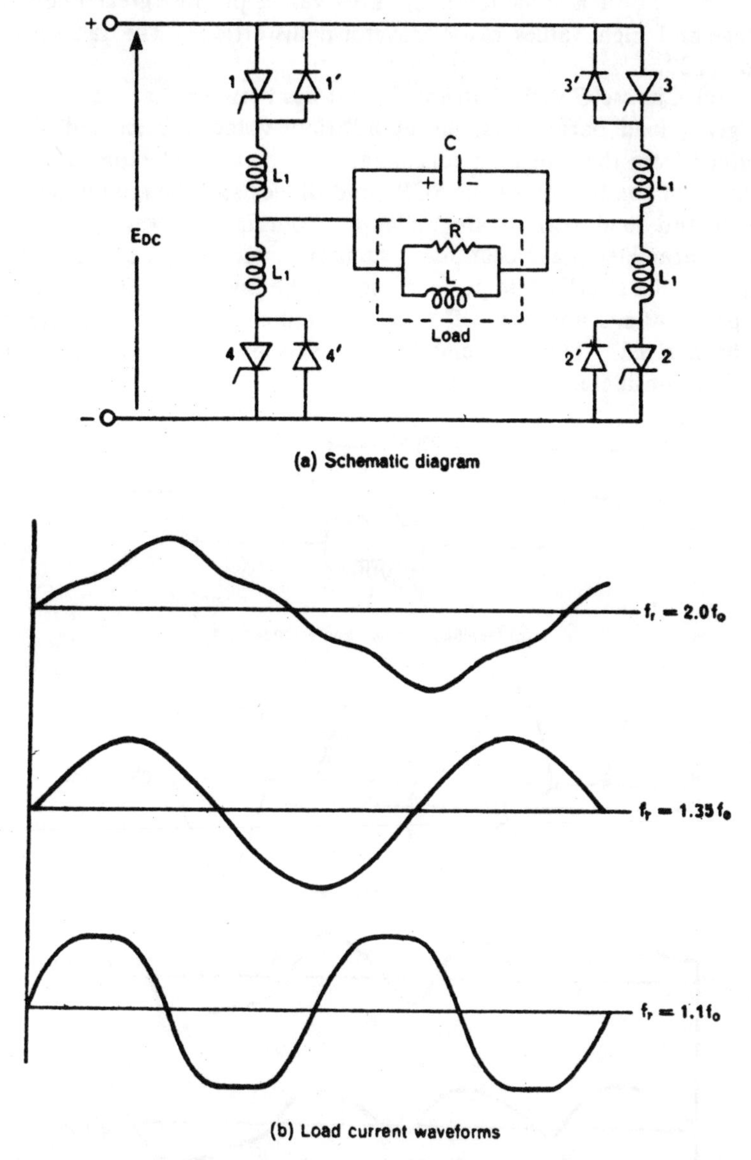

(a) Schematic diagram

(b) Load current waveforms

Fig. 8.16 Self-commutated inverter (*cont.*).

In Fig. 8.16a, the load is assumed to be inductive and is represented by a parallel connection of an inductor L and a resistor R. The steps for determining suitable values for the commutating components L_1 and C are as follows:

(a) Choose a proper ratio f_r/f_o. The ratio generally used is 1.35. Lower values give a shorter turn-off time and higher values produce waveform distortion.

(b) Choose a value for L/L_1. Low values produce greater component voltage and high values cause waveform distortion. The ratio generally used is 200.

(c) Choose C such that $R\sqrt{C/(2L_1)}$ lies between 3 and 5.

For given load parameters, the appropriate values for L_1 and C can be obtained from the foregoing considerations. Analytical expressions for the duration of conduction of the SCRs and diodes and the amplitudes of load voltage and load current are difficult to obtain. However, once all the circuit parameters have been fixed, numerical techniques can be employed to deduce the steady-state performance of the inverter. From such study, the peak ratings and turn-off time requirements of the SCRs and diodes can be derived. The mathematical analysis of a simplified form of this circuit is considered in Section 8.7.2.

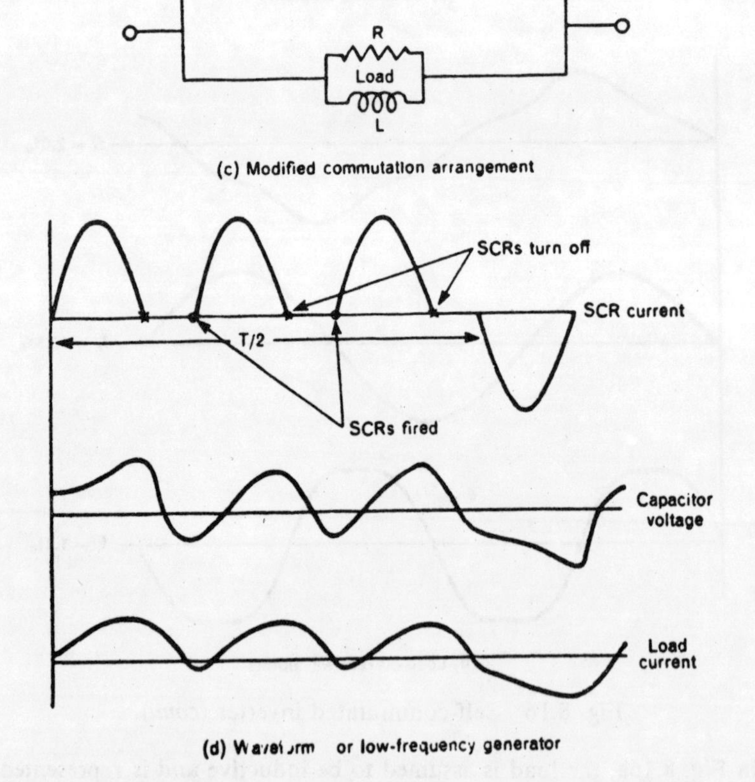

(c) Modified commutation arrangement

(d) Wavei orm or low-frequency generator

Fig. 8.16 Self-commutated inverter.

Figure 8.16c shows another commutation arrangement. Here, only one commutating inductor is required. The disadvantage of this arrangement is that the SCRs are subjected to a dv/dt larger than that when the inductors are placed between the SCRs. A capacitor alone, placed in parallel with the load, will act as a high-pass filter and make the load current more sinusoidal, as shown in Fig. 8.16b. The presence of an inductor in series with the capacitor, as in Fig. 8.16c, will produce waveform distortion. This circuit is often used as a high-frequency generator for induction heating.

The circuit in Fig. 8.16a can be used also as a low-frequency generator with an output voltage waveform similar to that for cycloconverters (discussed in Chapter 7). For this operation, feedback diodes 1′, 2′, 3′, and 4′ are removed and inductor L_1 is placed as shown in Fig. 8.16c. The design of commutating components L_1 and C, for given load parameters, is the same as that given earlier in this section, except that the ratio f_r/f_o is made equal to the number of pulses required in each half-cycle of the output. Figure 8.16d shows the voltage waveform across the load. SCRs 1 and 2 are repeatedly fired in the positive half-cycle. Each time, the SCRs will turn off when the current becomes zero. During the off-period, the load current will be supplied by the capacitor. Figure 8.16d also shows the waveforms for the capacitor voltage and the SCR current. It is also necessary that the load circuit, comprising L_1, C, R, and L, is underdamped.

The inverter configuration shown in Fig. 8.16a is usually referred to as a *bridge circuit*. In Chapter 9, the performance of various types of bridge inverters will be discussed in detail.

8.7.2 Mathematical Analysis

Figure 8.17a shows one form of the self-commutated inverter that uses only a single SCR. This circuit is chosen for our analysis because of its simplicity. Its operation is similar to that shown in Fig. 8.16a. The commutating components L and C are so chosen as to make the forward current through the SCR go through a zero value in order that the SCR undergoes self commutation. The load voltage waveform shown in Fig. 8.17b has a large DC value. In some applications, this may not be a serious disadvantage. By varying the triggering frequency of the SCR, the output frequency can be changed.

The SCR is fired at a (Fig. 8.17b). Current i_1 starts from zero, reaches a peak value at b, and again becomes zero at c, turning off the SCR. Then, the capacitor will discharge through the load, and the voltage across the capacitor will fall off. At d, the SCR will be turned on again and the same operations will be repeated. Let v be the voltage across the capacitor. The circuit operation is described by

$$E - v = L\frac{di_1}{dt},$$
$$i_1 = \frac{v}{R} + C\frac{dv}{dt}. \tag{8.18}$$

Eliminating v, the differential equation in i_1 will be

$$\frac{d^2i_1}{dt^2} + \frac{1}{RC}\frac{di_1}{dt} + \frac{1}{LC}i_1 = \frac{E}{RLC}. \tag{8.19}$$

Therefore,

$$i_1(t) = \frac{E}{R} + Ae^{[-1/(2RC)]t} \sin(\omega t + \alpha), \tag{8.20}$$

where

$$\omega = \sqrt{\frac{1}{LC} - \frac{1}{4R^2C^2}}.$$

The triggering frequency of the SCR must be less than ω. Assume that the peak value of the sinusoidal component of current is $2E/R$ to ensure

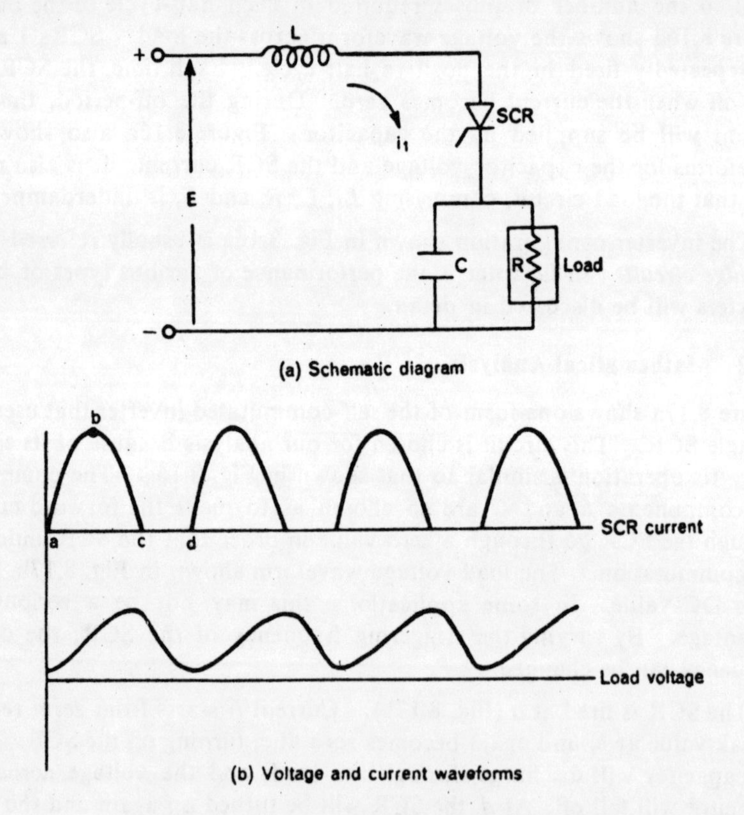

Fig. 8.17 Simple self-commutated inverter.

that the SCR current goes to zero. Since $i = 0$ at $t = 0$, the value of α in Eq. (8.20) is equal to $-\pi/6$. Therefore,

$$i_1(t) = \frac{E}{R}[1 + 2e^{-t/(2RC)} \sin(\omega t - \pi/6)]. \tag{8.21}$$

It is assumed that the current will again touch zero value when the sinusoidal component attains the negative peak value, that is, when

$$\omega t - \pi/6 = 3\pi/2.$$

Therefore,

$$0 = \frac{E}{R}\{1 - 2e^{[-1/(2RC)][5\pi/(3\omega)]}\}.$$

Hence,

$$\frac{5\pi}{6\omega RC} = \ln 2. \qquad (8.22)$$

For a given value of R, Eqs. (8.20) and (8.22) can be used to obtain proper values for L and C. After the SCR is turned off, the capacitor discharges through the load. The load current will then be

$$i_L(t) = \frac{v}{R} e^{-t/(RC)}. \qquad (8.23)$$

Equation (8.23) is valid from the instant the SCR is turned off to the instant the SCR is retriggered. For a given triggering frequency, Eqs. (8.18) and (8.23) can be solved using numerical methods, and the capacitor voltage $v(t)$ in the steady state can be determined. Figure 8.17b shows the steady-state load voltage waveform.

8.7.3 Example

For the circuit shown in Fig. 8.17a, obtain the appropriate values of L and C if the supply voltage $E = 100$ V, load resistance $R = 10\ \Omega$, and the triggering frequency of the SCR is 7 kHz.

Assume that the ringing frequency of the underdamped circuit is made 15 kHz. Using Eq. (8.22), we have

$$C = \frac{5\pi}{6 \times 2\pi \times 15 \times 10^3 \times 10 \times \ln 2} \times 10^6$$
$$\approx 4\ \mu\text{F}.$$

From Eq. (8.20),

$$\omega^2 = \frac{1}{LC} - \frac{1}{4R^2C^2}.$$

Therefore,

$$\frac{1}{L} = C[4\pi^2 \times 225 \times 10^6 + \frac{1}{400C^2}]$$
$$= 366 \times 10^2,$$
$$L = 0.028\ \text{mH}.$$

REFERENCES

Adams, R. D., Fox, R. S., Several modulation techniques for PWM inverter, IEEE Conference Record (IGA), 1970, p. 687.

Bedford, R. E., Nene, V. D., Analysis and performance of a three-phase ring inverter, *IEEE Trans.* (IGA), 1970, p. 488.

Corey, P. D., Methods for optimising the wave form of a stepped wave static inverter, AIEE Summer Power Meeting, 1962, Paper No. 62–1147.

Dewan, S. B., Duff, D. L., Practical considerations in the design of commutation circuits for choppers and inverters, IEEE Conference Record (IGA), 1969, p. 469.

Guggi, W. B., Practical design considerations for regulated sine wave inverter, IEEE Conference Record (IGA), 1971, p. 869.

Hauas, G., Sommer, R. A., A high frequency power supply for induction heating and melting, *IEEE Trans.* (IECI), 1970, p. 321.

Humphrey, A. J., Inverter commutation circuit, *IEEE Trans.* (IGA), 1968, p. 104.

Mapham, N., The classification of inverter circuits, IEEE International Convention Record, Part 4, 1964, p. 99.

Mapham, N., An SCR inverter with good regulation and sine wave output, *IEEE Trans.* (IGA), 1967, p. 176.

Mapham, N., A low cost ultrasonic frequency inverter using single SCR, *IEEE Trans.* (IGA), 1967, p. 378.

Ott, R. R., A filter for SCR commutation and harmonic attenuation in high power inverters, *AIEE Trans.* (Communication and Electronics), 1963, p. 259.

Pelly, B. R., Latest developments in static high frequency power sources for induction heating, *IEEE Trans.* (IECI), 1970, p. 297.

Penkowski, L. J., Pruzinsky, K. E., Fundamentals of a pulse width modulated power circuit, IEEE Conference Record (IGA), 1970, p. 669.

Ramamoorty, M., A simple high frequency static inverter, *IEEE Trans.* (IECI), 1976, p. 103.

Ramamoorty, M., Joshi, P. K., Variable frequency DC to AC parallel inverter circuit using silicon controlled rectifier, *JIE* (India), 1970, p. 6.

Robertson, S. D. T., Hebbar, K. M., A variable low frequency inverter using thyristors, *IEEE Trans.* (IGA), 1968, p. 501.

9

Bridge Inverters

9.1 PRINCIPLES OF OPERATION

Of the three types of inverters that use forced commutation, two, namely, parallel and series inverters, have been examined in detail in Chapter 8. The third type, known as the *bridge inverter*, will be considered here. The bridge configuration is widely used for controlled rectification (Chapter 6) and also for line-commutated inverters (Chapter 7). The main features of such a configuration are: (a) for rectifiers, there is no residual flux in the input transformer core, and so the problems resulting from magnetic saturation do not arise, and (b) for inverters, the output transformer is not essential.

Figure 9.1 gives the basic arrangement of a single-phase bridge inverter; the commutation circuits are not shown. When SCRs 1 and 2 conduct, the load voltage will be positive; when SCRs 3 and 4 conduct, the output

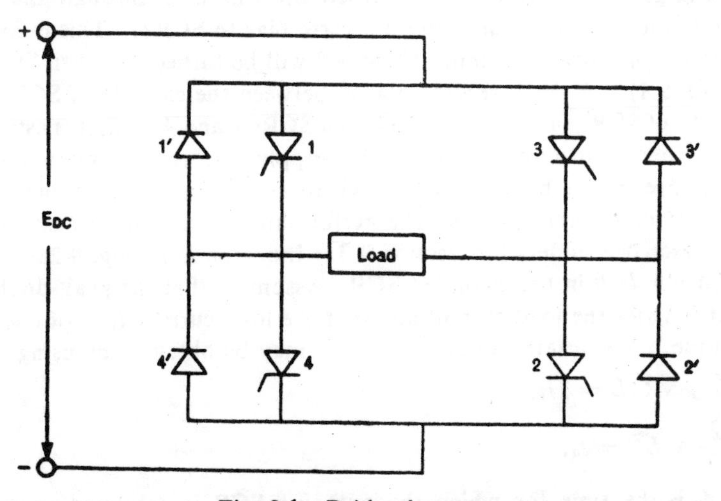

Fig. 9.1 Bridge inverter.

voltage will be negative. Thus, the output frequency is dependent on the triggering frequency of the SCRs. Diodes 1′ to 4′ serve to feed the load reactive power back to the DC supply. The load voltage waveform is fairly rectangular (as in parallel inverter circuits), and is not affected by the nature of the load. This is an advantage of the bridge inverter over the series inverter. Commutation between SCR pairs 1, 2 and 3, 4 can be achieved in two ways. One is by resonant commutation, where the load and commutating components are so designed that the SCR forward current goes to zero before the half-period is over. Such circuits use class A type of commutation; their operation and performance have been discussed in Chapter 8. The main drawbacks of this type of inverter are the load-dependent voltage and current waveforms, poor voltage regulation, and the limitations on output frequency. In the second method, class C or class D type of forced commutation is used. Bridge inverters employing such commutation will be considered in this chapter. The major advantages of these inverters are: (a) good voltage regulation, (b) wide range of control for output frequency, and (c) suitable for output voltage control and polyphase output. A description of various methods of commutation of SCRs in bridge inverters follows.

9.1.1 Commutation Circuits

Figure 9.2 illustrates three types of connection for the commutating components L and C in a bridge inverter. Only that portion of the circuit required for explaining the principle of commutation is shown. When SCRs 1 and 2 are conducting (Fig. 9.1), the commutating capacitor C, which is in parallel with the load, will get charged to full DC voltage with the polarities as shown in Fig. 9.2a. When SCR3 is fired (Fig. 9.2a) the capacitor will discharge through SCRs 1 and 3 and turn off SCR1. The excess charge on C after SCR1 is turned off will flow through the free-wheeling diode 1′, and thus apply a reverse bias to SCR1. This is known as *current commutation*. Similarly, SCR2 will be turned off when SCR4 is fired (Fig. 9.1). A small reactor placed between the cathode of SCR3 and the anode of SCR2 (and similarly between SCRs 1 and 4) will, it is assumed, prevent direct short-circuits on the DC supply. Current commutation is used for three-phase bridge inverters where commutation takes place between SCRs of different phases. The performance of the three-phase circuits will be described in detail in Section 9.3. Inductor L in Fig. 9.2a is used for reducing di/dt in the incoming SCR. Assuming that the peak discharge current is twice the load current and that the load current I_L is of constant amplitude, appropriate values of L and C may be obtained by using

$$\begin{aligned} E_{DC}\sqrt{C/L} &= 2I_L, \\ \frac{2\pi}{3}\sqrt{LC} &= t_q, \end{aligned} \tag{9.1}$$

where t_q is the time for which the outgoing SCR is subjected to reverse voltage, and is equal to the duration for which the diode conducts. The value of t_q must be more than the specified turn-off time of the SCR.

In Fig. 9.2b, the firing of SCR4 will turn off SCR1 as follows. Neglecting the drop in voltage across inductor L_1, capacitor C_4 will get charged to the full DC voltage when SCR1 is conducting. Capacitor C_1 will then be in the discharged state. If SCR4 is fired at the end of the half-period, voltage E_{DC} will appear across inductor L_4, which is closely coupled to inductor L_1. So, the cathode of the conducting SCR1 will be raised to a

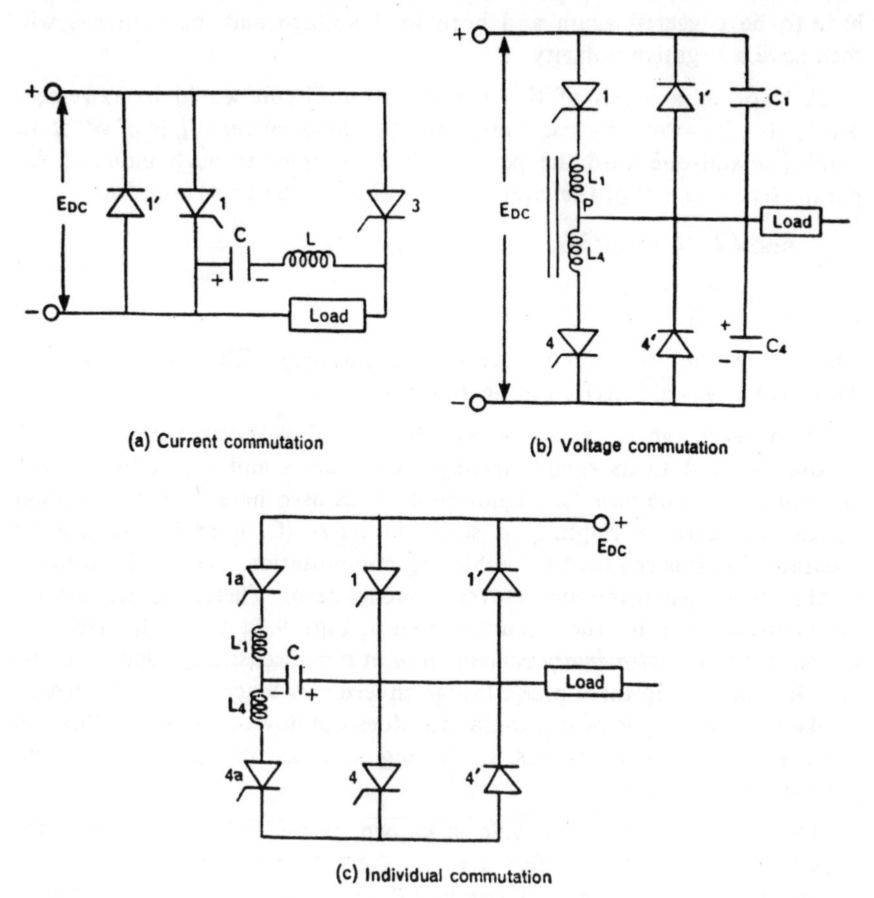

Fig. 9.2 Commutation circuits.

potential of $2E_{DC}$, and will therefore be turned off. Similarly, the firing of SCR3 will turn off SCR2 (Fig. 9.1). This is known as *voltage commutation.* After SCR1 has turned off, the load current will be shared equally by the two capacitors. The load current which was flowing in inductor L_1 will now flow in the same direction in inductor L_4 and thus maintain the same ampere-turns in the magnetic circuit. The current in L_4 is supplied by the two capacitors. When the potential of point P falls to $E_{DC}/2$, the cathode potential of SCR1 will be E_{DC} and SCR1 will begin to get forward-biased. Finally, when capacitor C_4 is discharged, capacitor C_1 will be fully charged. Free-wheeling diode 4′ will now get forward-biased and the load current will be taken up by the diode. When diode 4′ conducts, SCR4 will be turned off after the commutating energy in L_4 is completely dissipated. The load voltage will reverse its polarity and, if the load is inductive, the current will still be in the positive direction. The reactive energy in the load will be fed back to the DC supply through free-wheeling diodes 4′ and 3′ (Fig. 9.1). When the current polarity reverses, SCRs 3 and 4 (Fig. 9.1) will have to be triggered again and both load voltage and load current will then have a negative polarity.

A rigorous analysis of this commutation circuit would be extremely involved. However, by assuming that the load current I_L is of constant amplitude and one-third the peak resonant current through inductor L_4, parameters L and C of the commutating circuit may be obtained as

$$E_{DC}\sqrt{2C/L} = 2\sqrt{3}I_L,$$
$$t_q = \frac{\pi}{3\sqrt{2}}\sqrt{LC}, \tag{9.2}$$

where t_q is the duration for which the outgoing SCR is reverse-biased. These equations are derived in Section 9.1.2.

Both the foregoing types of connection use class D commutation, which is also referred to as *complementary commutation* and sometimes as *self commutation*. The term 'self commutation' as used here is in the context of resonant turn-off applied to series inverters (Chapter 8), because no separate circuit is required for achieving commutation. The SCRs turn off by themselves when the load current touches zero. Therefore, the turn-off mechanisms used by the circuits shown in Figs. 9.2a and 9.2b have been designated as *complementary commutation* in our discussion. Such circuits are often applied in three-phase bridge inverters. Since the load current, unlike in class A type of commutation, does not flow continuously through the commutating components, the rating and size of the inductor and capacitor will be small.

The circuit shown in Fig. 9.2c is known as the *individual commutation circuit* and is often also referred to as the *auxiliary commutation circuit*. It pertains to class C type of commutation. Here, an auxiliary SCR is used for turning off every SCR. For example, SCR1 in Fig. 9.2c will be turned off when SCR1a is fired. The operation of the circuit will be as follows. When SCRs 1 and 4a are fired, capacitor C will be charged to voltage $2E_{DC}$ because inductor L_4 overcharges the capacitor. SCR4a will be turned off due to natural commutation when the charging current becomes zero. SCRs 1 and 2 (Fig. 9.1) will then be turned on to apply positive voltage to the load. Capacitor C is already charged with polarities as shown in Fig. 9.2c. SCR1a is fired when SCR1 is to be turned off. Capacitor C will discharge through SCR1 as long as it is conducting the load current, but when the discharge current becomes equal to the load current, SCR1 will be turned off and diode 1′ will begin to conduct the load current and discharge current. The load voltage will remain positive as long as SCR1 or diode 1′ conducts. SCR1 will be reverse-biased during the conduction period of diode 1′. When the capacitor discharge current again becomes equal to the load current (the former is sinusoidal and the latter is assumed to be of constant amplitude), diode 1′ will stop conducting and diode 4′ will get forward-biased. Since SCR2 (Fig. 9.1) conducts continuously, the load voltage will become zero when diode 4′ begins to conduct. It is assumed that during this period the inductive energy in the load provides for continuous conduction of current. Capacitor C will continuously get charged and SCR1a will be turned off when the capacitor voltage becomes approximately equal to $2E_{DC}$ in a direction opposite to that shown in Fig. 9.2c. If SCR1a is to have control on the commutation of SCR1, the voltage across the capacitor will have to be reversed before the latter can be fired again to apply positive voltage to the load. This voltage reversal is achieved by firing SCR4a, and the conducting diode 4′ will help in reversing the charge on the capacitor. When SCR1 is fired, diode 4′ will be reverse-biased and the load current will again flow through SCRs 1 and 2 (Fig. 9.1). SCR1 can be turned off again, if required, to make the load voltage zero. Thus, the load voltage can be made zero a number of times during each half-cycle. This is known as *multiple commutation* and constitutes a special feature of the auxiliary commutation circuit. Similarly, the other load-carrying SCRs can be turned off by firing the corresponding auxiliary SCRs. Only one SCR needs to be turned off when the load voltage is to be made zero. If a reversal of load voltage is required, both conducting SCRs should be turned off simultaneously. The same commutating capacitor may be used to turn off SCRs 1 and 4 (Fig. 9.1).

The peak voltage across the capacitor will be $2E_{DC}$. Making the same assumptions as in the foregoing discussion on other commutation circuits, the values of the commutating components can be determined by

$$\begin{aligned} E_{DC}\sqrt{C/L} &= I_L \\ \frac{2\pi}{3}\sqrt{LC} &= t_q, \end{aligned} \tag{9.3}$$

where t_q is so chosen as to be slightly more than the turn-off time of the SCR. An additional consideration for this circuit is the minimum permissible duration of the free-wheeling time for the load current during which the output load voltage will be zero (i.e., the time for which diode 4′ and SCR2 will conduct). SCR1a, as stated earlier, will undergo natural commutation when the charging current of the capacitor becomes zero, at which stage the capacitor will be charged to voltage $2E_{DC}$ in the reverse direction. SCR4a will then need to be fired to reverse the voltage. When this voltage becomes E_{DC}, SCR1a will get forward-biased. The minimum time for which SCR1a will be reverse-biased is given by

$$t_{min} = \frac{\pi}{3}\sqrt{LC}. \tag{9.4}$$

According to Eq. (9.3), t_{min} will not be enough to turn off SCR1a. So, SCR4a should not be fired immediately after SCR1a is turned off, nor should it be fired immediately after the capacitor reaches a potential of

$2E_{DC}$. Thus, the minimum time interval from the instant SCR4a is turned on to when SCR1a is turned off should be $(\pi/3)\sqrt{(LC)}$. To avoid any overcharging of the capacitor, SCR1 should be turned on only after SCR4a has been turned off. Then, diode 4′ will stop conducting and the free-wheeling period will be over. Therefore, the total minimum duration for the free-wheeling is given by

$$t_{off} = (4\pi/3)\sqrt{LC} + t_o, \tag{9.5}$$

where t_o is the time taken for SCR1a to stop conducting after diode 4′ has started conducting. For the circuit in Fig. 9.2c, the value of t_o will be

$$t_o \approx LI_L/(E_{DC} \times 0.866). \tag{9.6}$$

Equation (9.6) is based on the mean value of the rate of decay of current through SCR1a. The rate of decay is calculated when diode 1′ stops conducting and when SCR1a is turned off, and the mean of the two values is taken. Combining Eqs. (9.5) and (9.6), the minimum duration of the off-time can be determined. Similarly, the minimum time for which the voltage will be applied to the load is given by

$$t_{on} = \frac{5\pi}{6}\sqrt{LC}.$$

The sum $(t_{on} + t_{off})$ will determine the maximum possible frequency of commutation.

The main advantage of individual commutation is that during each half-cycle of the output voltage the load voltage can be kept at zero by turning off only one of the conducting SCRs. This is known as the *free-wheeling period.* By multiple commutation, this period can be produced any number of times in each half-cycle. Such a mode of operation is used for controlling the output voltage, and is often referred to as *pulse-width-modulation* (PWM) *control.* Complementary-commutated inverters (Figs. 9.2a and 9.2b) have usually to be operated from a variable DC voltage supply since, for many applications, such as speed controllers for AC motors, the ratio of the output RMS phase voltage to frequency has to be kept constant. Commutation for such inverters may not be successful at low output frequencies because the commutating capacitor gets charged to a lower voltage. This problem is not encountered in auxiliary-commutated inverters where the output RMS voltage can be varied by PWM, even when the DC input voltage is kept constant. Thus, a much wider range of frequency is possible when individual commutation is used for inverters employing auxiliary SCRs. It must, however, be mentioned that PWM will result in a larger harmonic content, even though certain lower harmonics can be eliminated or attenuated by suitably choosing the conducting intervals.

9.1.2 Example

(a) Derive the equations from which appropriate values can be determined for the commutating components required for providing voltage commutation in the circuit shown in Fig. 9.2b. Assumptions considered necessary for the derivation may be made.

Capacitor C supplies one-half of the load current and one-half of the reactor current. The initial voltage across the capacitor is E_{DC} and the reactor current immediately after commutation is the load current I_L. Assuming the load current to be constant, the differential equation for the reactor current is

$$L\frac{d^2i}{dt^2}+\frac{i}{2C}=-\frac{I_L}{2C}.$$

Therefore,

$$i(\ )=A\sin(\omega t+\alpha)-I_L,$$

where

$$\omega=1/\sqrt{2LC}.$$

Since $i=I_L$ at $t=0$, and if the peak reactor current is assumed to be $3I_L$, A will be $4I_L$ and α will be $\pi/6$. Therefore,

$$i(t)=4I_L\sin(\omega t+\pi/6)-I_L.$$

The value of di/dt at $t=0$ is given by

$$E_{DC}/L=4I_L\omega\cos(\pi/6).$$

Therefore,

$$E_{DC}/2=I_L\omega L\sqrt{3}.$$

Substituting $1/\sqrt{2LC}$ for ω, we get

$$\frac{E_{DC}}{2}=\frac{I_L}{2C}\sqrt{3}\times\sqrt{2LC}.$$

The conducting SCR will be reverse-biased till the voltage across the commutating capacitor becomes equal to $E_{DC}/2$. The time required for this to take place is t_q (the circuit commutation turn-off time) and is given by

$$\frac{E_{DC}}{2}=E_{DC}-\left[\frac{2I_L}{C}\int_0^{t_q}\sin(\omega t+\pi/6)\,dt\right].$$

Therefore,

$$\frac{E_{DC}}{2}=\frac{2I_L}{C\omega}[-\cos(\omega t_q+\pi/6)+\cos(\pi/6)].$$

Substituting $[\sqrt{3}/(2C)]I_L\sqrt{2LC}$ for $E_{DC}/2$, the foregoing equation will be satisfied when

$$\cos(\omega t_q+\pi/6)\approx 0.5.$$

Therefore,

$$t_q=\frac{\pi}{6\omega}=\frac{\pi\sqrt{LC}}{4.24}.$$

(b) If $E_{DC} = 100$ V and $I_L = 10$ A, calculate the values of commutating components L and C for the circuits shown in Figs. 9.2a and 9.2b if the required turn-off time $t_q = 40$ μsec.

For the circuit in Fig. 9.2a, from Eq. (9.1) we get

$$\sqrt{C/L} = 2 \times 10/100 = 0.2,$$

$$\sqrt{LC} = 40 \times \frac{3}{2\pi} \times 10^{-6} \approx 20 \times 10^{-6}.$$

Therefore,

$$L = 100\ \mu\text{H},$$

$$C = 4\ \mu\text{F}.$$

For the circuit in Fig. 9.2b, from Eq. (9.2) we get

$$\sqrt{2C/L} = \frac{1.732 \times 20}{100} = 0.3464,$$

$$\sqrt{LC} = 40 \times 10^{-6} \times \frac{4.24}{\pi}.$$

Therefore,

$$L = 22.0\ \mu\text{H},$$

$$C = 13.2\ \mu\text{F}.$$

(c) For the circuit shown in Fig. 9.2c, if $L = 30$ μH and $C = 4$ μF, obtain the maximum possible number of multiple commutations in each half-cycle of the output when the supply voltage is 100 V, the load current is 10 A, and the output frequency of the inverter is 1000 Hz.

The value of t_{off} required for proper operation of the circuit [from Eq. (9.5)] is

$$t_{off} = \tfrac{4}{3}\pi\sqrt{LC} + t_o.$$

From Eq. (9.6),

$$t_o = \frac{30 \times 10^{-6} \times 10}{100 \times 0.866} = \frac{300}{86.6} \times 10^{-6} \approx 3\ \mu\text{sec}.$$

Therefore,

$$t_{off} = \tfrac{4}{3}\pi\sqrt{120 \times 10^{-6}} + 3 \times 10^{-6} \approx 49\ \mu\text{sec}.$$

The value of t_{on} will be

$$t_{on} = \frac{5\pi}{6}\sqrt{LC} \approx \frac{5\pi}{6} \times 11 \times 10^{-6} \approx 2.88\ \mu\text{sec}.$$

Hence, the maximum number n of multiple commutations will be $T/[2(t_{off} + t_{on})]$, where T is the period of the output. Therefore,

$$n = \frac{1}{2000} \times \frac{1}{52} \times 10^{6} \approx 10.$$

9.2 OPERATION OF A SINGLE-PHASE BRIDGE INVERTER

The operation of a single-phase bridge inverter (see Fig. 9.1) is explained here in detail. The load is assumed to be inductive. The output half-period is divided into four regions (Fig. 9.3a): (a) the active period *A*, during which power flows from the DC supply to the load (i.e., when SCRs 1 and 2 or 3 and 4 conduct), (b) the free-wheeling period *F*, when one SCR and one diode conduct (i.e., 1 and 3′, 3 and 1′, 4 and 2′, or 2′ and 4 conduct) and the load voltage becomes zero, (c) the recovery period *R*, when power flows from the load to the input (the load voltage and current will be of opposite polarity and the inductive energy stored in the load will be fed back to the supply; this will be possible only when the diametrically opposite diodes 1′ and 2′ or 3′ and 4′ conduct), and (d) the off-period *O*, during which neither the SCRs nor the diodes conduct, and the load voltage and the current are zero [this operation will take place when both the quality factor for the load *Q* (=ratio of reactance to resistance) and the output frequency are small].

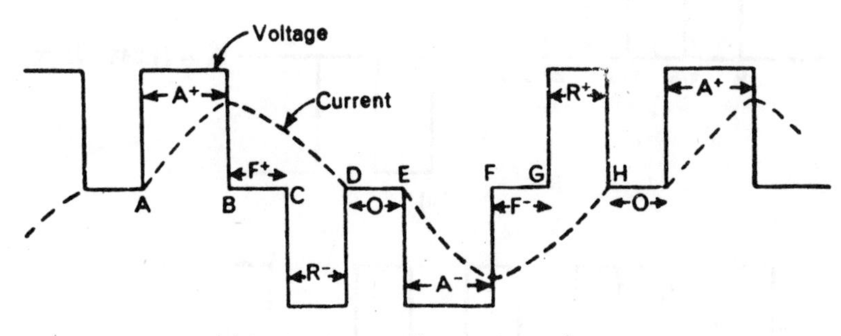

(a) Load voltage and current waveforms

Fig. 9.3 Output voltage and current waveforms for an inverter (*cont.*).

Figure 9.3 shows typical load voltage and current waveforms for a bridge inverter. Since free-wheeling periods are included, auxiliary commutation is assumed. During the *A* period, SCRs 1 and 2 of the circuit shown in Fig. 9.1 will conduct. At *B*, SCR1 will be turned off by firing SCR1a (individual commutation is assumed). After the commutation transients, diode 4′ will conduct the load current. Since SCR2 and diode 4′ conduct, the load voltage will be zero between *B* and *C*. This is the free-wheeling period (F^+). At *C*, SCR2 will be turned off by its auxiliary SCR2a. Now, diode 3′ will begin to conduct and the load voltage will reverse. The load inductance will, until all the reactive power is fed back, maintain the conduction of diodes 4′ and 3′ from *C* to *D*. This is the recovery period (R^-). At *D*, the current will become zero. Between *D* and *E*, the load will be open-circuited and both the voltage and the current will be zero. This is the off-period (*O*). At *E*, SCRs 3 and 4 are fired, the negative half-cycle starts, and the same sequence of operations as before will repeat.

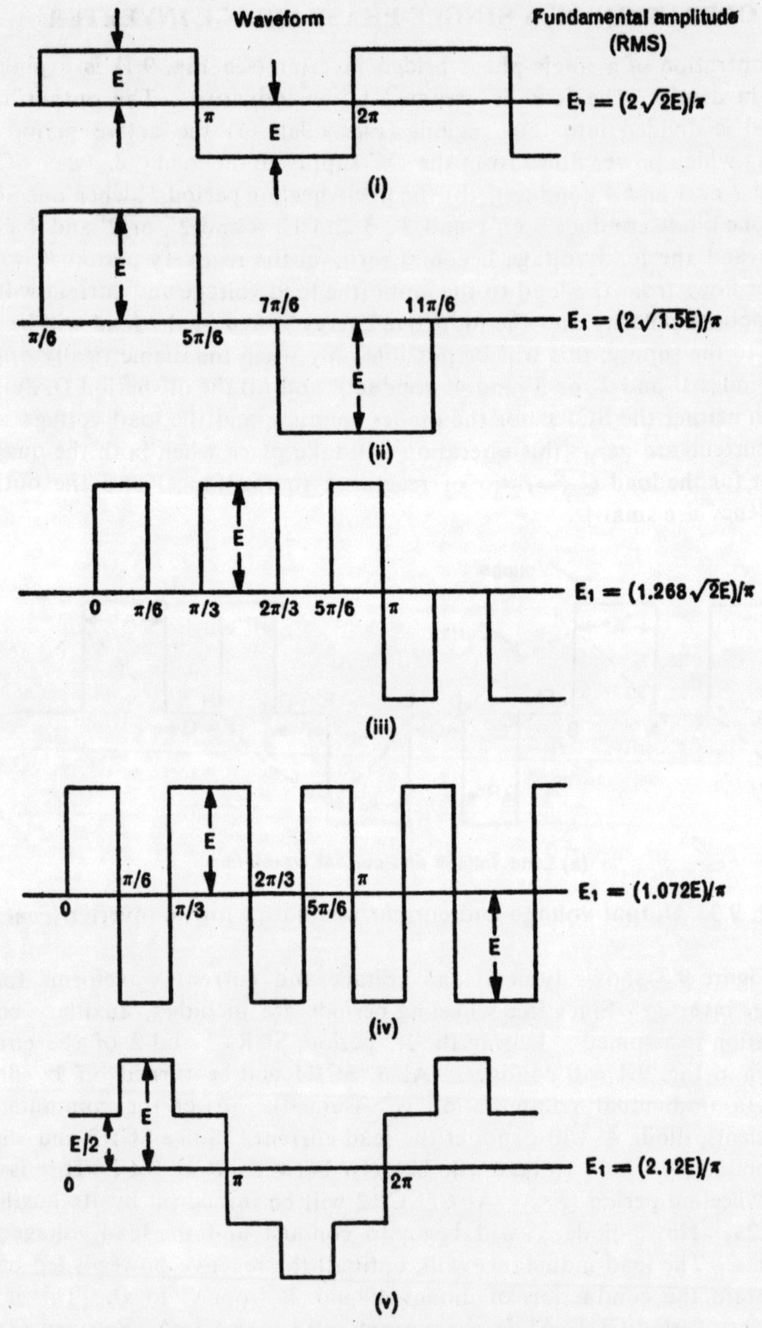

(b) Typical pulse-width-modulated and stepped voltage waveforms

Fig. 9.3 Output voltage and current waveforms for an inverter.

It has already been mentioned in Section 9.1 that the output voltage can be controlled by means of multiple commutation or pulse-width modulation. This is done by varying the number of commutations or duration of the four regions in the output half-period. Figure 9.3b shows the output voltage waveforms obtained from an auxiliary-commutated inverter using single and multiple commutation in each half-cycle of the output; the RMS value of the fundamental component of each waveform is also given.

Another way of controlling the output voltage of a bridge inverter (Fig. 9.1) is simultaneously to commutate the two conducting SCRs and fire the other two SCRs so that the load voltage is reversed. Figure 9.3b(iv) shows the typical output voltage resulting from such an operation. This method can be applied to both auxiliary- and complementary-commutated inverters. However, in view of the severe harmonic distortion it produces, this method of output voltage control is not often used.

9.2.1 Example

A single-phase bridge inverter produces the pulse-width-modulated output voltage waveform shown in Fig. 9.3b(ii). For a normal DC input voltage of 100 V, the pulse width is made $2\pi/3$ at the desired output frequency so that the RMS value of the output voltage can be controlled for both positive and negative changes in the supply voltage. What will be the pulse width if this input voltage is increased by 15 per cent ? Find the minimum input voltage required to keep the RMS output voltage constant. Describe briefly the relevant control scheme.

The RMS output voltage for normal input is

$$V_{\text{RMS}} = \sqrt{100^2 \times \frac{2\pi}{3} \times \frac{1}{\pi}} = \sqrt{\tfrac{2}{3}} \times 100 = 82 \text{ V}.$$

Let V_L be the minimum input voltage. The pulse width for this voltage will then be π and the RMS value of output will be equal to V_L. Therefore, the minimum value of input voltage will be 82 V. If the supply voltage becomes 115 V, then the required pulse width θ will be

$$115\sqrt{\frac{\theta}{\pi}} = 100\sqrt{\tfrac{2}{3}},$$

$$\theta = \tfrac{2}{3} \times \frac{1}{1.15^2} \times \frac{180\pi}{\pi} = 90°.$$

The comparator described in Chapter 8 for PWM control can be used for comparing the input voltage error with the reference signal having a triangular waveform of twice the input frequency. The error signal (i.e., the input voltage error) is obtained by subtracting V_L (=82 V) from the DC input voltage. Thus, when the DC input level is equal to V_L, the error voltage will be zero and the output pulse width will be equal to π. The slope of the reference triangular voltage should be so adjusted that when the DC voltage is equal to 100 V, the output pulse width will be $2\pi/3$. The triggering pulses are obtained from the comparator output.

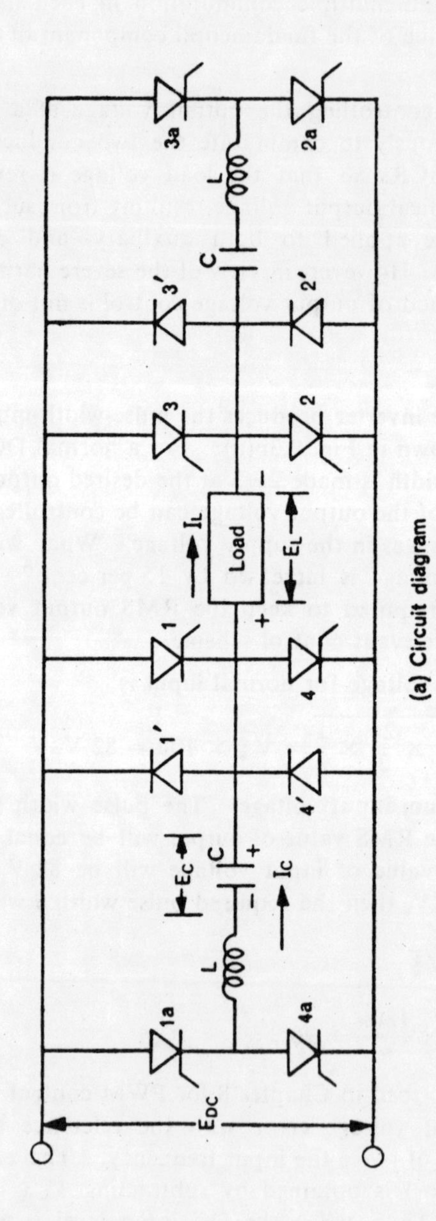

(a) Circuit diagram

Fig. 9.4 Auxiliary-commutated single-phase bridge inverter (*cont.*).

9.2.2 Auxiliary-Commutated Single-Phase Bridge Inverter

Figure 9.4a details the circuit of a single-phase bridge inverter, which has auxiliary commutation. The circuit consists of the main SCRs 1, 2, 3, and 4; the free-wheeling diodes 1′, 2′, 3′, and 4′; the auxiliary SCRs 1a, 2a, 3a, and 4a; and the commutating components L and C. The method of commutation (with reference to Fig. 9.2c) and the various modes of operation of this circuit have already been explained. Equation (9.3) can be used for obtaining appropriate values of the commutating components.

In Fig. 9.4b, SCR1a will be turned on at O. SCR1 will turn off at P and diode 1′ will be reverse-biased at Q. Diode 4′ will begin to conduct at Q, and the load voltage will either become zero (if the free-wheeling mode is used and only SCR1a is turned on at O) or reverse (if both SCRs 1 and 2 are turned off). At R, the capacitor discharge current will become zero and the capacitor will be charged to $2E_{DC}$ in the reverse direction. The commutating inductor L can be used as shown in Fig. 9.4a, or in two halves as in Fig. 9.2c. The circuit in Fig. 9.4a can be used as a building block for polyphase output. The load voltage waveform will be rectangular and independent of the type of load.

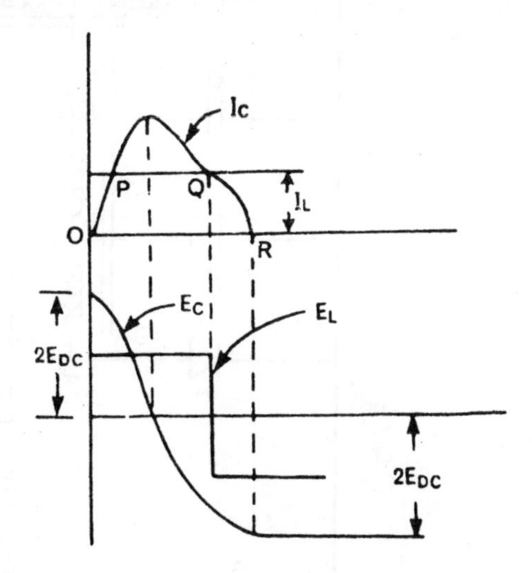

(b) Current and voltage waveforms

Fig. 9.4 Auxiliary-commutated single-phase bridge inverter.

9.2.3 Inverter Circuit with Complementary Commutation

Figure 9.5a shows the circuit for a single-phase bridge inverter with complementary commutation. The number of SCRs required here is less than that for the circuit discussed in Section 9.2 2. The method of commutation has already been explained with reference to Fig. 9.2b. In the positive half-cycle, SCRs 1 and 2 will conduct. At the end of the half-period, the firing of SCRs 3 and 4 will turn off SCRs 1 and 2. Figure 9.5b shows the load voltage waveform and the various conduction periods; transients during commutation are not given. SCRs 1 and 2 will conduct during period AB. At B, SCRs 3 and 4 will be turned on, and SCRs 1 and 2, as explained in Section 9.1 with reference to Fig. 9.2b, will turn off because of the reverse bias applied. Capacitor C_4 will supply the load current and capacitor C_3 will get charged by the load current. Therefore, the potential of P will go down and that of Q will go up, resulting in a

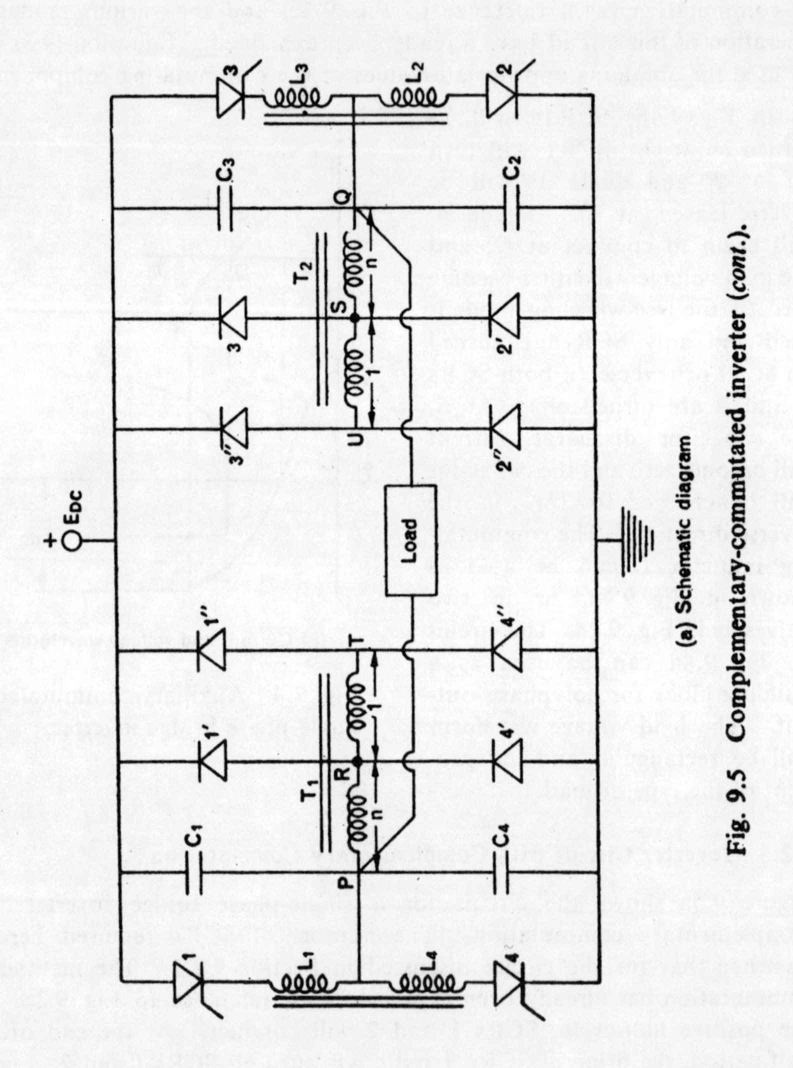

(a) Schematic diagram

Fig. 9.5 Complementary-commutated inverter (cont.).

decrease in the load voltage. SCRs 3 and 4 will turn off after the commutation transients are over. If the load current is assumed to be constant, then the load voltage will reduce approximately linearly (neglecting commutation transients), as shown in Fig. 9.5b.

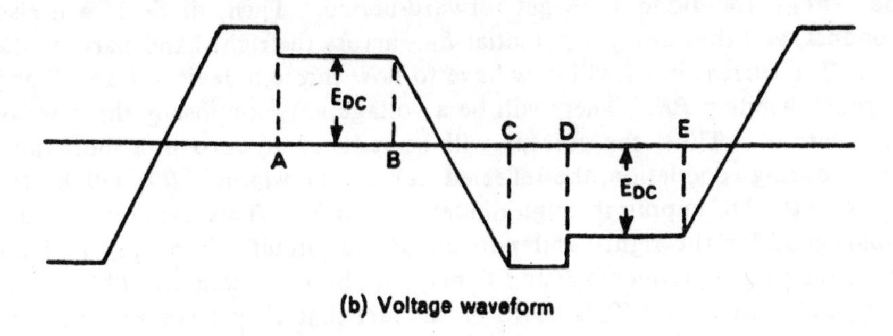

(b) Voltage waveform

Fig. 9.5 Complementary-commutated inverter.

In the circuit in Fig. 9.5a, the free-wheeling diodes 1′ and 4′ are not connected as in Fig. 9.2b to the common points (P and Q) of the SCRs but to the tap points R and S on coils T_1 and T_2. Therefore, the free-wheeling diode 4′ will not conduct from the moment the voltage across capacitor C_4 becomes zero (as explained for the circuit in Fig. 9.2b), but will start conducting only when the potential of point P becomes $-nE_{DC}$. Similarly, diode 3′ will not conduct unless the potential of $Q = E_{DC} + nE_{DC}$. When these diodes begin conducting, the potential of R and S, as also of P and Q, will be fixed. So, the capacitor discharge current will become zero and the load current will be supplied through these diodes. The load voltage during the period the free-wheeling diodes conduct will be $(2nE_{DC} + E_{DC})$. This duration is designated CD (in Fig. 9.5b), where the load current is positive and the load voltage negative. Hence, power will flow from the load to the supply. After this recovery period, i.e., when all the load inductive energy has been fed back, the current will go to zero. Since SCRs 3 and 4 are in the off-state, they will have to be retriggered to start the negative cycle. These SCRs will conduct during the period DE. At E, SCRs 1 and 2 will be fired and similar operations as just described will take place. By connecting the diodes to points R and S as in Fig. 9.5a, a part of the inductive energy trapped during commutation in inductors L_4 and L_3 will be returned to the DC supply through diodes 4′, 1″, and 3′, 2″. The increased load voltage during the recovery period following commutation also will absorb a part of this energy.

In spite of the fact that the load voltage is distorted, the efficiency of this circuit is better than that of the one shown in Fig. 9.2b. The reason for this is as follows. Consider only the left-hand side of the circuit in Fig. 9.5a. If diode 4′ is connected to P, then it will get forward-biased when the potential of P becomes negative with the result that the current flowing through L_4 will free-wheel through 4′ and all the energy in the inductor will have to be dissipated as heat in the components and in devices 4 and 4′. On the other hand, if 4′ is connected to the tap point R on the transformer coil T_1 having a tap ratio $n:1$, then the potential of P will have to be $-nE_{DC}$ for diode 4′ to get forward-biased. Then, diode 1″ will also conduct, and thus apply a potential E_{DC} across the right-hand part of the coil. The current in L_4 will now have to flow through devices 4 and 4′ and through winding PR. There will be a voltage nE_{DC} opposing the flow of this current. Thus, the current will be reduced to zero in a short time. Also, during conduction, the reflected current in winding RT will be fed back to the DC supply through diodes 4′ and 1″. This explanation also holds good for the right-hand portion of the circuit. The design of the commutating components L and C may be obtained from Eq. (9.2). The design of coils T_1 and T_2 is based on the fact that they have to carry the load current at voltage $(1 + n)E_{DC}$ during the free-wheeling period.

The half-bridge version of the MacMurray-Bedford bridge inverter is obtained by slightly modifying the scheme just discussed. Only the left-half portion of this circuit is used. Diodes 1″ and 4″ are disconnected and point T (see Fig. 9.5a) is connected to the centre tap of the power supply. The load and capacitors C_1 and C_4 are connected to points P and T, in parallel with coil T_1. The method of commutation and the operation of the circuit are similar to those described for the circuit in the preceding paragraph.

9.3 THREE-PHASE BRIDGE INVERTERS

The bridge configuration shown in Fig. 9.6a is ideally suited for generating polyphase output. Output transformers, such as those provided for the parallel or ring inverters described in Chapter 8, are not required for this

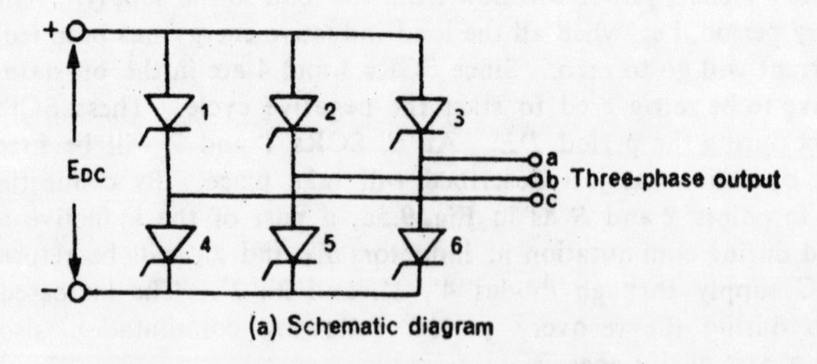

(a) Schematic diagram

Fig. 9.6 Three-phase bridge inverter (*cont.*).

circuit. Basically, there are two schemes for three-phase bridge inverters. In one scheme, a maximum of two SCRs conduct at any instant. This is called the *120°-mode operation* because each SCR conducts for $2\pi/3$ radians in every cycle of the output. It produces a stepped phase-to-phase voltage waveform. The conducting sequence and the phase-to-neutral output voltage waveforms for a three-phase balanced resistive load are as shown in Fig. 9.6b. The parallel-capacitor-commutation scheme shown in Fig. 9.2a is suitable for this mode of operation.

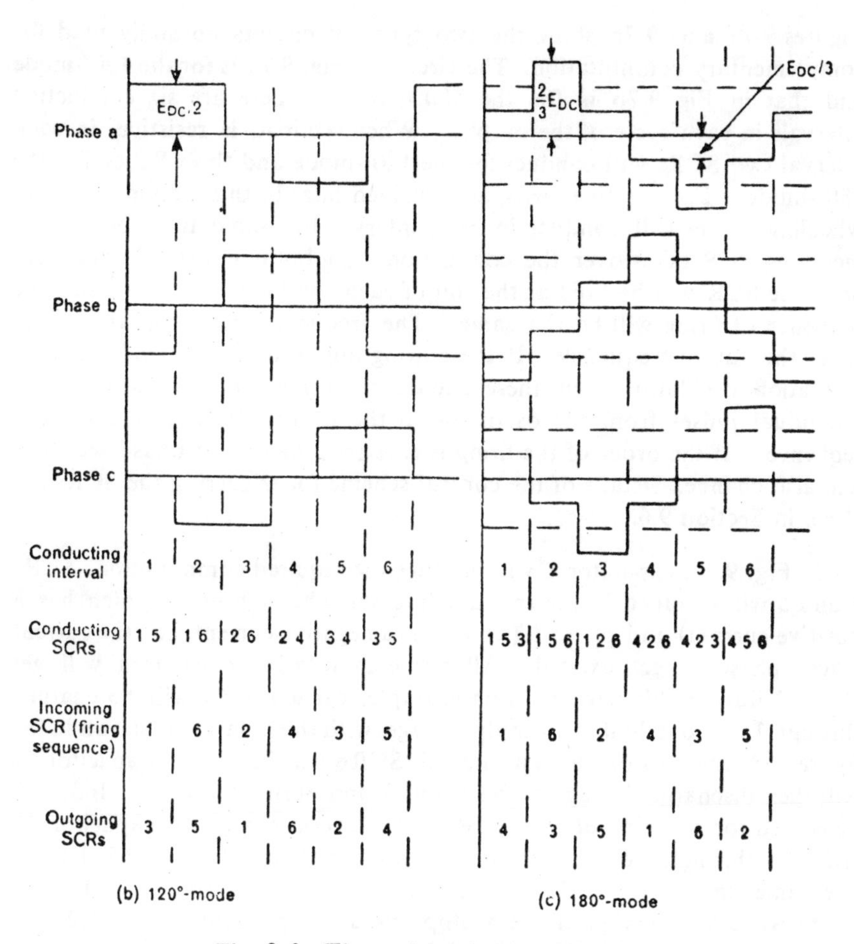

(b) 120°-mode

(c) 180°-mode

Fig. 9.6 Three-phase bridge inverter.

In the other scheme, called the *180°-mode operation*, a maximum of three SCRs conduct at any instant and each SCR conducts for π radians in every cycle of the output. The utilisation factor will be more for SCRs operating in this scheme. The conducting sequence and the voltage waveforms are as shown in Fig. 9.6c. Complementary commutation as illustrated in Fig. 9.2b is suitable for this mode of operation.

For both these schemes, class D type of forced commutation is used. The commutation circuits appear as shown in Fig. 9.7. The triggering frequency of the SCRs will decide the output frequency. Output voltage control is obtained by either varying the DC input voltage or having a tap-changing transformer at the output. The former may result in commutation failures at low levels of DC input voltage.

9.3.1 Commutation Circuits

Figures 9.7a and 9.7b show the two types of circuits normally used for complementary commutation. The circuit in Fig. 9.7a is for the 120°-mode and that in Fig. 9.7b is for the 180°-mode. There are six conducting intervals in each cycle of the output. When the load is resistive, in each interval two SCRs will conduct for the 120°-mode and three SCRs for the 180°-mode. For reactive loads, as explained later in this section, the free-wheeling diodes will conduct in each interval for some time before the incoming SCR takes over the conduction. In both modes of operation, only one SCR will be fired at the end of each conducting interval, and the sequence of firing will be the same. The frequency of firing will be six times the output frequency. If these firing pulses are generated by a UJT relaxation oscillator, then there should be a logic circuit to guide the individual pulses from the oscillator to the proper SCR in a particular sequence. If the order of the firing is reversed, the output phase sequence will also change. Details of the control scheme for triggering the SCRs are given in Section 9.6.

In Fig. 9.7a, capacitor C will produce the required commutation. SCRs 1 and 5 will conduct during the first interval when phase a (which has a positive current) and phase b (which has a negative current, i.e., the current leaves phase b) get excited. All the commutating capacitors will get charged during this period. For example, capacitor C_6 will be charged through C_4 to one-half the supply voltage, with the polarity as shown in the figure. At the end of the first interval, SCR6 will be fired. Capacitor C_6 will then discharge through SCRs 6 and 5 and turn off SCR5. Inductor L_6 is used for reducing di/dt in SCR6. The excess charge on capacitor C_6 will flow through diode 5′, which will apply a reverse bias to SCR5. During the same period, diode 5′ will also conduct phase b current. SCR5 will get forward-biased when diode 5′ stops conducting. Diodes 7′ to 12′ are used for preventing the discharge of the capacitors through the load. After SCR5 is turned off, capacitor C_6 will finally get charged in the opposite direction through capacitor C_5 to one-half the DC voltage. If the load is reactive, phase b current, which was flowing through SCR5, will not become zero immediately after SCR5 is turned off and diode 5′ stops conducting. Instead, the current will flow through diode 11′ into capacitor C_6, charging it more rapidly. When the left-hand terminal of capacitor C_6 attains the potential E_{DC}, diode 2′ will conduct and return the load reactive energy to the supply. There will be no further charging of C_6, and SCR6 will carry the phase c current. When diode 2′ conducts, the phase-a-to-phase-b voltage will be zero. Thus, it will be seen that with reactive loads the commutating capacitor will get charged to voltage E_{DC}. When diode 5′ stops conducting, SCR5 will experience a sudden forward voltage. To

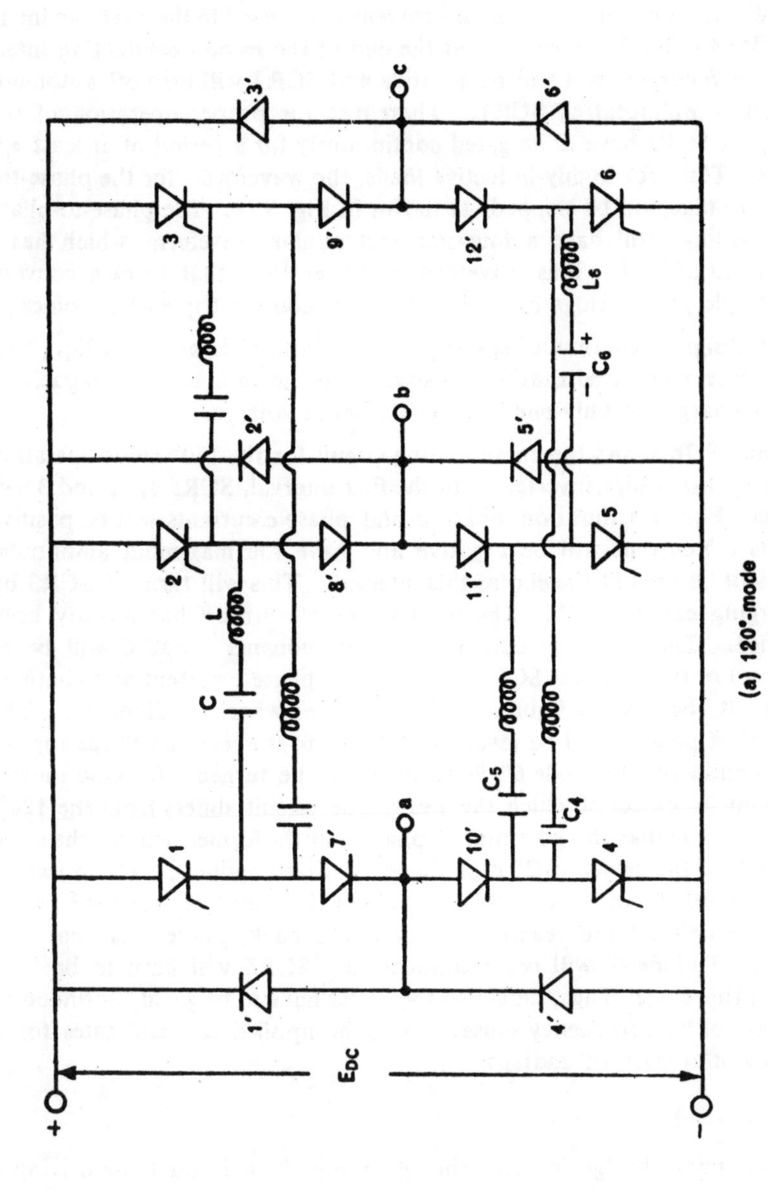

(a) 120°-mode

Fig. 9.7 Commutation circuits for three-phase inverters (*cont.*).

prevent the SCR from turning on due to a large *dv/dt*, proper snubber circuits must be used.

For highly-inductive loads, the maximum duration of conduction of the free-wheeling diodes is about $\pi/2$. Thus, after SCR5 has been turned off, diode 2′ can conduct till the phase current is reversed in the next two intervals. When SCR2 is turned on at the end of the second conducting interval, phase *b* current will still be negative and SCR2 will turn off automatically after commutating SCR1. Therefore, for proper operation of the circuit, the SCRs have to be gated continuously for a period of at least $\pi/6$ radians. Thus, for highly-inductive loads, the waveforms for the phase-to-neutral voltage will be stepped, as shown in Fig. 9.6c. The phase-to-phase output voltage will have a four-step rectangular waveform which has a pulse width of $2\pi/3$. This waveform is better than that from a conventional single-phase bridge because it does not contain triplen harmonics.

The design of commutating components *L* and *C* is based on Eq. (9.1). It must be remembered that for resistive loads the commutation capacitors here are charged to only one-half the DC input voltage.

Figure 9.7b shows the commutating circuit for the 180°-mode operation of three-phase bridge inverters. In the first interval, SCRs 1, 5, and 3 will conduct. For this duration, phase *a* and phase *c* currents will be positive and phase *b* current will be negative and have the maximum amplitude. SCR6 will be fired at the end of this interval. This will turn off SCR3 by discharging capacitor C_6. The mechanism of turn-off has already been explained. The values of commutating components *L* and *C* will be as given by Eq. (9.2). After SCR3 is turned off, phase *c* current will continue to flow (if the load is inductive) through free-wheeling diode 6′. The potential of phase *c* will be reversed and, due to the reverse voltage applied by the conduction of diode 6′, SCR6 will also be turned off. One important point in respect of which the 180°-mode circuit differs from the 120°-mode circuit is that the potential of phase *c* in the former remains the same regardless of the device (SCR6 or diode 6′) that conducts. The duration of conduction of the free-wheeling diode will depend on the load power factor. When all the reactive energy is fed back, phase *c* current will reverse and diode 6′ will cease conduction. SCR6 will have to be fired again at this stage. Thus, here also the SCRs have to be gated continuously or a train of high-frequency pulses has to be applied to their gates for a duration of at least $\pi/2$ radians.

9.3.2 Example

The three-phase bridge inverter shown in Fig. 9.7b is used for driving a three-phase induction motor. The normal rating of the motor is 400 V, 10 A, 1450 rpm. The maximum to minimum speed ratio desired is 10 to 1. Obtain the minimum and maximum DC input voltage required for the inverter. If this voltage is obtained from a fully-controlled three-phase bridge with three-phase 400 V input, calculate the required firing angles.

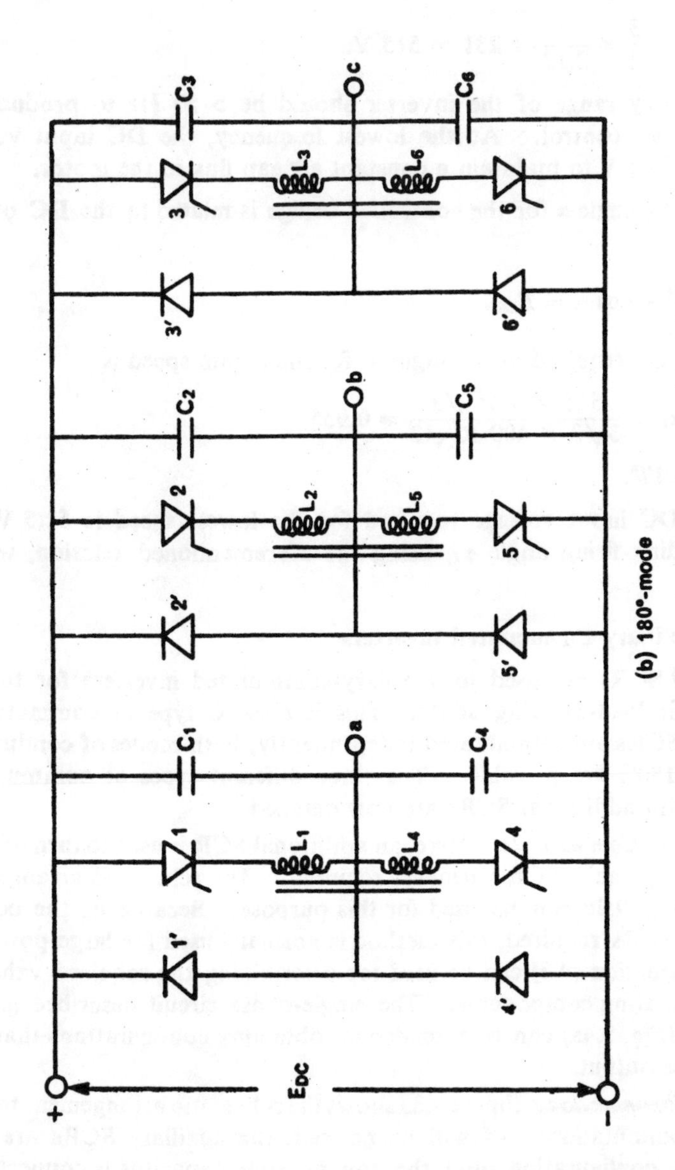

(b) 180°-mode

Fig. 9.7 Commutation circuits for three-phase inverters.

At normal operation, the synchronous speed for the motor will be 1500 rpm at 50 Hz. The motor has four poles. The input fundamental frequency RMS voltage is $400/\sqrt{3} = 231$ V. The inverter will have a stepped output voltage waveform as shown in Fig. 9.6c. From Fig. 9.3b(v), the DC input voltage is given by

$$E_{DC} = \frac{3}{2} \times \frac{\pi}{2.12} \times 231 = 515 \text{ V}.$$

The frequency range of the inverter should be 5–50 Hz to produce the desired speed control. At the lowest frequency, the DC input voltage should be 51.5 V to maintain a constant air gap flux in the motor.

The firing angle α for the controlled bridge is related to the DC output voltage by

$$\frac{3\sqrt{3}E_m}{\pi} \cos \alpha = E_{DC}.$$

Therefore, the required firing angle α_1 for maximum speed is

$$\cos \alpha_1 = \frac{515 \times \pi \times \sqrt{3}}{3\sqrt{3} \times 400 \times \sqrt{2}} = 0.955,$$

$$\alpha_1 = 17°.$$

Since the DC input voltage required for the lowest speed is 51.5 V, the corresponding firing angle α_2, using the aforementioned relation, will be 84.5°.

9.3.3 Auxiliary-Commutated Inverters

Additional SCRs are used in auxiliary-commutated inverters for turning off the main load-carrying SCRs. This is class C type of commutation. Since the SCRs are commutated independently, both modes of conduction, 120° and 180°, are possible. The three different types of commutation possible with additional SCRs are now detailed.

Individual Commutation Here, an additional SCR is used to turn off each load-carrying main SCR whenever required. The schematic arrangement shown in Fig. 9.2c can be used for this purpose. Because of the cost of the extra SCRs required, this method is normally used for large power inverters. Equation (9.3) can be used for determining the required values of the commutating components. The single-phase circuit described in Section 9.2.2 (Fig. 9.4a) can be extended for obtaining configurations that give three-phase output.

Half-Commutation Figure 9.8a shows the schematic arrangement for this type of commutation. As will be noticed, the auxiliary SCRs are used in a bridge configuration, and the commutating capacitor is connected to the midpoints of this bridge. The term half-commutation is used because the external commutation circuit turns off—depending on the voltage polarity of the capacitor during each commutation—either all the top SCRs (1, 2, and 3) (see Fig. 9.7a), or all the bottom SCRs (4, 5, and 6).

The advantage in this scheme is that a lesser number of commutating components is required than in individual commutation. This circuit works as a chopper on DC input. A detailed design of various forms of chopper circuits will be considered in Chapter 10.

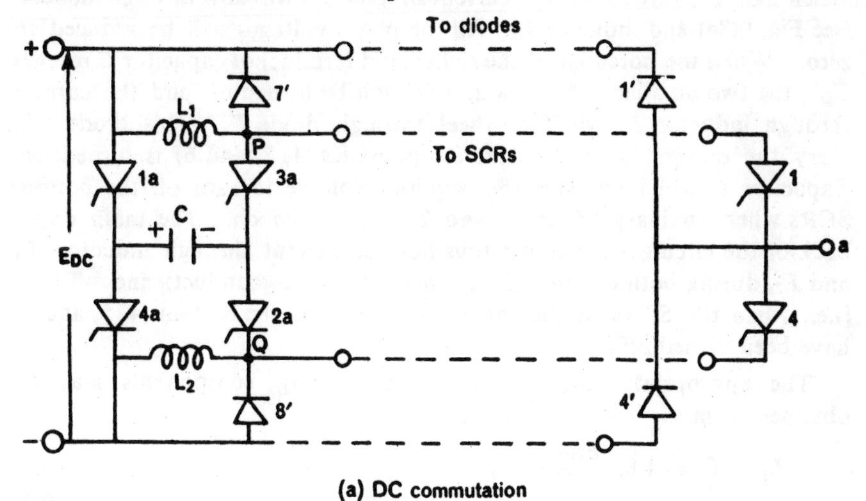

(a) DC commutation

(b) AC commutation

Fig. 9.8 Auxiliary-commutated circuits (*cont.*)

The method of commutation for the circuit in Fig. 9.8a is as follows. Assume that capacitor C is charged to voltage E_{DC} from the preceding commutation process, with the polarity as shown in Fig. 9.8a. Let SCRs 1, 5, and 6 be conducting in the second conducting interval for the 180°-mode (Fig. 9.6c). At the end of this conducting interval, auxiliary SCRs 3a and 4a will be fired, bringing the potential of point P to $-E_{DC}$. So, all the top SCRs will be reverse-biased and turned off. This is the voltage commutation. The current flowing through L_1 will be diverted through auxiliary SCRs 3a and 4a, and the capacitor will get charged to $+E_{DC}$ in the reverse direction. Diode 7′ will prevent overcharging of the capacitor. When SCR1 is turned off, the current of phase *a* will flow through diode 4′ (see Fig. 9.8a) and inductor L_2. All the phase voltages will be reduced to zero. When the potential of the right-hand terminal of capacitor *C* reaches E_{DC}, the two auxiliary SCRs 3a and 4a will be turned off and the current through inductor L_1 will free-wheel through diode 7′. This diode will carry the current until the next set of SCRs (1, 2, and 6) is turned on. Capacitor *C* will now have the required voltage to turn off the bottom SCRs when auxiliary SCRs 1a and 2a are turned on. The main drawback of the circuit is the continuous flow of current through inductors L_1 and L_2, during both on-time (i.e., when the SCRs conduct) and off-time [i.e., when the SCRs at the top (1, 2, and 3) or the bottom (4, 5, and 6) have been turned off].

The appropriate values of the commutating components may be obtained from

$$L_1 = L_2 = 1.82\,\frac{E_{DC}}{I_L},$$
$$C = \frac{1.47 I_L t_q}{E_{DC}}, \qquad (9.7)$$

where t_q is the turn-off time of the SCRs. Turn-off is achieved by reducing the potential of point *P* or by increasing the potential of point *Q*, thereby reverse-biasing the top or bottom SCRs, respectively. Since points *P* and *Q* are the DC input terminals, this type of commutation is also known as *DC commutation.*

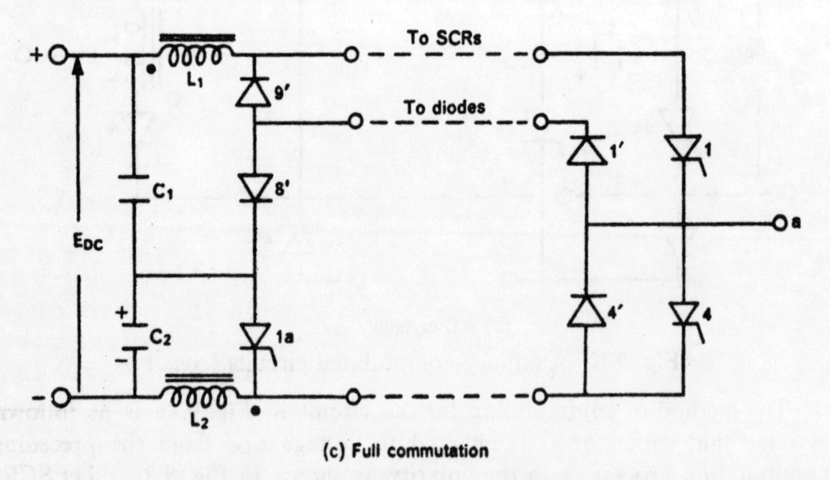

(c) Full commutation

Fig. 9.8 Auxiliary-commutated circuits.

In the process known as *AC commutation,* the action that takes place is the reverse of that in DC commutation because *P* and *Q* are on the AC side. The top SCRs 1, 2, and 3 are turned off by raising the potential of P and the bottom SCRs 4, 5, and 6 are switched off by decreasing the potential of Q. The commutation scheme is shown in Fig. 9.8b. Just as the inductors on the input side in DC commutation are required to rapidly change the potentials of P and Q, in AC commutation the inductors between the cathodes of the top SCRs and the anodes of the bottom SCRs are required to sustain the voltages applied between P and Q. Capacitor C will be charged to potential $+E_{DC}$ when SCRs 1, 2, and 6 are conducting. In the first interval, SCRs 1 and 2 will be turned off by firing auxiliary SCRs 1a and 2a, because of the capacitor discharge current. This is current commutation. The load current of phases a and b will flow through diodes 4′ and 5′, and will charge the capacitor in the reverse direction. When this voltage becomes E_{DC}, diode 7′ will get forward-biased and the load current will be diverted to it. SCRs 1a and 2a will then be turned off and the phase-to-phase voltages of the output will become zero. In the next interval, SCRs 2, 4, and 6 will be fired; diode 5′ will be reverse-biased and the current in phase b will flow through SCR2. The current in phase a will then quickly reverse due to the reverse voltage applied. When diode 4′ stops conducting, SCR4 will be turned on. Thus, in each interval, the SCRs have to be gated continuously for a period of $\pi/2$ radians for the 180°-mode and $\pi/6$ radians for the 120°-mode of operation. At the end of the second interval, auxiliary SCRs 3a and 4a will be fired. Consequently, the bottom SCRs will be turned off, and an operation identical to that just described will take place. The capacitor will again be charged to E_{DC} with the appropriate polarity for the next commutation.

The values of L and C may be obtained from

$$\frac{I_L t_q}{C} = \frac{\sqrt{3}}{2} E_{DC},$$
$$E_{DC}\sqrt{\frac{2C}{L}} = 2I_L, \tag{9.8}$$

where I_L is the maximum phase current and t_q is the SCR turn-off time. The derivation of these equations is given in Section 9.3.4.

Full Commutation Figure 9.8c shows the circuit for this type of commutation. Here, all the conducting SCRs are turned off by diverting the input current. This is a DC commutation because during commutation the potential difference of the input terminals is reversed by providing reactors L_1 and L_2 on the input side. These reactors are closely coupled, with the polarity markings as indicated in Fig. 9.8c. Capacitor C_2 is charged to voltage $2E_{DC}$ with the polarity as shown in the figure. When commutation is desired, SCR1a is fired. Capacitor C_2 will then discharge through SCR1a and L_2. The right-hand terminal of L_2 will become positive and by induction an equal voltage, opposing the applied voltage, will be induced in L_1. Then, capacitor C_1 is connected to the DC supply and it begins to get charged. The induced current in L_1 will reduce the input current to zero and turn off all SCRs in the bridge. The induced voltage in L_1 will reverse-bias 8′ and keep it in the blocking state while commutation is in progress. When capacitor C_2 is completely discharged, capacitor C_1 will be fully charged to the DC supply. SCR1a will be maintained in the conducting state by the inductive load currents flowing through the free-wheeling diodes (2′, 4′, and 6′ for the first conducting interval in Fig. 9.6c), diode 8′, and SCR1a. During this period, the phase-to-phase voltages will be zero. In the next conducting interval when the other set of three SCRs is triggered, C_1 will discharge through reactor L_1, the conducting SCRs, the free-wheeling diodes, and diode 8′. The presence of this current in L_1 will induce an opposite current in L_2 which will flow to SCR1a and turn it off. When the discharge current becomes zero, capacitor C_1 will be charged to E_{DC} in the reverse direction and the voltage across capacitor C_2 will be $2E_{DC}$, preparing capacitor C_2 for the next commutation.

Suitable values of the commutating components may be determined from

$$C = \frac{20}{\pi} I_L t_q / E_{DC},$$

$$L = \frac{20}{\pi} t_q E_{DC} / I_L, \tag{9.9}$$

where t_q is the required turn-off time for the SCRs.

For all the circuits with auxiliary commutation discussed in this section, the DC input voltage is fixed. It is possible to have multiple commutation in each conducting interval, as explained for the single-phase inverter in Section 9.2, and thereby control the output voltage without reducing the DC input voltage. Therefore, these circuits are best suited where the ratio of the RMS output phase voltage to frequency has to be kept constant, as in the speed control of AC motors.

9.3.4 Example

(a) Using all types of auxiliary commutation, obtain suitable values of the commutating components L and C for a three-phase bridge inverter. The DC input voltage is 100 V, the commutating current is 10 A, and the SCR turn-off time is 40 μsec.

Individual commutation Using Eq. (9.3),

$$\sqrt{\frac{C}{L}} = \frac{10}{100} = 0.1,$$

$$\sqrt{LC} = 40 \times 10^{-6} \times \frac{3}{2\pi}.$$

Therefore,

$$C = \frac{4 \times 3 \times 10^{-6}}{2\pi} = 1.9\ \mu\text{F},$$

$L = 190\ \mu H.$

Half-commutation (DC commutation) Using Eq. (9.7),

$$L = \frac{1.82 \times 100 \times 40 \times 10^{-6}}{10} = 650\ \mu H,$$

$$C = \frac{1.47 \times 10 \times 40 \times 10^{-6}}{100} = 5.9\ \mu F.$$

Half-commutation (AC commutation) Using Eq. (9.8),

$$C = \frac{10 \times 40 \times 10^{-6}}{0.866 \times 100} = 4.6\ \mu F,$$

$$\frac{2C}{L} = 4 \times \frac{100}{10^4} = 0.04,$$

$$L = \frac{9.2}{0.04} = 230\ \mu H.$$

Full commutation (DC commutation) Using Eq. (9.9),

$$C = \frac{20}{\pi} \times 10 \times \frac{40 \times 10^{-6}}{100} = 25.5\ \mu F,$$

$$L = \frac{20}{\pi} \times 40 \times 10^{-6} \times \frac{100}{10} = 2.55\ mH.$$

It can be observed that the size of the commutating components increases considerably when full commutation is used. For individual commutation, a large number of commutating components of lower rating is required. Hence, whenever pulse-width modulation is desired in bridge inverters, half-commutation (AC or DC) is preferred.

(b) Obtain equations for designing commutating components L and C for the commutation circuit in Fig. 9.8b.

The commutating capacitor is charged to voltage E_{DC}. For the given voltage polarity, when SCRs 1a and 2a are fired the main SCRs 1 and 2 will be turned off at the end of the third conducting interval (Fig. 9.6c). Therefore, the effective commutating inductance in the discharge path will be $L/2$. The capacitor discharge current is given by $i_C = A \sin \omega t$, where $\omega = \sqrt{2/(LC)}$. The peak value of this current $E_{DC}\sqrt{2C/L}$ is taken to be twice the maximum phase current I_L. Then, the SCRs will be turned off at an angle $\alpha = \omega t = \pi/6$. When SCRs 1 and 2 are turned off, the load current will flow through capacitor C and will charge it in the opposite direction. The voltage across the capacitor, when the main SCRs are turned off, will be $E_{DC} \cos (\pi/6) = 0.866 E_{DC}$. Assuming the load current I_L to be constant, the voltage across the capacitor will become zero in time t_q. Therefore,

$$\frac{I_L t_q}{C} = 0.866 E_{DC}.$$

The turned-off SCRs will be reverse-biased for duration t_q. Hence, t_q must be more than the turn-off time of the SCRs. The other constraints on L and C are obtained from the assumption made earlier for the peak discharge current, i.e., $E_{DC}\sqrt{2C/L} = 2I_L$. Using these two equations, the required values of L and C can be obtained. Equations (9.7) and (9.9) can be derived in a similar manner.

9.4 CURRENT-SOURCE INVERTER

Most inverters used for AC motor drives or for generating variable frequencies are voltage driven, i.e., the input is a DC source with a small resistance. The advantages provided by current-source inverters, namely, greater simplicity, better controllability, higher regenerative capability, and ease of protection, are now widely recognised. Here, the source has a large reactor with high impedance which maintains a constant current.

Figure 9.9 shows the schematic arrangement for a three-phase inverter. The firing sequence and the method of commutation are the same as those for the complementary-commutated inverters discussed in Section 9.3 (Fig. 9.7a). If, in the 120°-mode operation discussed here, the DC is obtained

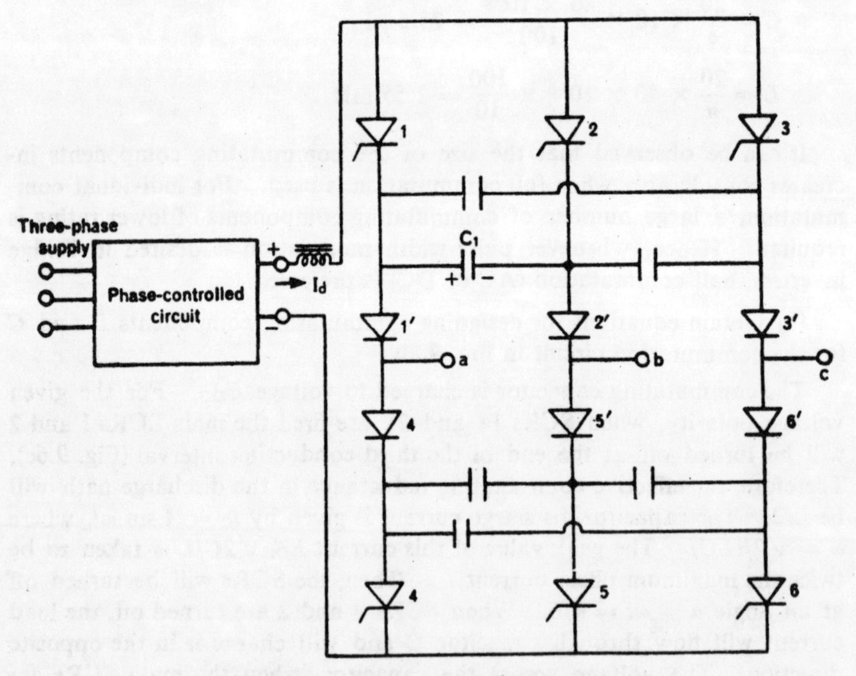

Fig. 9.9 Current-source inverter.

from a phase-controlled circuit, only a large inductance, without any filter capacitance, is used at the output. The series inductance will feed ripple-free current to the inverter. This current can be supplied to different phases of the load by properly gating the SCRs. The output voltage waveform of the inverter depends on the nature of the load. Only two SCRs will be on at any one time, and the conducting SCR will be turned off by firing the SCR adjacent to it. For example, if SCRs 1 and 6 are conducting, phase *a* will get positive current while the current flows out of phase *c*. Capacitor C_1 will get charged to one-half the input voltage with the polarity as shown in the figure. At the end of this conducting interval (60°), SCR2 will be fired. Capacitor C_1 will then discharge through SCR1 and turn it off. Since the input current is maintained constant, it will flow through SCR2 and capacitor C_1 into phase *a*, and thereby linearly charge the capacitor in the opposite direction. When diode 2′ is properly biased, phase *b* will begin conducting and the current in phase *a* will decrease. When phase *a* current goes to zero, phase *b* will carry the full current. The phase current waveform will be identical to the voltage waveform shown in Fig. 9.6b, except that its edges will have an exponential rise and decay.. Diodes 1′ to 6′ are used for preventing the capacitors from discharging through the load. This circuit can be operated also in the 180°-mode by using the appropriate commutation circuits.

One main difference between voltage- and current-source inverters is the absence of free-wheeling diodes in the latter. The direction of current in the input is the same even when the load is reactive. In the case of single-phase circuits (discussed in Section 9.5), the voltage at the DC terminals of the inverter automatically adjusts itself so as to properly feed back the reactive energy of the load. Then, the inverter input voltage will become negative and power will be fed from the load to the input. In three-phase circuits (Fig. 9.9), it is the phase voltage that reverses to absorb the reactive power. It is possible to feed the power also into the three-phase input supply of the controlled rectifier. This is generally known as *regenerative action*. The direction of the output current I_d will be the same as that shown in Fig. 9.9, but the firing angle of the controlled bridge will be advanced such that its DC output voltage will be negative. The circuit will operate as a line-commutated inverter (see Chapter 7). Thus, current-source inverters are ideally suited for incorporating regenerative control. Capacitor C is chosen on the basis of the time for which the reverse voltage is applied to the SCRs after these have been turned off. The required equation is

$$t_q = \frac{CE_d}{2I_d}, \tag{9.10}$$

where t_q is the turn-off time, E_d is the maximum input voltage, and I_d is the input current of the inverter. It is assumed here that the commutating capacitor will be charged to one-half the peak phase-to-phase voltage (which is equal to the maximum value of the input DC voltage), and that phase *b* will begin conducting only after capacitor C_1 is completely discharged.

9.4.1 High-Frequency Inverter with Forced Commutation

The turn-off time for inverter-grade SCRs is usually around 10 μsec to

15 μsec. This sets a limit on the maximum possible frequency of operation for SCR controlled inverters. One technique for improving the operating range of frequency is to connect a capacitor in parallel with the load so that the effective load current leads the voltage. The SCRs consequently turn off before the load voltage reverses. Reverse current will then flow through the free-wheeling diodes, which apply a reverse voltage across the turned-off SCRs. Known as *self commutation*, this technique is used for the series inverter discussed in Chapter 8; a modified version of this inverter (based on self-commutation) for high-frequency operation has also been explained there. Operation at still higher frequencies is possible by using forced commutation in which the turn-off time is reduced by the application of a large reverse voltage.

Figure 9.10 shows a circuit with a current source, a compensated load, and forced commutation for obtaining a high-frequency output. This circuit is used for induction heating or as a variable frequency source. A description of its operation follows.

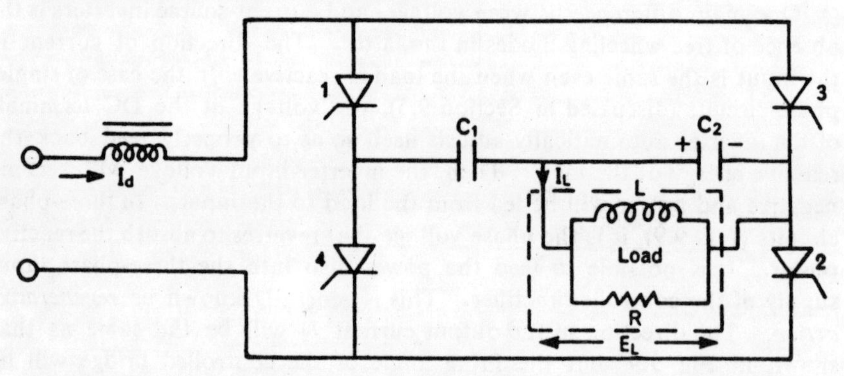

Fig. 9.10 High-frequency inverter with forced commutation.

The large input reactor maintains a constant current. This current can be switched into the load in opposite directions by firing opposite pairs of SCRs. The current waveform will then be rectangular. The parallel combination of L and R in the circuit is the equivalent representation of the load. Capacitor C_2 is used to make the effective load power factor leading at the operating frequency. This is achieved by keeping the ringing frequency of the compensated load slightly lower than the triggering frequency. If the load parameters change (as they do in induction heating, depending on the state of the metal in the furnace), the triggering frequency cannot be maintained constant for optimum operation. If the triggering frequency is kept much above the resonance frequency, the compensated load power factor will again be poor. So, the triggering frequency is adjusted by monitoring the load voltage and initiating the firing of the SCRs before the zero crossing of the voltage. When SCRs 1 and 2 conduct, capacitor C_2 will be charged as shown. When the other pair of SCRs is fired, capacitor C_2 will turn off SCRs 1 and 2. Even though the waveform for the net load current I_d is rectangular, the actual load current I_L and voltage E_L will be fairly sinusoidal since all the higher harmonics will pass through capacitor C_2. At the time of switching the inverter, capacitor C_2 will remain uncharged and, at the end of the first reversal of current, will be unable to provide sufficient reverse voltage to commutate the pair of conducting SCRs. This means that with only one parallel compensating capacitor C_2, starting the inverter by merely firing the SCRs at the chosen frequency will not be possible. Therefore, additional energy for commutation will be required when switching on the inverter. This is provided by capacitor C_1. The voltage across C_1 being current-dependent is enough to produce successful commutation and the inverter can therefore be started from the cold state at the highest operating frequency.

From the foregoing description, it is evident that capacitor C_1 essentially produces the required energy for commutation and capacitor C_2 primarily compensates for the load reactance; the latter serves also as a high-pass filter so that the load current is fairly sinusoidal. Because of the energy supplied by capacitor C_2 to the load, the load current I_L will be many times the input current I_d. The Q-factor ($\omega L/R$) of the load coil will determine the approximate ratio I_d/I_L. The load power can be controlled in two ways. One is by adjusting the level of the input current I_d. Because of the large source reactance, the effective time constant for any change in I_d will also be large; therefore, this method of control will be slow. The other method is by varying the frequency. This changes the current distribution and thereby controls the load power for a given I_d. For example, if the triggering frequency is increased, capacitor C_2 will take a larger share of the total current, I_L will decrease, and the power factor will become more leading. A small change in frequency is enough to produce a large variation in the current distribution, and the associated time constant will be negligible.

The effect obtained by varying the triggering frequency provides an open loop control for power. However, it is necessary that this frequency be automatically adjusted whenever the load parameters change so that the net effective power factor is leading, and there is sufficient voltage across the commutating capacitor to turn off the conducting SCRs. If the frequency is increased, the voltage across the capacitor at the time of commutation will rise and the duration for which the conducting SCR is reverse-biased will increase. Thus, for controlling the frequency automatically, the voltage across the bridge has to be measured. This voltage will have negative excursions. The duration of this negative voltage must be more than the required turn-off time of the SCRs. A comparator compares the average duration of the negative cycle of the input voltage with the required turn-off time, and its output is used for properly adjusting the triggering frequency. Since the load parameters do not change quickly (the change being due mainly to temperature), it is desirable to use the average of the negative-cycle durations so that unnecessary drifts in the frequency are avoided.

9.4.2 Example

A coil for an induction furnace has an effective Q-factor of 10 and an inductance 100 μH at a frequency 3 kHz. Calculate the value of the capacitance that should be connected across the coil to provide power factor compensation and commutation. The turn-off time of the SCRs is 40 μsec and the input current I_d is 10 A.

Since the input current I_d is 10 A and the Q-factor for the load at the triggering frequency is 10, the fundamental frequency current amplitude I_{max} through the load will be $(4 \times 10 \times 10)/\pi = 127$ A. It is assumed that all high-frequency currents will be shorted by the capacitor and that the load voltage will be sinusoidal. The input current must have a phase lead of at least ωt_q, where $t_q = 40$ μsec, to provide effective commutation. Therefore, the voltage across the capacitor at the time of commutation will be

$$V_C = \omega L I_{max} \sin \omega t_q$$
$$= 2\pi \times 3 \times 10^3 \times 100 \times 10^{-6} \times 127 \times 0.7 = 167 \text{ V}.$$

During commutation, the capacitor discharge current will be equal to the sum of the load current and input current. It is assumed to be constant and equal to $I_d(Q + 1)$, and will reduce the capacitor voltage to zero in time t_q. Therefore,

$$\frac{1}{C} \times 10 \times 11 \times 40 \times 10^{-6} = 167,$$

$$C = \frac{110 \times 40}{167} = 26.3 \text{ μF}.$$

This value is smaller than 31 μF required to provide a fundamental phase angle lead of $\omega t_q = 45°$. Therefore, the latter value is used.

9.5 INVERTER OUTPUT VOLTAGE AND WAVEFORM CONTROL

Since the phase-to-phase voltage of the output of the inverter is equal to the DC input voltage, the output voltage can be easily controlled by varying the DC input voltage. If the DC is obtained from controlled rectification of a three-phase supply (see Chapter 7), a wide variation in DC voltage can be obtained. If the DC is obtained from a battery supply, then a chopper can be used to regulate the input to the inverter. In both cases, the output voltage waveform will not be affected. For resistive loads, the phase-to-neutral voltage will be a stepped wave for the 180°-mode operation (Fig. 9.6c) or a pulse of width $2\pi/3$ radians for the 120°-mode operation (Fig. 9.6b). The main problem with this type of control is that commutation failure may take place at low DC input voltage. The parallel inverter discussed in Chapter 8 also suffers from the same drawback. This problem can be overcome by using auxiliary-commutated inverters. Here, the AC output voltage can be varied by pulse-width modulation, keeping the DC input voltage constant. The resulting output voltage waveform will get distorted and its harmonic content will increase due to multiple commutation in each half-cycle (see Fig. 9.3b). Howevèr, it is possible to eliminate certain harmonics by suitably choosing

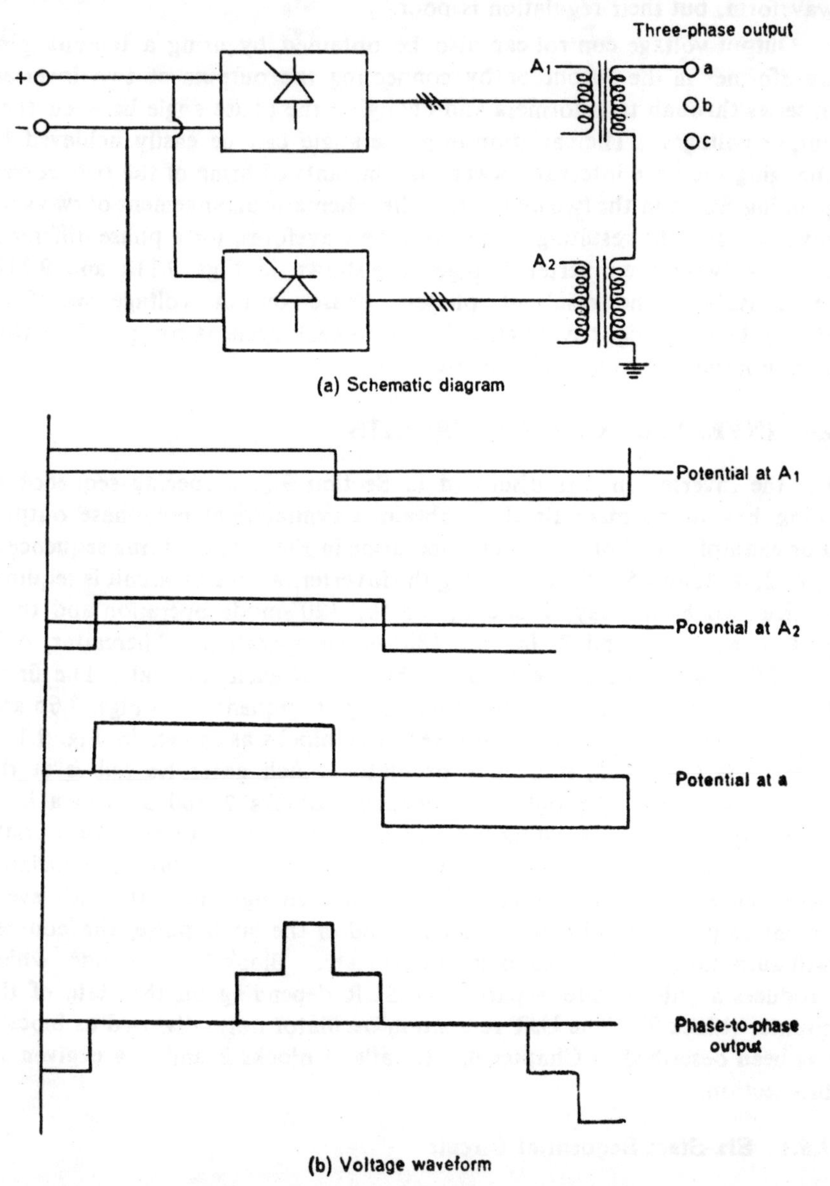

(a) Schematic diagram

(b) Voltage waveform

Fig. 9.11 Inverter voltage control.

the on-off intervals. When sinusoidal output waveforms are desired, filters must be connected to obtain only the fundamental frequency component. (The design of one type of *LC* filter has been discussed in Chapter 8.) The stepped waveform obtained from a 180°-mode inverter—for instance, the six-step voltage waveform in Fig. 9.6c—is sufficient for normal applications, such as in the speed control of AC motors. Inverters with series-capacitor commutation usually provide a better voltage waveform, but their regulation is poor.

Output voltage control can also be obtained by using a tap-changing transformer in the output or by connecting the outputs of two inverters in series through transformers and changing the phase angle between their output voltages. The variation in phase angle can be easily achieved by changing the time interval between the instants of firing of the two corresponding SCRs in the two inverters. The schematic arrangement of two such inverters and the resulting output voltage waveform, for a phase difference of $\pi/6$ between the inverter voltages, are shown in Figs. 9.11a and 9.11b, respectively. The resultant phase-to-phase output voltage waveform (Fig. 9.11b) has eight symmetrical steps in each cycle as compared to four steps obtained from a single inverter.

9.6 INVERTER CONTROL CIRCUITS

For the inverter circuits discussed in Section 9.3, a specific sequence of firing has to be maintained to obtain a symmetrical polyphase output. For example, for both modes of operation in Fig. 9.6, the firing sequence is 1, 6, 2, 4, 3, and 5. When starting the inverter, a control circuit is required to fire two SCRs, say, 1 and 5, for the 120°-mode operation and three SCRs, say, 1, 5, and 3, for the 180°-mode operation. Thereafter, only one SCR will need to be fired at the end of each interval. The firing frequency will be six times the desired output frequency (see Figs. 9.6b and 9.6c). The control circuit comprises three blocks as shown in Fig. 9.12a. Block 1 is a variable frequency oscillator which generates pulses at the rate of six times the output frequency. Blocks 2 and 3 form a logic circuit which applies a gating pulse at the right instant to the appropriate SCR. Block 2 is a six-state binary counter. Each state is associated with one SCR in the bridge. The counter changes its state with every incoming pulse from block 1. At the end of the sixth pulse, the counter will automatically be reset to its initial state. Block 3 is a decoder which produces a pulse to fire a particular SCR depending on the state of the counter (block 2). The UJT relaxation oscillator normally used as block 1 has been described in Chapter 4. Details of blocks 2 and 3 are given in this section.

9.6.1 Six-State Sequential Circuit

A six-state binary counter is shown in Fig. 9.12b. It consists of three *JK* flip-flops. Each flip-flop has two input terminals (*J* and *K*) and two output terminals. The latter are designated X and $\overline{X}$, Y and $\overline{Y}$, Z and $\overline{Z}$. The state of the flip-flop is given by the voltage at the output terminals.

For example, if X is at a high voltage, then $\overline{X}$ will be at a low voltage, i.e., X and $\overline{X}$ are complementary output terminals. A similar relation will

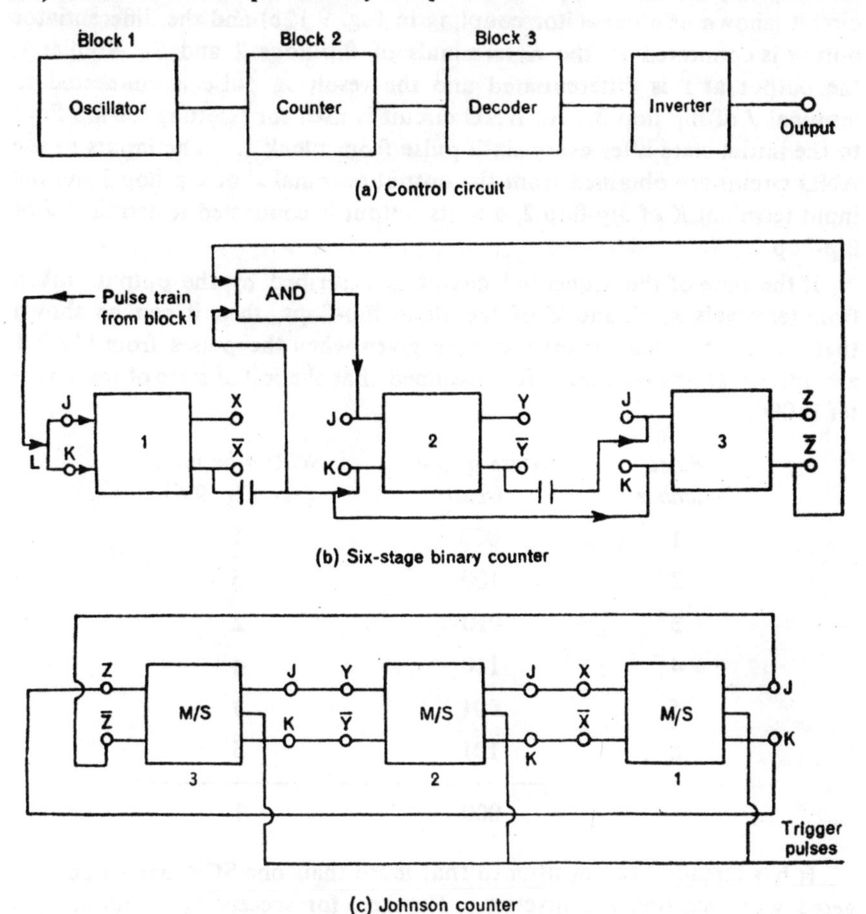

Fig. 9.12 Inverter control circuit.

hold for terminals Y and $\overline{Y}$, and Z and $\overline{Z}$. The high- and low-voltage states are designated 1 and 0, respectively. The truth table that follows describes the performance of the JK flip-flop. The inputs to terminals J and K are in the form of pulses which have positive amplitudes.

Input at	*Present state* X	$\overline{X}$	*Future state* X	$\overline{X}$
J	0	1	1	0
J	1	0	1	0
K	1	0	0	1
K	0	1	0	1
J and K	1	0	0	1
J and K	0	1	1	0

The pulses from block 1 are connected to both input terminals J and K of flip-flop 1. The output terminal $\bar{X}$ is connected to a differentiating circuit (shown as a capacitor coupling in Fig. 9.12b) and the differentiator output is connected to the K terminals of flip-flops 2 and 3. Similarly, the output at $\bar{Y}$ is differentiated and the resulting pulse is connected to terminal J of flip-flop 3. An AND circuit is used for resetting the flip-flops to the initial state after every sixth pulse from block 1. The inputs to the AND circuit are obtained from the output terminal $\bar{Z}$ of flip-flop 3 and the input terminal K of flip-flop 2, and its output is connected to terminal J of flip-flop 2.

If the state of the sequential circuit is described by the outputs taken from terminals X, Y, and Z of the three flip-flops, then it can be shown that the states occur in the sequence given when the pulses from block 1 are applied to the counter. It is assumed that the initial state of the counter is 000.

Pulse number	*State of the circuit*	*SCR to be fired (see Fig. 9.6)*
1	000	1
2	100	6
3	010	2
4	110	4
5	001	3
6	101	5
1	000	1

It has already been mentioned that more than one SCR has to be triggered when starting the inverter. Further, for succeeding commutations only one SCR has to be turned on according to the sequence just given. Thus, a separate firing circuit is required, which will come into operation only when the inverter is started. Similarly, when the initial state of the counter is kept at 000, the remaining states follow in the order specified. There are many forms of the six-stage counter. The binary counter considered here is better than the six-stage ring counter in so far as it requires a lesser number of components. Certain counters have a built-in feature for counting the states from 1 to 6, or in the reverse order. These are called *up down counters*. By using them to reverse the firing order of the SCRs, the phase sequence of the output voltages can be altered. Such an operation is essential when speed reversal is required for inverter-driven motors.

With three flip-flops, the total number of independent states is eight. The counter shown in Fig. 9.12b uses only six of these eight states. Hence, it is called a *modified counter*. If the counter goes into either of the remaining two states (111 and 011), maloperation of the firing controller will take place. Additional logic circuits have therefore to be used to bring the counter back into one of the six usable states. The counter with simple *JK* flip-flops described here has another drawback known as *racing,* which also leads to maloperation. This problem can be overcome by edge-triggered master-slave (M/S) flip-flops. Details of such flip-flops can be obtained from any book on digital electronics.

Figure 9.12c shows an improved six-stage counter (Johnson counter) which uses a twisted feedback from the last stage. Here, the trigger pulses from block 1 are applied simultaneously to all three flip-flops. When this is done, the state of the flip-flop output will change depending on the inputs to terminals J and K. If the outputs are taken from terminals X, Y, and Z, the six states in this counter will be as follows:

Pulse number	X	Y	Z
1	1	0	0
2	1	1	0
3	1	1	1
4	0	1	1
5	0	0	1
6	0	0	0
1	1	0	0

Thus, it can be observed that the outputs at terminals X, Y, and Z form a three-phase voltage with a phase displacement of 120°. These signals can be directly used to trigger the required SCRs in the main power circuit. For the 180°-mode operation, the voltage at X is used for firing SCR1 and that at $\overline{X}$ for firing SCR4. Similarly, each terminal is associated with a specific SCR. For this counter, no separate starting or decoding circuit is required.

9.6.2 Decoding Circuit

Each of the six states in the counter described in Section 9.6.1 is associated with the firing of a particular SCR. If, for the 180°-mode, the inverter has to feed power to inductive loads, it is necessary that the SCRs be gated continuously for a maximum period of $\pi/2$ radians. This is so because the incoming SCR, after commutating one of the conducting SCRs, will itself be turned off by the flow of load current through the free-wheeling diode. After all the reactive power has been fed back, the same SCR will have to be fired again in order to reverse the load current within the phase. It is therefore advisable to use a decoding circuit which applies a train of pulses to the proper SCR for the required duration. This is generally known as *carrier frequency gating* (discussed in Section 8.4.1).

Figure 9.13 shows the decoding circuit using a diode matrix. Resistors R_1 and R_2 should be so arranged that the flip-flops in block 2 do not get overloaded and the voltage appearing across R_2 is enough to drive the pulse-stretching circuits whose output modulates the gating pulses applied

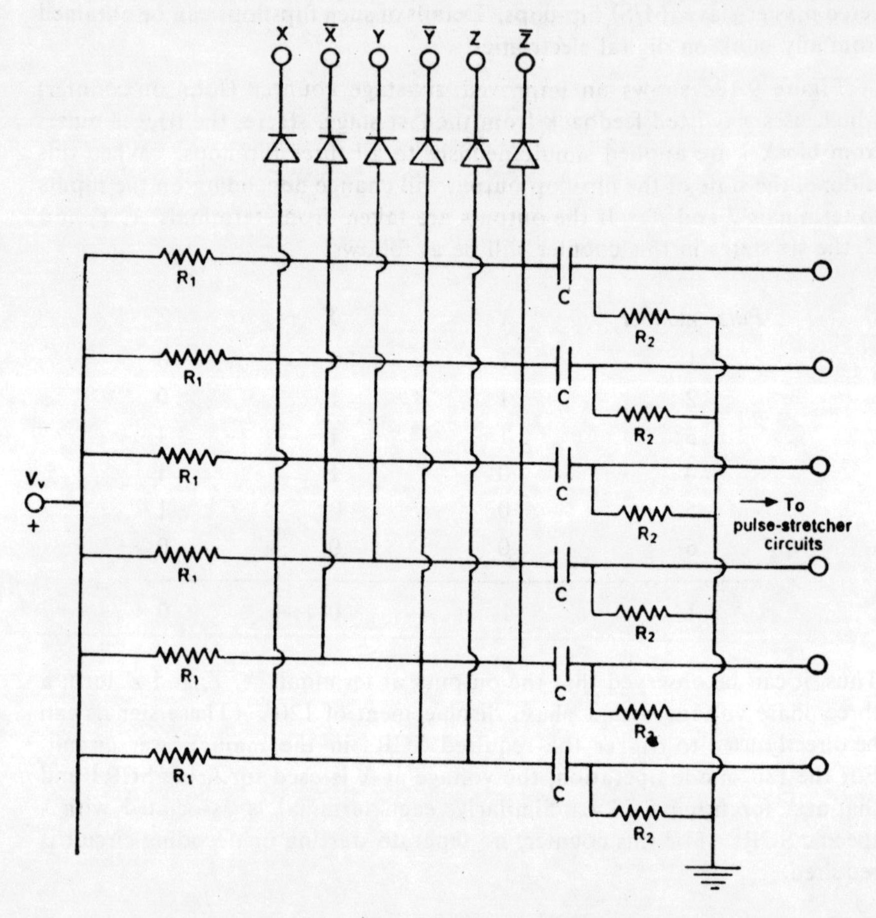

Fig. 9.13 Decoding circuit.

to the SCRs. The output of the decoding circuit is a positive pulse with a width $\pi/3$ radians. This output is differentiated and the positive pulse is applied to a pulse-stretching circuit (monostable) to provide at the desired frequency an output pulse with a width $\pi/2$ radians. Further details on counters, decoders, and logic circuits can be obtained from the references at the end of this chapter.

9.7 INVERTER APPLICATIONS

There are many applications of variable frequency power sources. The availability of high-power SCRs has made it possible to design power supplies suitable for induction heating. The circuit shown in Fig. 9.5 can be used for frequencies upto 10 kHz. For higher frequencies, the time sharing series-capacitor-commutated inverters, described briefly in Chapter 8, can be used. Inverters are widely used also in variable speed AC motor drives. There are several drawbacks in regulating the speed of induction motors by voltage control, e.g., their efficiency and pull-out torque decrease with speed. For rotor on-off control, even though high starting torques are possible, the efficiency will decrease with increase in slip. Variable frequency supply can be effectively used for the speed control of reluctance, synchronous, and induction motors. If the RMS value of the AC output voltage can be made directly proportional to the frequency by some suitable control (either by pulse-width modulation or by controlling DC input voltage), the full load torque and the pull-out torque of the motor can be made fairly independent of speed. Magnetic saturation is avoided since the air gap flux is approximately constant.

Figure 9.14a shows the schematic arrangement of a typical control scheme for an inverter-fed induction motor. Here, the motor is run at a constant rotor frequency $(\omega_o - \omega)$ at all speeds by changing the triggering frequency of the inverter. The optimum rotor frequency may be chosen

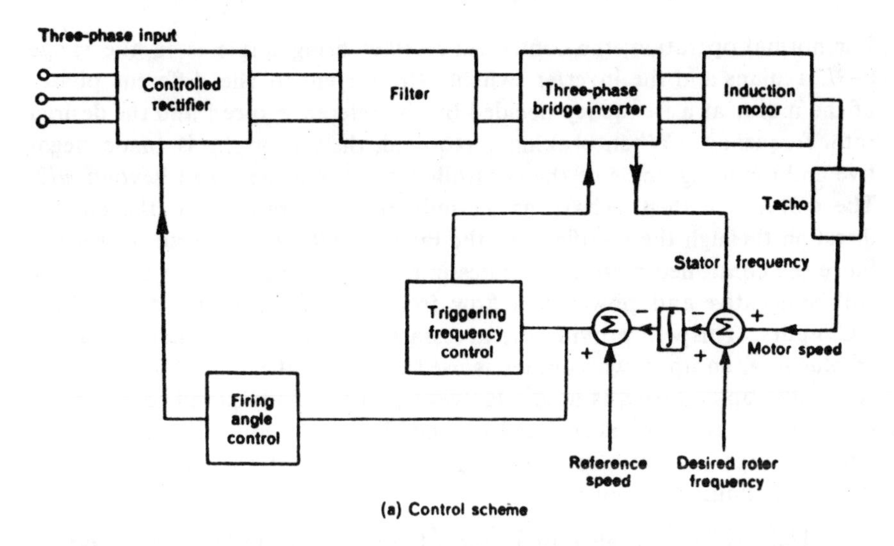

Fig. 9.14 Variable frequency control of induction motors (*cont.*).

on the basis of efficiency, power factor, and the ratio of full load torque to pull-out torque. At constant torque, variation in speed can be obtained by changing the reference signal. Figure 9.14b shows the torque-speed characteristics of an induction motor supplied from a variable frequency source. It will be observed that, by keeping the rotor frequency $(\omega_o - \omega)$ constant, the torque developed by the motor will be the same from zero speed to full speed. By slightly modifying this scheme, both regenerative and dynamic braking can be obtained. For regenerative braking, the inverter output frequency is made less than the rotor speed, and for dynamic braking the triggering sequence of SCRs is changed.

The current-source inverter described in Section 9.4 is ideally suited for variable speed control of induction motors if regenerative braking is desired.

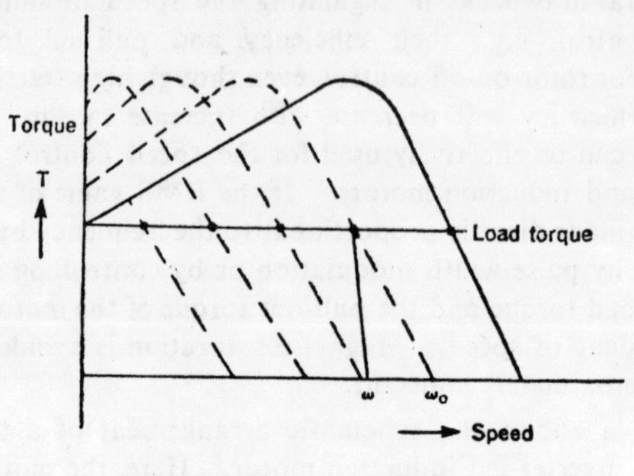

(b) Torque-speed characteristics

Fig. 9.14 Variable frequency control of induction motors.

For normal operation, the controlled rectifier firing angle is in the range 0–$\pi/2$ radians and the inverter switches the current to the different phases of the motor at a frequency decided by the reference speed and the desired rotor frequency. When braking is required, the motor slip is made negative and the firing angle of the controlled rectifier is increased beyond $\pi/2$. The motor will then behave as an induction generator, and the current direction through the rectifier and the inverter will be the same as before. Since the controlled rectifier operates in the inverting mode, the DC voltage will be negative and power will flow from the motor to the three-phase AC input. This will provide the necessary braking. For dynamic braking or plugging, an up-down counter is used to trigger the SCRs sequentially. Normally, up-counting is used. However, when it is required to brake the motor, down-counting is resorted to, and the phase sequence of the voltage applied to the motor is reversed. The frequency of the inverter is controlled by limiting the current.

The inverter is also used for obtaining an uninterrupted power supply. Such a supply is often required for process-control computers. Here, the AC supply to the load is taken directly from the lines and the inverter output is connected in parallel with the supply. The input to the inverter is from a battery which is charged by the same AC supply. As long as this supply is available, the battery will be fully charged and the inverter floating. If it fails because of a fault, the inverter will begin to supply power to the load at the appropriate voltage and frequency. This scheme is shown in Fig. 9.15.

One way to obtain phase conversion is by using SCRs to connect different portions of the input sine wave to the output terminals. An AC chopper (see Chapter 10) can be conveniently used for this purpose. A

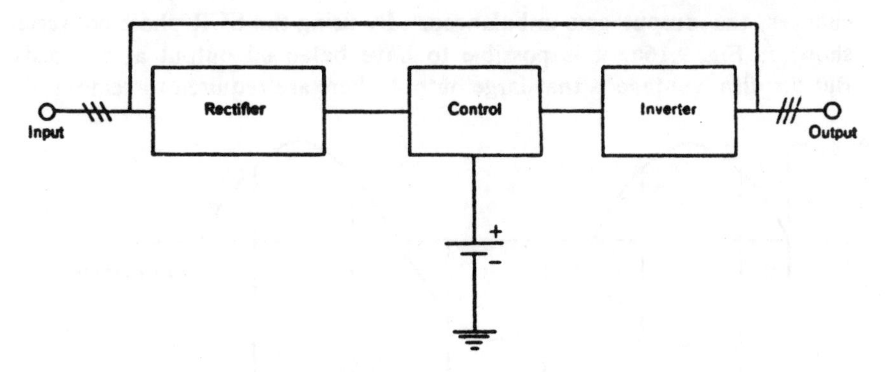

Fig. 9.15 Uninterrupted power supply.

block diagram of such a phase converter is given in Fig. 9.16a, and the output voltage waveform is shown in Fig. 9.16b. Even though it is possible to obtain a reasonably balanced three-phase output at the fundamental frequency (by adjusting the transformer ratios and the instants at which the gating of SCRs takes place), the output waveform will have a very high harmonic content. Unless suitable output filters are used, the three-

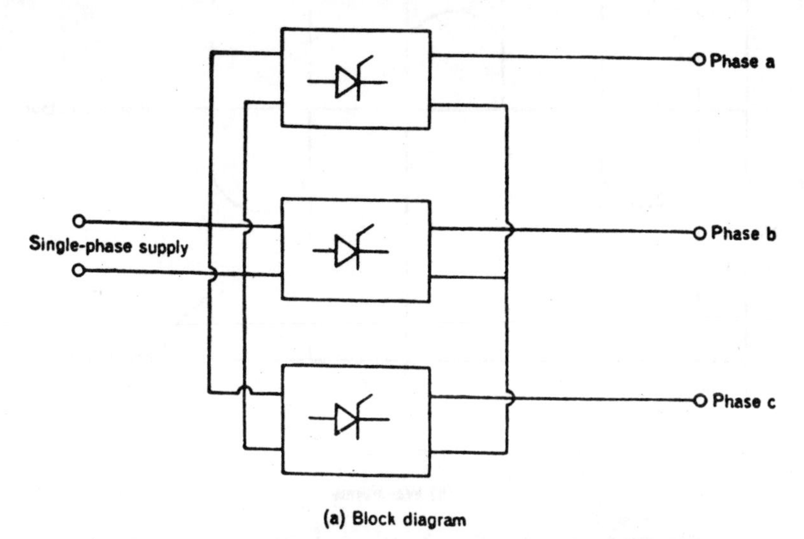

Fig. 9.16 Single-phase to three-phase converter (*cont.*).

phase output resulting from such a circuit, as shown in Fig. 9.16b, may not be very useful for any application.

Converters for transforming single-phase to three-phase are applied in several distribution systems in which a single-phase high-voltage line is used for transmission, and static phase-shifting circuits consisting of L and C are employed for obtaining three-phase output. For a given load, it is possible to choose the proper phase-shifting components. As the load changes, the output gets unbalanced. By using the SCR phase converter shown in Fig. 9.16a, it is possible to have balanced output at all loads. But the disadvantage is that large output filters are required to remove the

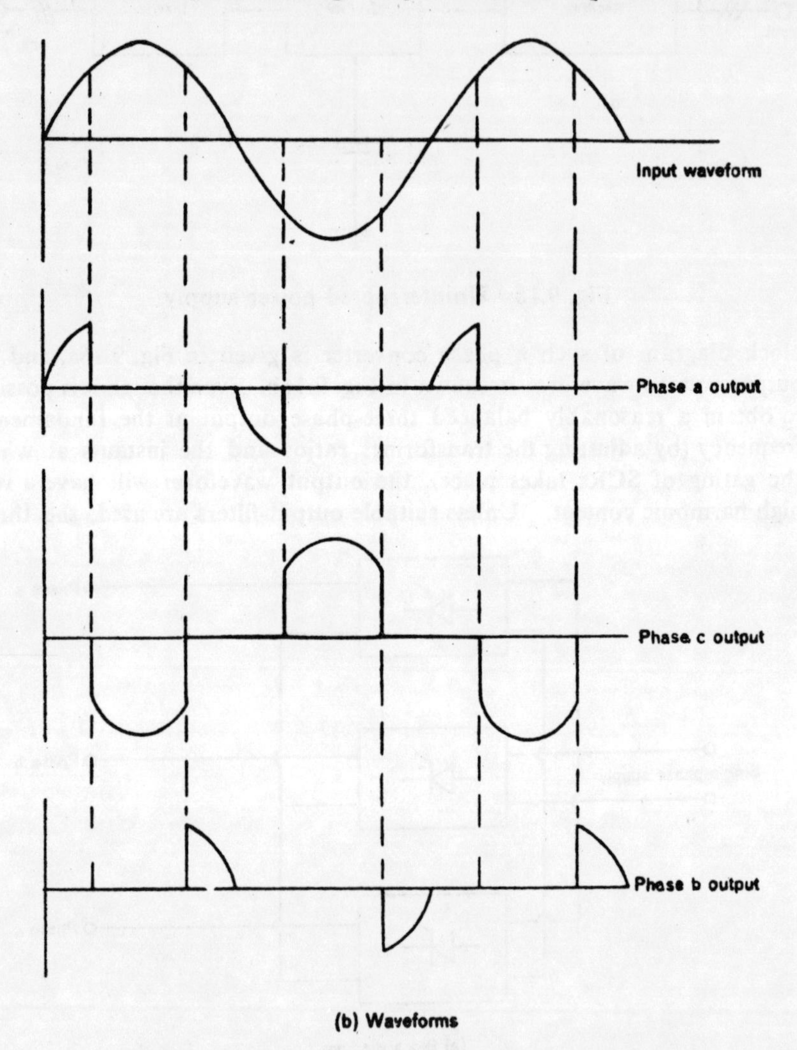

(b) Waveforms

Fig. 9.16 Single-phase to three-phase converter (*cont.*).

harmonics. Another phase converter, used today in electric locomotives, is of the rotating type and is known as the *Orno system*. For electric traction, single-phase supply is used. Some electric motors (other than traction motors) in locomotives require three-phase supply which is obtained by means of such a phase converter.

9.8 EXAMPLE

Design a suitable phase-shifting circuit consisting of passive elements to give a balanced three-phase output from a single-phase supply.

Figure 9.16c shows the required circuit. The single-phase input is connected to terminals *a* and *c*, and the centre tap of the input transformer

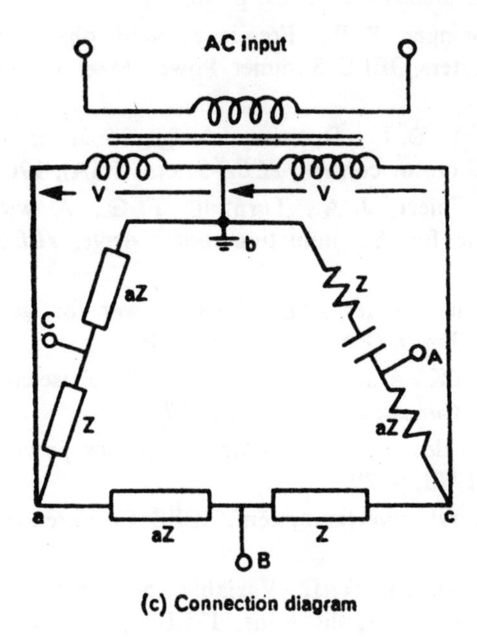

(c) Connection diagram

Fig. 9.16 Single-phase to three-phase converter.

secondary winding is connected to terminal *b* which is grounded. A symmetrical three-phase output will be obtained at terminals *A*, *B*, and *C*, the voltage across the terminals being

$$V_{AB} = -\frac{VaZ}{Z + aZ} = a^2V,$$

$$V_{BC} = -\frac{VZ}{Z + aZ} = aV,$$

$$V_{CA} = \frac{VaZ}{Z(1 + a)} + \frac{VZ}{Z(1 + a)} = V,$$

where V is the RMS value of the input voltage and $a = e^{j2\pi/3}$. If aZ is taken as resistance R, then

$$Z = R^{\angle -2\pi/3} = \frac{R}{2} - jR\frac{\sqrt{3}}{2},$$

and is obtained by using a resistor and capacitor in series. The resulting circuit will be useful for obtaining three-phase balanced output when the impedance of the load is high.

REFERENCES

Agarwal, P. D., The GM high performance induction motor drive system, *IEEE Trans.* (PAS), 1969, p. 86.

Dewan, S. B., Biringer, P. P., Frequency and phase conversion with thyristor converters, IEEE Summer Power Meeting, 1966, Paper No. 31, pp. 66–389.

Dewan, S. B., Duff, D. L., Optimum design of an input commutated inverter for AC motor control, *IEEE Trans.* (IGA), 1969, p. 699.

Espelage, P. M., Chierr, J. A., Turnbull, F. G., A wide range static inverter suitable for AC induction motor drive, *IEEE Trans.* (IGA), 1969, p. 438.

Farber, J. O., Static inverter standby AC power for generating station controls, *IEEE Trans.* (PAS), 1968, p. 1270.

Flairty, C. W., A 50KVA adjustable frequency 24-phase controlled rectifier inverter, *Direct Current* (UK), 1961, p. 278.

Frank, W. E., New developments in high frequency power sources, *IEEE Trans.* (IGA), 1970, p. 29.

Guggi, W., Sine wave inverter system, IEEE Conference Record (IGA), 1970, p. 517.

Johnston, R. W., Newill, W. J., Variable speed induction motor drive system for industrial applications, IEEE Conference Record (IGA), 1970, p. 581.

Klingshirn, E. A., Jordan, H. E., Polyphase induction motor performance and losses on nonsinusoidal voltage sources, *IEEE Trans.* (PAS), 1968, p. 624.

Maeno, T., Kobato, M., AC commutator-less and bush-less motor, IEEE Conference Record (IGA), 1971, p. 25.

Malvino, A. P., Leach, D. P., Digital Principles and Applications, Tata McGraw-Hill, New Delhi, 1975.

Mokrytzki, B., The controlled slip static inverter drive, *IEEE Trans.* (IGA), 1968, p. 312.

Nims, J. W., Static adjustable frequency drives, *IEEE Trans.* (IGA), 1963, p. 75.

Peterson, W. V., Yohman, E. J., Pulse width modulated inverters for UPS applications, *IEEE Trans.* (IECI), 1970, p. 339.

Philips, K. P., Current source inverter for AC motor drive, IEEE Conference Record (IGA), 1971, p. 385.

Ramamoorty, M., Thyristor controlled frequency doubler, *JIE* (India), 1971, p. 241.

Ramamoorty, M., Steady state analysis of inverter driven induction motors using harmonic equivalent circuits, IEEE Conference Record (IGA), 1973, p. 437.

Ramamoorty, M., Abrol, R. K., Solid state frequency doublers for power application, *JIE* (India), 1973, p. 178.

Ramamoorty, M., Ilango, B., A static single phase to three phase converter, *Proc. IEE*, 1971, p. 1288.

Rigberg, R. L., A wide speed range inverter fed induction motor drive, IEEE Conference Record (IGA), 1969, p. 629.

Robertson, S. D. T., Hebbar, K. M., Torque pulsations in induction motor with inverter drives, *IEEE Trans.* (IGA), 1971, p. 318.

Salihi, J. T., An inverter for traction applications, IEEE Conference Record (IGA), 1971, p. 393.

Segsworth, R. S., Dewan, S. B., Thyristor power units for induction heating and melting, IEEE Conference Record (IGA), 1967, p. 617.

Sleman, G. R., Forsythe, J. B., Dewan, S. B., Controlled power angle synchronous motor inverter drive system, IEEE Conference Record (IGA), 1970, p. 663.

Texas Instruments, Inc., Designing with TTL Integrated Circuits, McGraw-Hill, New York, 1971.

Turnbull, F. G., Wide range impulse commutated static inverter with a fixed commutation circuit, IEEE Conference Record (IGA), 1966, p. 475.

10

Choppers

10.1 ON-OFF CONTROL

In the on-off method of power control, voltage is applied to the load for a specified period called *on-time* (T_{on}), and the load is open-circuited or the applied voltage is removed for a duration known as *off-time* (T_{off}). The load power can be controlled by varying the on- and off-time. The ratio T_{on} to ($T_{on} + T_{off}$) is known as the *duty cycle*. This method can be applied to both AC and DC circuits. In AC circuits with direct on-off control, the load current is alternating, and therefore a triac can be very conveniently used.

Figure 10.1 shows a typical on-off scheme for illumination or temperature control. The main power circuit is shown in Fig. 10.1a and the load current in Fig. 10.1b. During on-time, the triac receives high-frequency gate pulses which turn it on in every half-cycle. These pulses are obtained from a UJT relaxation oscillator (UJT1 in Fig. 10.1c), whose output is connected to the gate through a logic circuit. The logic circuit during on-time applies the pulses to the gate; during off-time, the pulses are cut off. The logic gate, which is a transistor switch, is driven by a flip-flop. The on- and off-period correspond to the durations of the two states of the flip-flop. The state transition is attained by triggering the pulses obtained from a variable-pulse-width monostable circuit and another UJT relaxation oscillator (UJT2 in Fig. 10.1c). The control circuit is shown in Fig. 10.1c. The output frequency of UJT2 decides the (on + off)-time. By changing the pulse width of the monostable circuit, the on-time can be varied. This procedure can be automatically controlled by means of a suitable feedback scheme which regulates the output voltage. A similar scheme can be employed to regulate the speed of AC motors. For given on- and off-time, the RMS voltage applied to the stator is given by

$$V_{RMS} = \frac{V_m}{\sqrt{2}} \sqrt{\frac{T_{on}}{T_{on} + T_{off}}}, \tag{10.1}$$

where V_m is the peak amplitude of the AC voltage. By varying the on- and off-time (or the duty cycle), the value of V_{RMS} can be changed.

Using the steady-state equivalent circuit of the induction motor (Fig. 10.2a), and neglecting the effect of harmonics, the developed torque can be computed for any applied voltage at a given slip. The power dissipated in resistance R_r'/S gives the torque (in synchronous watts). In the equivalent circuit, R_r'/S and X_r' are respectively the rotor resistance and rotor reactance per phase referred to the stator at slip S. R_s and X_s are the

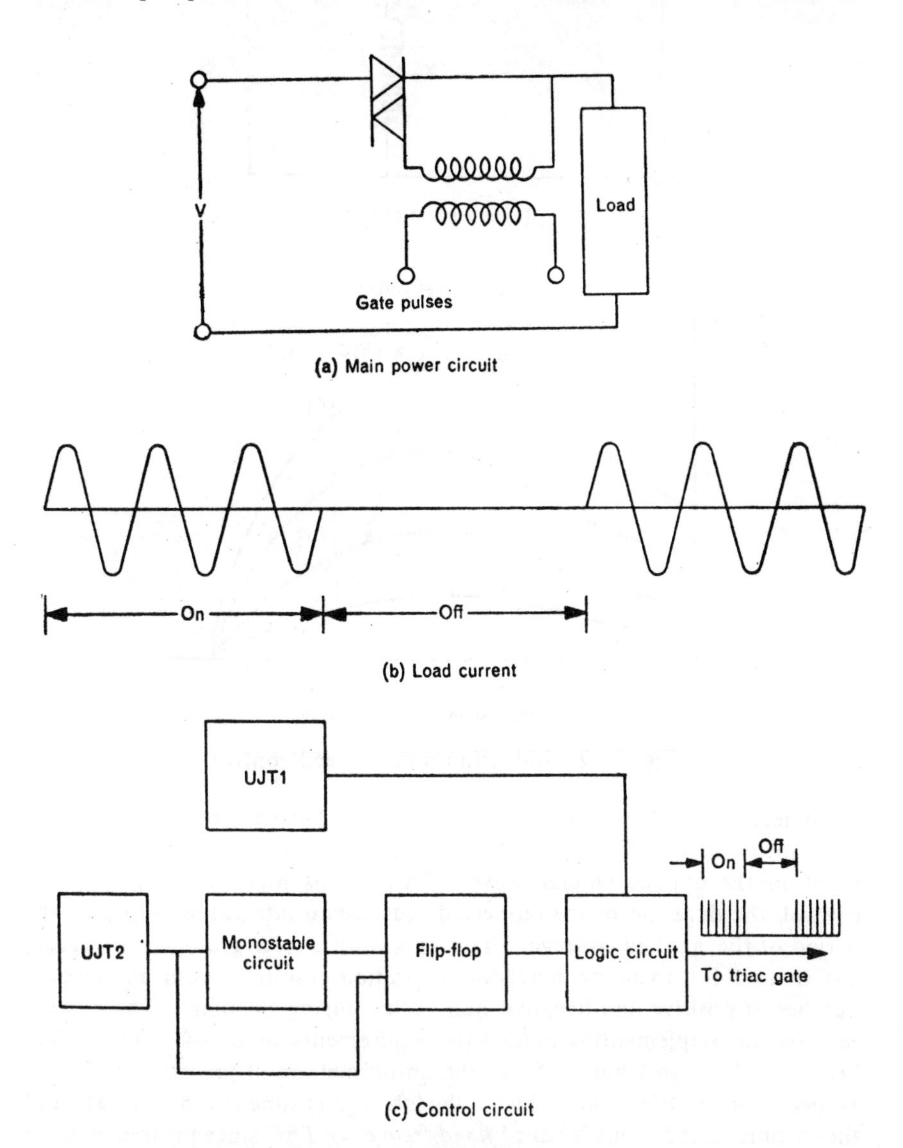

Fig. 10.1 On-off scheme for temperature.

stator parameters and X_m is the magnetising reactance. Figure 10.2b shows the torque-speed characteristics of an induction motor for several applied voltages. For a given load torque, the speed can be changed by varying the applied voltage. Thus, the characteristics are similar to those obtained by phase control (discussed in Chapter 6).

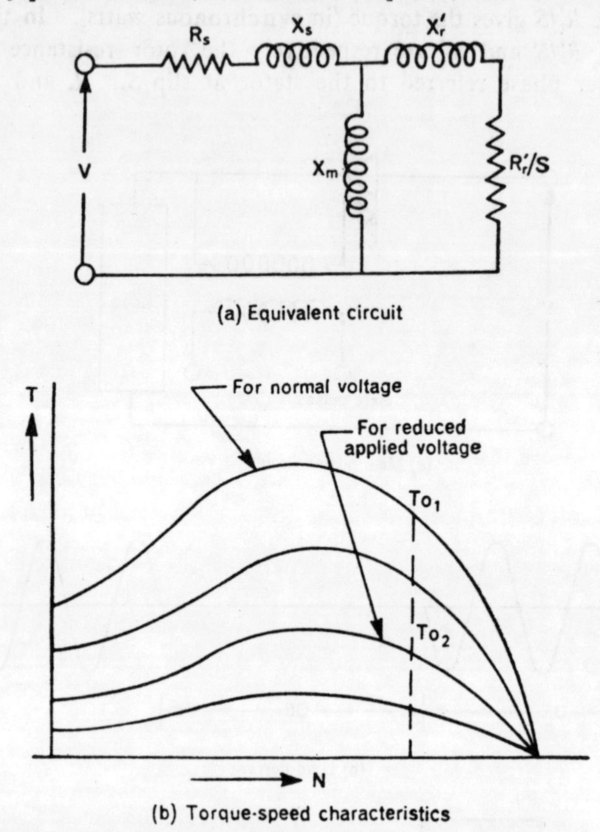

Fig. 10.2 Induction motor speed control.

When the load is resistive, it is advisable to have a zero-crossing switch built into the on-off control, so that on-time always starts at the zero point of the applied voltage wave. This avoids high *di/dt*. For motor control, the duration of the on-period must be an integral multiple of the period of the applied voltage. This is known as *integral-cycle triggering* and is used for avoiding magnetic saturation resulting from an unequal number of positive and negative half-cycles during on-time. The circuits required for implementing these two requirements in on-off control have been considered in Chapter 5. In the on-off method of power control, also known as *time-ratio control* (TRC), the (on + off)-time is kept constant and the on-time varied. This is called *fixed-frequency TRC* since the frequency of application of voltage is constant. In a modified version of TRC, the on-time is maintained constant and power variation is obtained by changing the (on + off)-time. This is known as *variable-frequency TRC*.

Another convenient method of voltage control in AC circuits with chopper which uses forced commutation is explained in Section 10.9.

10.1.1 Example

A single-phase circuit for temperature regulation uses on-off control. The AC input is at 230 V and 50 Hz. The circuit has a constant on-time and a variable (on + off)-time. If the input voltage goes up by 10 per cent, calculate the per cent change required in the triggering frequency of the chopper.

Since the furnace temperature is to be kept constant, the RMS value of the output voltage must remain the same. Therefore,

$$V_{RMS} = 230\sqrt{\frac{T_{on}}{T_{on} + T_{off}}} = 230\sqrt{T_{on}} \times \sqrt{f}$$

$$= 230 \times 1.1\sqrt{T_{on}}\sqrt{f_1},$$

where f and f_1 are the triggering frequencies. Hence,

$$\frac{f_1}{f} = \frac{1}{1.21},$$

i.e., the triggering frequency must be reduced by 17.4 per cent.

10.2 ROTOR ON-OFF CONTROL

Speed control of slip-ring induction motors can also be achieved by using on-off control on the rotor side. In this scheme, the rotor windings are periodically subjected to open- and short-circuit, and the voltage is continuously applied to the stator. During the period the rotor windings are open-circuited (off-time), the developed torque is zero. In the remaining period (on-time), the torque developed can be obtained (neglecting the effect of harmonics) from the equivalent circuit of the motor. By controlling the on- and off-period, the average torque can be changed. This will produce the required speed change for a given load torque. The average torque-speed characteristic resulting from rotor on-off control is similar to that resulting from stator voltage control when the on-off or phase-control method is used. If in Fig. 10.2b To_1 is the developed torque under normal operation, and To_2 the torque with rotor on-off control, then

$$To_2 = To_1 \frac{T_{on}}{T_{on} + T_{off}}. \tag{10.2}$$

To_1 is computed by using the equivalent circuit shown in Fig. 10.2a.

A chopper is used to open- and short-circuit the rotor winding. The performance of various chopper circuits will be discussed in Section 10.4. Figure 10.3a shows the scheme for on-off control with a chopper. The slip-frequency voltages in the rotor windings are rectified by the bridge rectifier and applied to a filter which provides a steady DC output voltage and avoids the sudden chopping of the current in the rotor windings at the beginning of the off-period. R_1 is a large resistance.

When the rotor chopper is off (i.e., when switch S_1 is open), the effective resistance of the rotor will be high and the torque-speed characteristic will be given by curve 1 in Fig. 10.3b. During on-time, switch S_1 will be

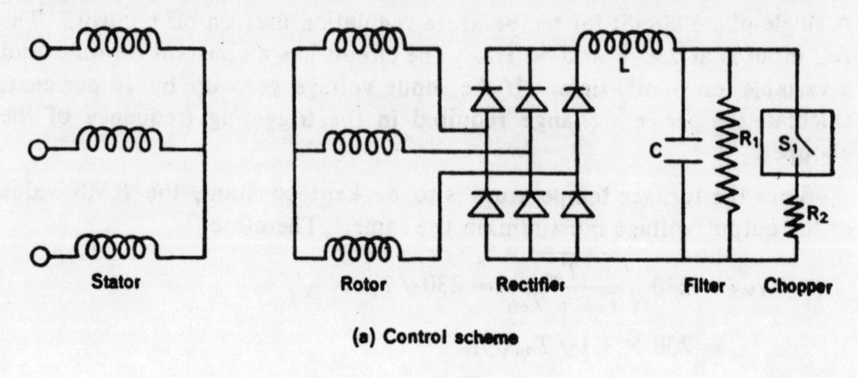

(a) Control scheme

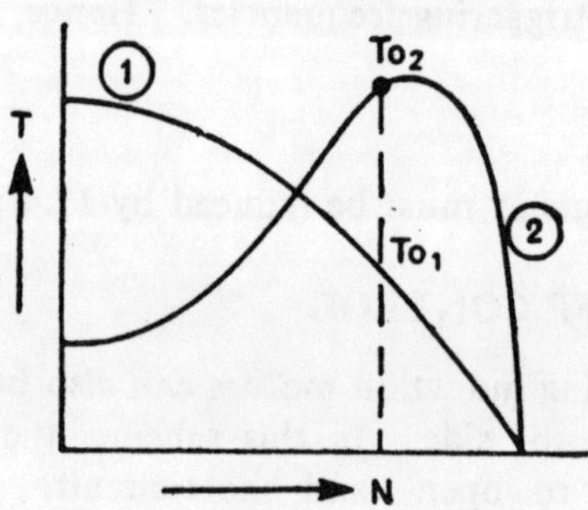

(b) Torque-speed characteristic

Fig. 10.3 Rotor on-off control.

closed and the effective resistance of the rotor will be very low since R_2 is a small resistance which shunts R_1. The torque-speed characteristic in this case will be as shown by curve 2 in Fig. 10.3b. For a particular speed of operation, let To_1 and To_2 be the torques corresponding to the off- and on-time. Then, the net average torque will be

$$T_{net} = \frac{To_1 T_{off} + To_2 T_{on}}{T_{on} + T_{off}}. \tag{10.3}$$

Thus, by varying the on- and off-time the average torque for any speed can be changed. This scheme is similar to the conventional rotor-rheostat control. The advantage it has over other on-off schemes used to control the stator voltage is its wider range of speed variation. The on-off methods applied in controlling the stator voltage or rotor impedance of the motor are less efficient at high values of slip, since the power dissipation in the rotor is, approximately, given by (slip × power input), and therefore the efficiency (neglecting stator and mechanical losses) will be

$$\frac{\text{mechanical power output}}{\text{power input}} = 1 - S. \tag{10.4}$$

The characteristics and disadvantages of the scheme in which the phase-controlled resistance is used, for speed control, on the rotor side (see Section 6.6) of a slip-ring induction motor are similar to those of the on-off control scheme discussed here. These speed control methods are therefore normally applied to small motors where the cost of the control scheme has to be minimised. The inverter-fed drive for induction motors (discussed in Chapter 9) requires more thyristors and other components than do the on-off control circuits described here. It must, however, be noted that inverter drives are more efficient, have better speed-torque characteristics, and can be applied to all types of AC motors.

10.2.1 Example

A three-phase slip-ring induction motor uses rotor on-off control for variable speed operation. The effective rotor resistance is increased ten times during the off-period. If the motor develops 0.4 per unit (pu) torque at a slip of 1 per cent for normal operation, calculate the average torque developed at the same slip for 50 per cent duty cycle of the chopper.

At low values of slip, the torque developed by the induction motor will be

$$T_d \approx \frac{V^2}{R_r'} S,$$

where V is the applied voltage, R_r' is the effective resistance of the rotor, and S is the slip. Therefore,

$$T_d \text{ (normal)} = \left(\frac{V^2}{R_r'}\right) \times 0.01 = 0.4,$$

$$T_d \text{ (with chopper)} = 0.5\left[\left(\frac{V^2}{R_r'} \times 0.01\right) + \left(\frac{V^2}{10R_r'} \times 0.01\right)\right]$$

$$= 0.5 \times 0.4 \times 1.1 = 0.22.$$

10.3 CONTROL OF DC MOTORS

The speed control of DC motors driven by a rectified AC input through a phase-controlled SCR bridge has been discussed in Chapter 6. If the available power supply is DC, say, a battery, then the variation in applied voltage to the motor armature is obtained through an on-off control. This type of control is generally employed for battery-operated vehicles where a DC series motor is used for the main drive. Speed reversal is obtained by interchanging the armature terminals, keeping the field polarity the same. Since the input voltage is DC, the on-off control is achieved through forced commutation of the SCR, unlike that in on-off control for an AC circuit where the SCR or triac is turned off at the natural zero value of the alternating current. A chopper is used for on-off control in DC circuits. There are many variations of chopper circuits but only a few types are discussed in the sections that follow.

The schematic diagram of a chopper control circuit is given in Fig. 10.4a. The output voltage of the DC chopper can be approximated from the waveforms shown in Fig. 10.4b. During the on-time of the chopper

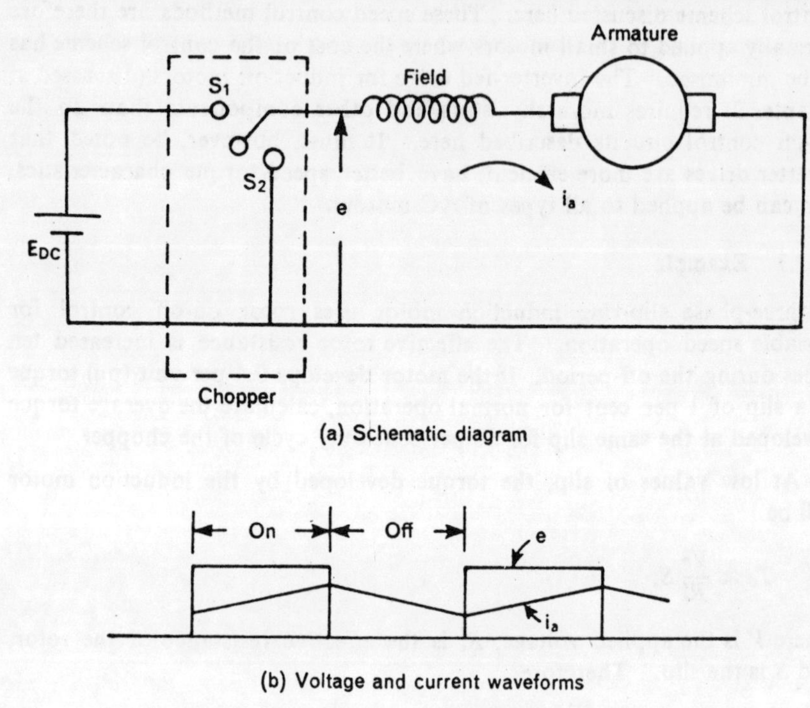

Fig. 10.4 Chopper control for DC series motors.

(i.e., when switch S_1 is closed and switch S_2 is open), the supply voltage E_{DC} is applied to the motor. During the off-period (i.e., when switch S_2 is closed and switch S_1 is open), the applied voltage to the motor is zero.

In the steady state, neglecting speed variation and assuming constant mutual inductance between the field and armature circuits, the differential equation which relates applied voltage e and armature current i_a is given by

$$e = (L_f + L_a)\frac{di_a}{dt} + i_a(R_f + R_a) + M_{af}i_a\omega_m, \tag{10.5}$$

where L_f and L_a are the field and armature inductances, R_f and R_a are their corresponding resistances, M_{af} is the mutual inductance, and ω_m is the speed. Voltage e is equal to E_{DC} during on-time and is zero during off-time. Since the average current, voltage, and speed of the motor are of particular interest in the steady-state analysis, the terms in Eq. (10.5) are integrated over one complete (on + off)-period and divided by the corresponding time to get the equation in terms of the average values of the variables e_a, i_a, and ω_m:

$$E_{av} = I_{av}(R_f + R_a) + M_{af}I_{av}\omega_{av}, \quad (10$$

where

$$E_{av} = E_{DC}(\frac{T_{on}}{T_{on} + T_{off}}).$$

Similarly, the steady-state average torque can be obtained in terms of average values as

$$T_{av} = M_{af} I_{av}^2 k^2, \tag{10.7}$$

where k is the form factor for the current waveform i_a shown in Fig. 10.4b. The value of k is approximately equal to unity. Equations (10.6) and (10.7) can be used for obtaining the steady-state torque-speed characteristics of the DC motor for various on- and off-times. The motor speed can be kept constant by a proper feedback from a tacho-generator, which adjusts the ratio $T_{on}/(T_{on} + T_{off})$ for the chopper.

10.3.1 Example

A fixed-frequency TRC system is employed for the chopper control of a battery-driven vehicle which uses a series motor. The on- and off-time are so adjusted that the average motor voltage is 110 V. Calculate the motor speed if the average torque developed is 10 N-m. Neglect the voltage drops in the field and armature resistances and take mutual inductance M_{af} to be 0.1. Assume $k = 1$ for the armature current.

The average armature current from Eq. (10.7) is given by

$$I_{av} = \sqrt{\frac{10}{0.1}} = 10 \text{ A}.$$

From Eq. (10.6)

$$\omega_{av} = 110/(0.1 \times 10) = 110 \text{ rad/sec}.$$

10.4 CHOPPER CIRCUITS

The circuit shown in Fig. 5.3 can be used as a chopper circuit. The voltage across the load will be as shown in Fig. 10.4b. The turn-off of SCR1 (Fig. 5.3) is achieved by class C type of forced commutation (discussed in Chapter 8).

Figure 10.5a shows a widely used chopper circuit. This circuit has class B type of forced commutation in which the conducting SCR is turned off by the capacitor discharge current passing through it. Hence, it is also called an *oscillating chopper*. L_2 is a large inductor which makes the load current ripple-free. This inductor may be the field winding of a DC series motor or of any load. In the analysis that follows, the load current is assumed to be steady and continuous. Voltage e will be equal to the supply voltage when SCR1 is conducting (on-time) and zero when diode D_1 is conducting (off-time). Here, the on-time is fixed by the natural period of the resonant circuit L_1C_1. SCR1 is triggered by a UJT relaxation oscillator. The triggering frequency determines the (on + off)-time

So, this chopper produces a variable-frequency TRC. The current and

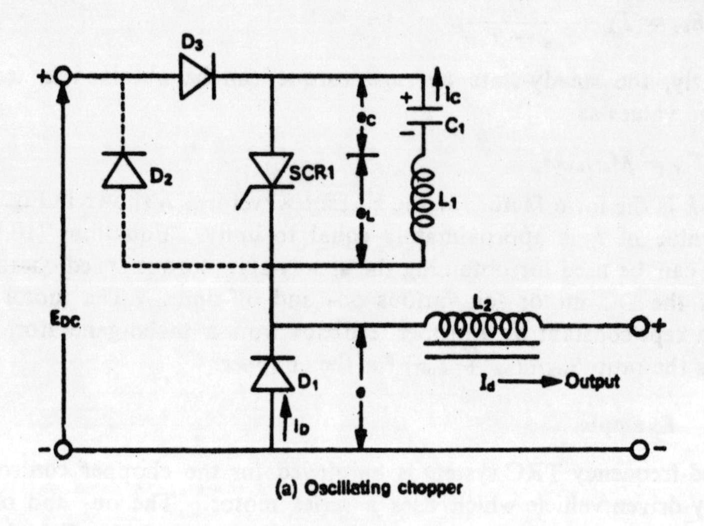

(a) Oscillating chopper

Fig. 10.5 Chopper circuit (*cont.*).

voltage waveforms across the various elements in the chopper circuit are shown in Fig. 10.5b.

10.4.1 Circuit Performance

Let capacitor C_1 in Fig. 10.5a be initially charged to voltage E_1 when SCR1 is triggered. This instant corresponds to point *a* in Fig. 10.5b. Then, the output voltage *e* will be equal to E_{DC}. The capacitor will discharge through SCR1 and L_1. The discharge current i_C will flow from the anode to the cathode, along with the load current I_d. The voltage waveforms across the capacitor and the inductor (e_C and e_L) for this period are shown in the figure. Diode D_1 will be reverse-biased as long as SCR1 is conducting. At the end of the positive half-cycle of discharge current (point *b*), the capacitor voltage will reverse, and since SCR1 will still be conducting, this discharge current will also reverse and now flow from the cathode to the anode. At point *c*, the load current will equal the discharge current. Then, the net current will be zero and SCR1 will be turned off. Since the load current I_d is made continuous by inductor L_2, this current which is of constant amplitude will begin to flow through capacitor C_1 and inductor L_1. Its duration will be from point *c* to point *d*. SCR1 will remain reverse-biased until the capacitor voltage becomes zero (point *f*). In the interval *cd*, the capacitor voltage will vary linearly because the capacitor is charged by a steady current, and the inductor voltage e_L will be zero. At point *d*, the capacitor voltage will become equal to E_{DC}, and diode D_1 will begin to get forward-biased. A part of the load current will then be taken by this diode (i_D) and the output voltage *e* will be zero. At point *e*, the load current will be completely taken over by diode D_1, and the capacitor current will be zero. Voltage

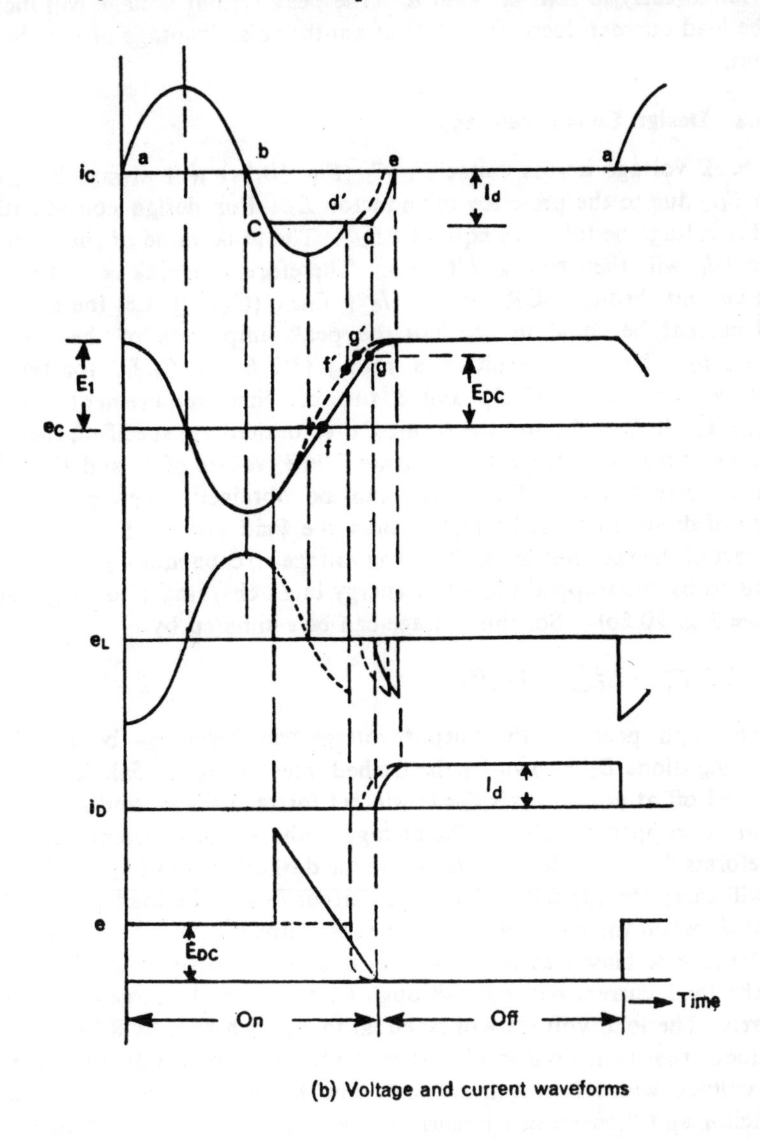

(b) Voltage and current waveforms

Fig. 10.5 Chopper circuit.

e_C across the capacitor will be E_1, which will be more than the supply voltage E_{DC}. The capacitor will retain this overcharge because of diode D_3. The duration *ae* is the on-time of the chopper; it is dependent mainly on the duration of the capacitor discharge circuit and the magnitude of the load current. As the load current increases, the overall on-time will be reduced since the capacitor will get charged faster. This is one of the disadvantages of the chopper circuit. When the load current switches from SCR1 to the capacitor, the output voltage will suddenly rise (point *c*) and fall linearly to zero at point *d*. The peak output voltage will increase as the load current decreases. This is another disadvantage of the chopper circuit.

10.4.2 Design Considerations

The peak voltage across capacitor C_1 (Fig. 10.5a) will always be greater than E_{DC} due to the presence of inductor L_1. For design considerations, let this voltage be taken as equal to E_{DC}. The peak value of the discharge current i_C will then be $E_{DC}\sqrt{(C_1/L_1)}$. Therefore, the peak repetitive transient current through SCR1 will be $[I_d + E_{DC}\sqrt{(C_1/L_1)}]$. Let the maximum load current be equal to one-half the peak amplitude of the discharge current i_C. This will result in a rating $(3/2)E_{DC}\sqrt{(C_1/L_1)}$ for the peak repetitive current in SCR1, and a turn-off time requirement $(C_1\sqrt{3}/2) \times E_{DC}/I_d$. If E_{DC}, I_d, and the turn-off time desired are specified, then using the criteria for peak current and turn-off time, values of L_1 and C_1 and the proper current rating for SCR1 can be obtained. The peak current rating of diode D_1 must be higher than the load current I_d. Inductor L_2 is a part of the specified load. The overvoltage on capacitor C_1 $(=E_1 - E_{DC})$ is caused by the trapped inductive energy in L_1 charging it during interval *de* (see Fig. 10.5b). So, this voltage can be estimated by

$$\tfrac{1}{2}C_1(E_1^2 - E_{DC}^2) = \tfrac{1}{2}L_1 I_d^2. \tag{10.8}$$

The high peaks in the output voltage waveform can be avoided by providing diode D_2 (shown by the dashed lines in Fig. 10.5a). When SCR1 is turned off at point *c*, this diode will get forward-biased and load voltage *e* will be clipped to E_{DC}. The change in the various current and voltage waveforms due to diode D_2 is shown by the dashed lines in Fig. 10.5b. Diode D_2 will carry the capacitor discharge current i_C and the load current I_d. At point d', when these two currents are equal, diode D_2 will be cut off. SCR1 will be reverse-biased as long as D_2 is conducting (from *c* to f'). Beyond f', the load current will pass through C_1 and L_1 and charge the capacitor linearly. The load voltage will be equal to E_{DC} whether SCR1 or diode D_2 conducts, that is, from *a* to f'. After diode D_1 stops conducting, the output voltage will fall instantaneously due to the positive voltage on the capacitor and then reduce linearly to zero (see the waveform for output voltage *e* shown by the dashed line in Fig. 10.5b). The capacitor voltage will become equal to the supply voltage at g' when diode D_1 begins to conduct. When diode D_2 is connected the load voltage will not suddenly rise. However, it can be seen that the capacitor will get overcharged and the actual on-time will still be dependent on the load current. Figure 10.6 shows a modified chopper circuit in which such overcharging of the capacitor can be avoided. Diode D_1 will clip the voltage across capacitor C to a maximum of E_{DC}. The operation of the circuit will be similar to that of the circuit shown in Fig. 10.5a.

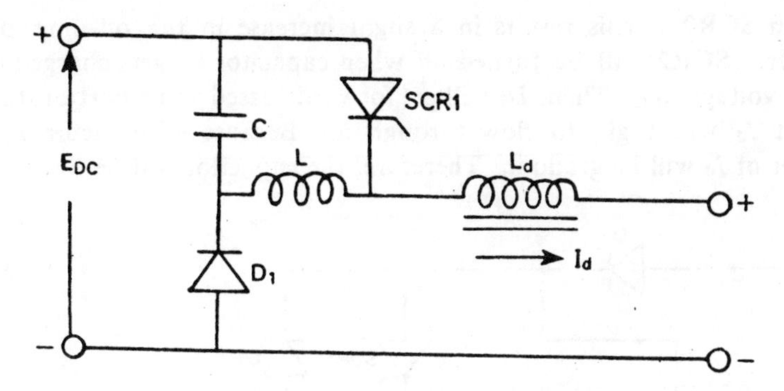

Fig. 10.6 Modified chopper circuit.

10.4.3 Example

Obtain appropriate values of the commutating components L_1 and C_1 for the chopper shown in Fig. 10.5a if the chopper input voltage is 100 V and the average load current is 10 A. Make necessary assumptions. The required turn-off time for the SCR is 40 μsec, and the peak intermittent current rating I_p is 15 A.

Assuming that the peak discharge current of the capacitor is twice the load current, the following relationships can be obtained:

$$I_p = 15 = \frac{3}{2} \times 100\sqrt{\frac{C_1}{L_1}},$$

$$t_q = 40 \times 10^{-6} = \frac{\sqrt{3}}{2} \times 100 \times \frac{C_1}{10}.$$

Therefore,

$$C_1 = \frac{8}{1.732} = 5 \ \mu\text{F},$$

$$L_1 = 0.5 \ \text{mH}.$$

10.5 IMPROVED CHOPPER CIRCUITS

Figure 10.7 shows an improved version of the oscillating chopper. Here, since the on- and off-time are independently controllable, a longer on-time is possible. The operation of this circuit is the same as that of the one covered in Section 10.4, except that capacitor C_1 cannot discharge through SCR1 in the reverse direction due to the presence of diode D_2. Only when the auxiliary SCR2 is fired will reverse voltage appear across SCR1 and turn it off. If diode D_3 is not used, then the load current will begin to flow through C_1, L_3, and SCR2 after SCR1 is turned off. Under these conditions, the load voltage e will experience a spike. The use of diode D_3 will reduce this spike. Inductor L_3 is employed for limiting the discharge current amplitude when SCR1 is off, and for restricting the initial value of *di/dt* in SCR2. This results in a slight increase in the on-time of the chopper. SCR2 will be turned off when capacitor C_1 gets charged to the supply voltage E_{DC}. Then, D_1 will get forward-biased and a part of the load current I_d will begin to flow through it. Because of inductor L_3, this transfer of I_d will be gradual. Therefore, the capacitor will be overcharged

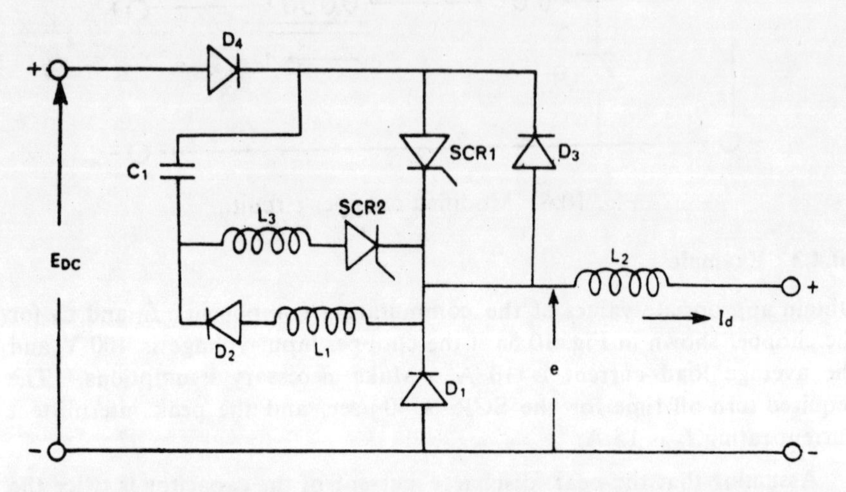

Fig. 10.7 Improved chopper circuit.

as in the circuit described earlier. Diode D_4 will retain this overcharge. Here also, the total on-time will change slightly with the load current. However, commutation is initiated by the firing of SCR2, and therefore the on- and off-time can be independently controlled.

The spikes in the load voltage waveform can be controlled also by connecting the cathodes of SCRs 1 and 2 to taps on inductor L_1, removing at the same time inductor L_3 and diode D_4. The resulting circuit is known as the *Jones chopper*. Another well known circuit for on-off control is the *Morgan chopper*. Here, inductor L (Fig. 10.6) is a saturable-core reactor. As long as the core is not saturated, it will offer a very high impedance, and the discharge current of capacitor C_1 will be zero. Only when the core is saturated will this capacitor discharge through the saturated inductor L_1. Thus, the time taken for the core flux to move from $-B_s$ to $+B_s$ (where B_s is the saturation flux density) will determine the time for which the chopper will be on.

The choppers described here are used for the control of DC motors, for regulated DC power supplies, and for on-off power control. The speed of the motor, the average output voltage, and the load current can be kept constant by means of a suitable feedback (see Section 6.7). The feedback signals appropriately vary the on- and off-time by changing the triggering frequencies of the SCRs shown in Figs. 10.5 and 10.7. Many electric traction systems employing DC series motors use this type of chopper control.

10.6 STEP-UP CHOPPER

The chopper circuits discussed so far are called *step-down choppers* because the maximum output voltage is equal to the input voltage, and any change in the duty cycle $T_{on}/(T_{on} + T_{off})$ only reduces the output voltage. A similar principle can be applied also in stepping up the DC voltage. Here, a *step-up chopper* is used to make the minimum output voltage equal to the input voltage and to increase the output voltage by changing the duty cycle. The circuit for a step-up chopper appears as shown in Fig. 10.8. Switch S_1 functions as the conventional chopper circuit described in Section 10.4.

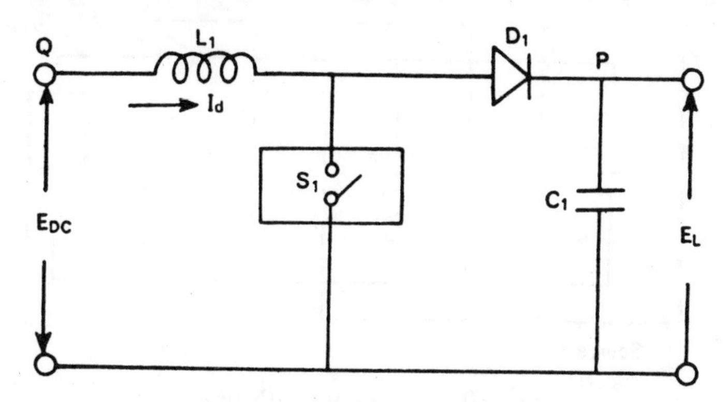

Fig. 10.8 Step-up chopper.

When S_1 is closed (i.e., when the chopper is on), current will build up in inductor L_1. At the end of on-time, S_1 will be opened (i.e., the chopper will be turned off) and current I_d will flow through diode D_1 and charge capacitor C_1. The capacitor will retain this charge and provide a continuous output voltage. Diode D_1 will permit the potential of P to rise above that of Q. Assuming that current I_d remains constant, the energy input E_i to the inductor during on-time will be

$$E_i = E_{DC} I_d T_{on}. \tag{10.9}$$

The energy E_0 delivered by the inductor to the load will be

$$E_0 = (E_L - E_{DC}) I_d T_{off}, \tag{10.10}$$

where E_L is the output load voltage (Fig. 10.8). In a system that has no loss, E_i is equal to E_0 for the steady state. Therefore,

$$E_L = E_{DC}\left(\frac{T_{on} + T_{off}}{T_{off}}\right). \tag{10.11}$$

Thus, voltage E_L can be changed by controlling the on- and off-time.

10.7 MULTIPHASE CHOPPER CIRCUITS

In all the chopper circuits thus far described, only during on-time does the DC source supply power to the load, making the source current equal to the load current. Therefore, the DC source should be capable of supplying the peak load current. This requirement poses a serious constraint on the source, particularly if the chopper has a short duty cycle and is used for the speed control of a motor which, because it operates at low speeds, requires a high ratio of peak current to average current (see Fig. 10.4b). The peak current demand on the source can be reduced by providing a filter between the DC source and the chopper, as shown in Fig. 10.9. The filter consists of an *RC* network. During off-time, the capacitor will get charged, and when the chopper is on, the load current will be supplied partly by the source and partly by the capacitor.

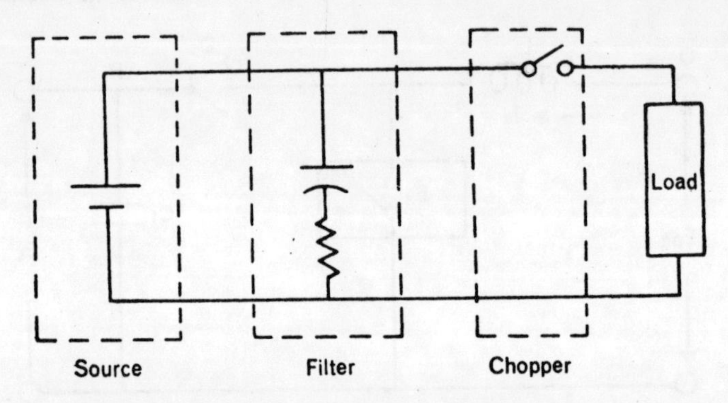

Fig. 10.9 Chopper with filter.

Thus, the ratio of peak current to average current can be very much reduced and an almost continuous flow of power from the source obtained. The only drawbacks of this scheme, with particular reference to traction, are: (a) a large capacitor (which adds to the weight) is required, and (b) the power loss associated with the charging resistance is high. These shortcomings can be overcome by using a multiphase chopper circuit shown in Fig. 10.10. This circuit has two identical choppers (of the type shown in Fig. 10.5) driving a common load. The control is such that when one chopper is on, the other is off. This is known as a *two-phase sequentially-switched chopper*. It can be seen from Fig. 10.10 that current I_2 will free-wheel through the load (armature) and diode D_2 when chopper 2 is in the off-state, and that current I_1 will flow from the source to the load when chopper 1 is in the on-state. The total load current will be $(I_1 + I_2)$. In the next cycle, when the states of the choppers get reversed, the source will supply current I_1 through chopper 2. Thus, there will be continuous flow of power from the source and the peak current demand will be reduced.

For controlling battery-driven vehicles by using choppers, both multiphase circuits and input filters are used. To brake the vehicle, a chopper-controlled resistor is connected across the armature, and its series field is connected across the DC supply. Then, the motor behaves as a separately-excited DC generator and its kinetic energy is dissipated as heat in the resistor. The chopper is used for controlling the dissipation of this power and thereby for maintaining a suitable braking torque on the vehicle. Further information on traction control is provided by the references at the end of this chapter.

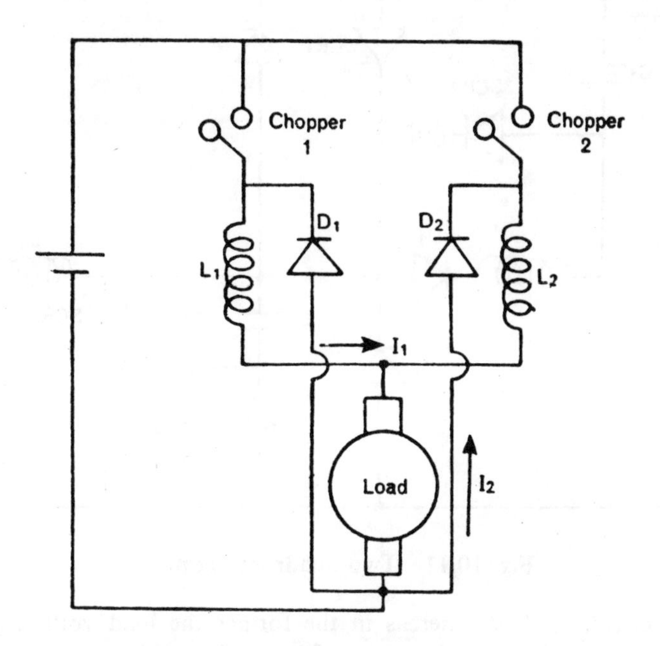

Fig. 10.10 Two-phase chopper.

10.8 TWO-QUADRANT CHOPPER

In a two-quadrant chopper, the polarity of the load current can be reversed, that is, the direction of power flow between the normal input and output terminals can be reversed. Figure 10.11 shows the circuit for the two-quadrant operation. The principle of the step-up chopper (discussed in Section 10.6) is applied for reversing the flow of power. For normal operation, SCR1 is triggered to apply voltage to the load. When SCR2 is gated, it will turn off SCR1, making the load voltage zero. In this mode of operation, power flows from the DC supply to the load. If the load is a DC motor with separate excitation, then regenerative braking can be provided if the power flow through the chopper can be reversed. This is done by SCRs 3 and 4. Initially, SCR4 is fired to charge capacitor C_2; it will turn off when the charging current becomes zero. When SCR3 is turned on, the armature will be short-circuited through L_d and current will build up in this inductor. If SCR4 is again fired, it will turn off SCR3 as it would in a normal oscillating chopper circuit. Inductor L_d will now force the current into the main supply E_{DC} through the feedback rectifier D_2, and thus return to the supply the inductive energy it has stored. By controlling the firing instants of SCRs 3 and 4, the required amount of power can be allowed to flow from the load to the supply.

The two-quadrant chopper is similar to the line-commutated inverter described in Chapter 6. The major difference between the operations of

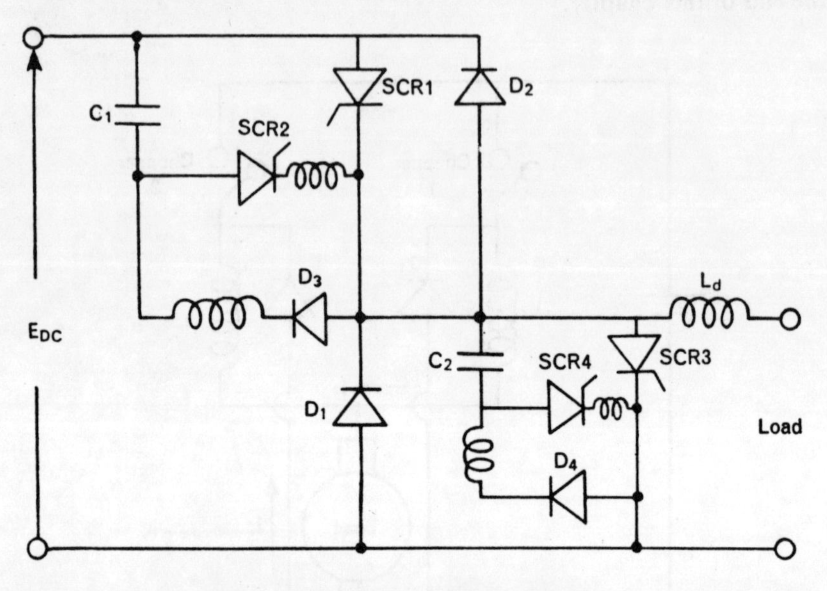

Fig. 10.11 Two-quadrant chopper.

the two circuits is that whereas in the former the load voltage polarity remains the same for both directions of power flow and the current polarity is reversed, in the latter, the load voltage polarity reverses and the current polarity remains the same. In either case, the load should contain a DC voltage source to supply the necessary amount of power. For regenerative braking through a line-commutated inverter, the armature terminals of the DC motor must be reversed and the field winding connected across the DC input. No reversal of armature connections is required when a two-quadrant chopper is used for the same purpose.

10.9 AC CHOPPERS

A simple and very convenient way of controlling power is by using line commutation (discussed in detail in Chapter 6) where the instant of firing the SCR is varied to change the RMS value of the load voltage. The main disadvantage of such a scheme is that the input power factor is poor, particularly with large firing angles. This drawback can be overcome by using forced commutation. The required circuit is shown in Fig. 10.12a, and the load voltage in each half-cycle in Fig. 10.12b. The load power can be changed by varying the pulse width β. The operation of the circuit is as follows. SCR1 is gated at instant t_1 and will be turned off by firing SCR3 at instant t_2. Similarly, SCRs 2 and 4 are used for providing the negative pulse. The voltage across capacitors C_1 and C_2 is used for the commutation of the SCRs. The main advantage of this scheme is that, whatever the pulse width β, the fundamental input power factor is always unity. This circuit is useful for obtaining a regulated AC output. The

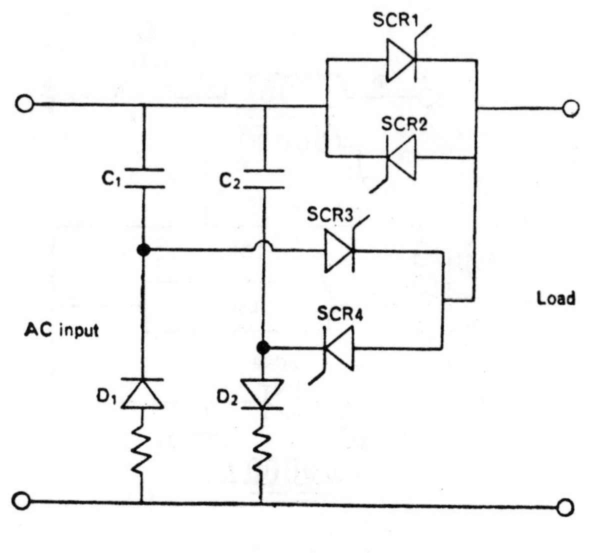

(a) Commutation circuit

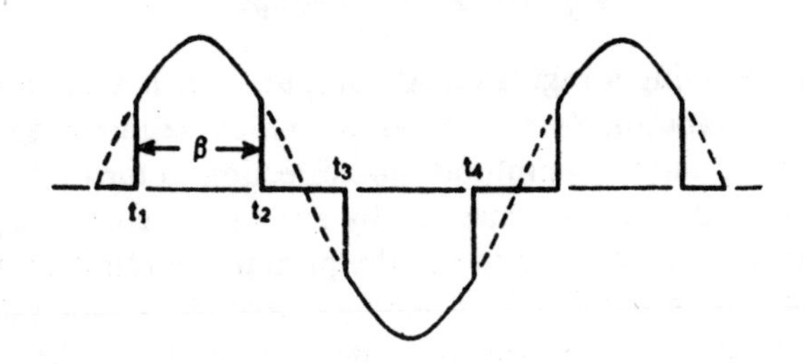

(b) Output voltage waveform

Fig. 10.12 AC chopper (*cont.*).

firing of SCRs 1 and 3 (and SCRs 2 and 4) must occur at angles α and $(\pi - \alpha)$, symmetrical to the peak value ($\alpha = \pi/2$) of the input voltage. This can be very easily achieved by comparing a sinusoidal voltage, in phase with and proportional to the input supply, with a DC control voltage in a comparator, such as the one described in Section 7.6.3. This control voltage is the error between the rectified AC output voltage and the reference voltage. To make the static error zero, the error signal is passed through a PI block before it is applied to the comparator.

The circuit shown in Fig. 10.12a is useful only if one forced commutation is desired in each half-cycle. In certain applications, multiple commutation in each half-cycle is required to speed up the response of the controlled system. The circuit to achieve such multiple commutation is given in Fig. 10.12c. Here, the output is DC. A similar circuit can be

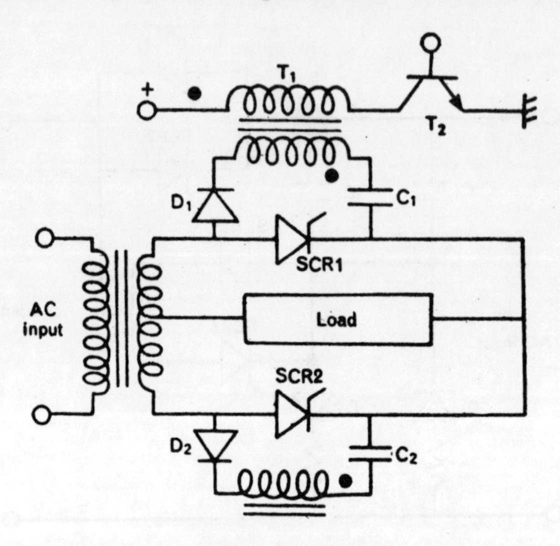

(c) Chopper with class E type of commutation

Fig. 10.12 AC chopper.

used also for obtaining a regulated AC output. Such a circuit uses class E type of commutation (see Chapter 8) which requires an externally-generated pulsc of suitable amplitude and direction to turn off the conducting SCR. The pulse can be generated by applying a control signal to the base of transistor T_2, which connects the primary winding of pulse transformer T_1 to the ground. This induces secondary voltage pulses of suitable amplitude in the secondary windings. One of these pulses will turn off the conducting SCR. A large number of commutations in each half-cycle is possible with this circuit. The scheme shown in Fig. 10.12c is useful for the speed control of DC motors and for regulated DC power supplies. Because of multiple pulses, the inductance required to maintain continuous current in the load is small as compared with that required for other schemes, and therefore the effective load time constant is reduced. This helps improve the dynamic response of the controlled system.

REFERENCES

Berman, B., Battery powered regenerative SCR drive, IEEE Conference Record (IGA), 1970, p. 657.

Franklin, P. W., Theory of the DC motor controlled by power pulses, Parts I and II, IEEE Conference Record (IGA), 1970, p. 59.

Morgan, R. E., Basic magnetic functions in converters and inverters including new soft commutation, *IEEE Trans.* (IGA), 1966, p. 58.

Ramamoorty, M., Battery driven series motor controlled by DC chopper using thyristors, *The Electrical Research* (India), 1972, p. 53.

Ramamoorty, M., Chopper controlled slipring induction motor, *JIE* (India), 1976, p. 206.

Ramamoorty, M., Parihar, J. S., Variable speed induction motor with on-off control using thyristors, IEEE Conference Record (IGA), 1972, p. 105.

Reimers, E., Design analysis of multi phase DC chopper motor drive, IEEE Conference Record (IGA), 1970, p. 587.

Schofield, J. R. G., Smith, G. A., The applications of thyristors to the control of DC machines, IEE Conference Publication, No. 17, 1965, p. 219.

Sen, P. C., Ma, K. H. J., Rotor chopper control for induction motor drive, *IEEE Trans.* (IA), 1975, p. 43.

Thompson, R., A thyristor alternating voltage regulator, *IEEE Trans.*. (IGA), 1968, p. 162.

Wiser, E. F., Thyristor chopper control system for transportation equipment, *IEEE Trans.* (IGA), 1969, p. 470.

Wouk, V., High efficiency high power load insensitive DC chopper for electronic automobile speed control, IEEE Conference Record (IGA), 1969, p. 393.

11

Reliability

11.1 THYRISTOR PROTECTION CIRCUITS

The operation of a thyristor is greatly affected by temperature. Its reliability is ensured only when it is subjected to voltages and currents within specified limits based on permissible junction temperatures. These limits are called the *thermal ratings* of the device. The different types of thyristor ratings, such as the continuous RMS or averge rating, the intermittent or recurrent peak rating, and the surge or nonrepetitive rating, have already been discussed in Chapter 2. These ratings relate to the anode-to-cathode main power circuit and the gate-to-cathode control circuit. If, for any reason, the device carries voltage and current greater than its ratings, the junction temperature may rise beyond the safety limit and permanently damage the device. Therefore, some form of protection against such hazards is essential.

11.2 GATE-CONTROL CIRCUIT

Protecting the gate-to-cathode circuit against overvoltage and overcurrent is not very difficult because of the low-power level of the control circuit. For instance, the gate-control circuits (discussed in Chapter 2) provide the required reliable protection. The more important problem here is that of shielding the control circuit from interference by external electric and magnetic fields which induce spurious signals in the gate-to-cathode circuit. Normally, a number of SCRs are mounted close together in many power-control applications. When any one of the SCRs is triggered, electromagnetic radiations are generated by the sudden collapse of the electric field caused by the SCR that is turned on, and the sudden collapse of the magnetic field caused by the SCR that is turned off. The induced voltages resulting from the rate of change of these fields produce appreciable voltage in the neighbouring SCR gate-control circuits; this may turn on the SCRs, causing maloperation of the entire control scheme which is generally known as *radio interference*. To shield the control circuits and also to minimise the radio interference of SCRs, an RF filter, such as the one shown in Fig. 11.1, may be used. This filter consists of a small inductor in series with the SCR (to slow down the rate of rise of forward current) and a shunt capacitor (to reduce the rate of decay of the forward voltage); the shielding required for the SCR and the gate-control circuit is also illustrated in the figure.

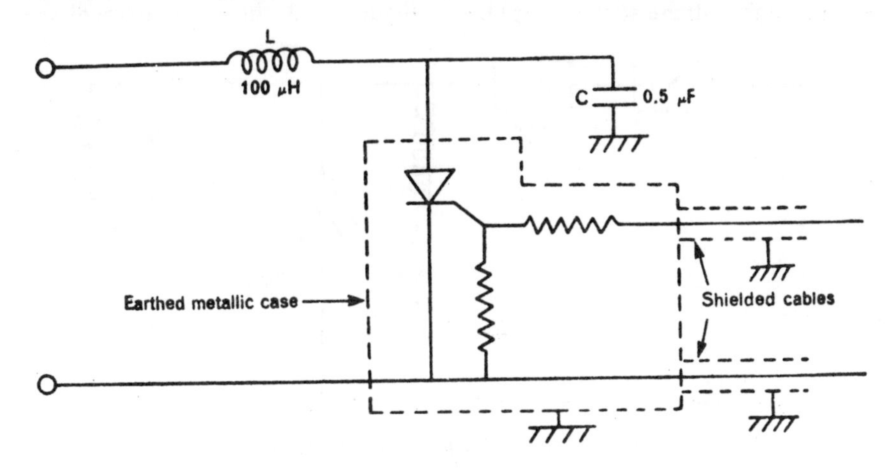

Fig. 11.1 RF filter.

Values of L and C that will produce the required attenuation in the radio frequency band are:

$$L = \frac{R_L}{2\pi f_0}, \qquad C = \frac{1}{2\pi R_L f_0}, \tag{11.1}$$

where R_L is the load resistance, and f_0 is the corner frequency, usually taken as 50 kHz, for the filter.

Another method for reducing radio interference is *zero voltage switching*. Here, the SCR is gated only at the zero crossing of the voltage waveform to reduce di/dt. The use of fast-recovery diodes increases radio interference due to a phenomenon known as *snap action*; this is the result of the very fast decay of the reverse recovery current. The use of such diodes can be avoided at least for low-frequency applications.

11.3 OVERVOLTAGE AND OVERCURRENT PROTECTION

The main power circuit can be subjected to large voltages and currents on two accounts, viz., internal and external factors. The internally-generated overvoltages and overcurrents may result from bad commutation, short-circuits, and inadvertent loading of the circuit. The externally-generated overvoltages and overcurrents may be caused by variations in the supply or by any switching that may take place in the external circuit. Sometimes, the harmonics produced by the control circuit generate overvoltages in the power circuit because of resonance. Switching of inductive loads may also result in overvoltages. The thyristor can be safeguarded against such overvoltages by using shunt-connected nonlinear resistance devices.

These protective devices will register a fall in resistance with increase in voltage, and therefore produce a virtual short-circuit across the thyristor when a high-voltage surge is applied. Figure 11.2 shows an overvoltage

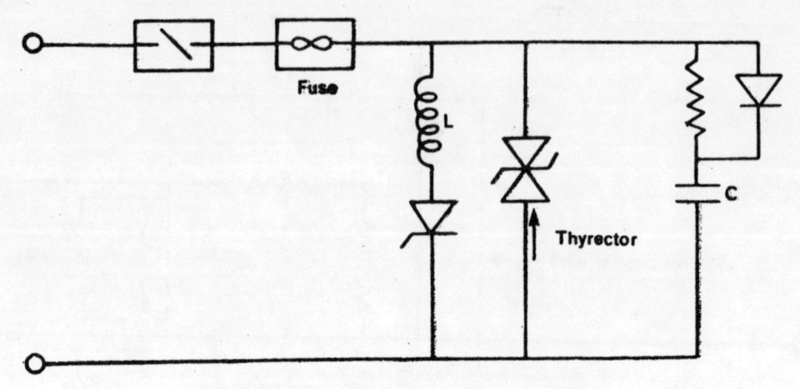

Fig. 11.2 SCR protection circuit.

protection circuit using the thyrector diode, which has a low resistance at high voltage and vice versa; inductor L and capacitor C protect the SCR against large di/dt and dv/dt. Incidentally, L and C will also work as RF filters. A large di/dt will produce a hot-spot temperature in the junction, and this damages the SCR. Sometimes, a saturable-core reactor is used in place of L for soft start. A large dv/dt can, without any gate current, turn on the device, thereby causing short-circuits and maloperation of the control scheme. The detailed design of the *snubber circuit* for protection against large dv/dt is given in Section 11.4.

11.3.1 Overcurrent Protection

Overcurrent protection can be provided by connecting a circuit breaker and a fuse in series with the thyristor, as shown in Fig. 11.2. Its small thermal capacity notwithstanding, a semiconductor device is capable of taking overloads for a limited period; surge and intermittent ratings are higher than continuous ratings for this reason. A circuit breaker, because of its longer tripping time, is generally used for protecting a semiconductor device against continuous overloads or against surge currents of long duration. When the circuit breaker is used in such situations, its tripping time has to be properly coordinated with the device rating. A fast-acting fuse can be used for protecting thyristors against large surge currents of very short duration, called *subcycle surge currents*. Here too coordination of the fusing time with the subcycle duration rating of the device is essential. For reliable protection of the SCR, the circuit breaker or the fuse must open the circuit before the SCR suffers any permanent damage due to large currents. Figure 11.3 shows the desired coordination between the ratings for SCRs and the circuit breakers or fuses.

The one-cycle surge rating $\hat{I}$ of an SCR is defined as the peak amplitude of the sinusoidal current which the SCR can carry for one half-cycle (10 msec on a 50-Hz basis). The subcycle surge current rating may be obtained from

$$\hat{I}_{\text{subcycle}\,(t)} = \sqrt{\frac{\hat{I}^2 \times 1/100}{t}}, \tag{11.2}$$

where $\hat{I}_{\text{subcycle}}$ is the peak amplitude of the sinusoid whose period is $2t$ and t is the duration of the surge in seconds. The fusing time for this circuit must be less than t to ensure reliable protection of the SCR. To facilitate the right choice of fuse, the manufacturers' data include the I^2t rating (square of the RMS value of the one-cycle surge current multiplied by the cycle duration) of the device. The fuse ratings give values of I^2t at different prospective fault currents and the corresponding peak let-through current.

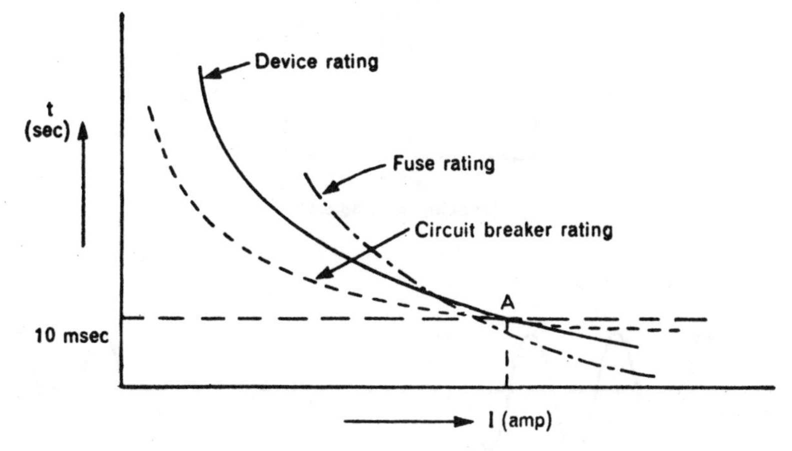

Fig. 11.3 Fuse and circuit breaker coordination.

From these data, the fusing time for a given subcycle surge current can be computed. Assuming, as is usual, that the fault current waveform is triangular, the fusing time t_c will be

$$t_c = 3\,\frac{I^2t}{I_p^2}, \tag{11.3}$$

where I_p is the peak magnitude of the let-through fault current.

The manufacturers' data on peak surge ratings of SCRs correspond to peak values of rectified sinusoidal waveforms in a half-wave circuit operating at 50 Hz or 60 Hz. The one-cycle point (*A* in Fig. 11.3) will therefore give peak values of a nonrecurrent half-sine wave of duration 10 msec. The fuse will provide protection when the current is higher than that corresponding to point *A*. For lower currents, the device will usually be protected by the circuit breaker.

11.3.2 Example

A circuit having a prospective fault current 1 kA is protected by a fuse with I^2t rating of 100 A^2sec on a 50-Hz basis. The faulted circuit is opened in 5 msec. Calculate the peak value of the fault current.

Assuming a triangular waveform for the fault current with a peak value of I_p, we have

$$t_c = 5 = 3I^2t/(I_p^2).$$

Therefore,

$$I_p^2 = 3 \times 100 \times 10^3/5,$$

$$I_p = \sqrt{6 \times 10^4} = 245 \text{ A}.$$

11.3.3 Intermittent Current Waveforms

When an SCR is subjected to an irregular or intermittent but periodic

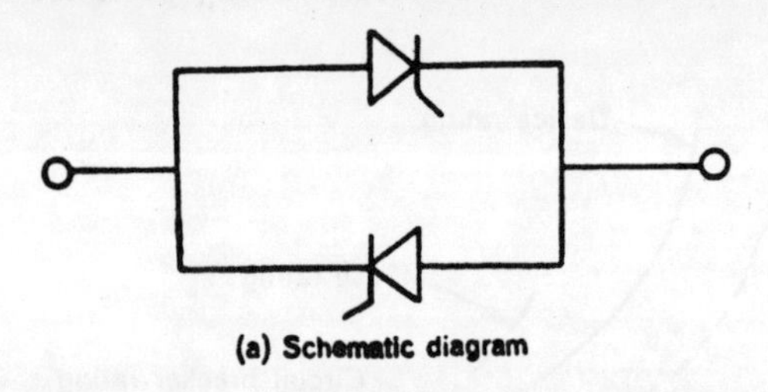

(a) Schematic diagram

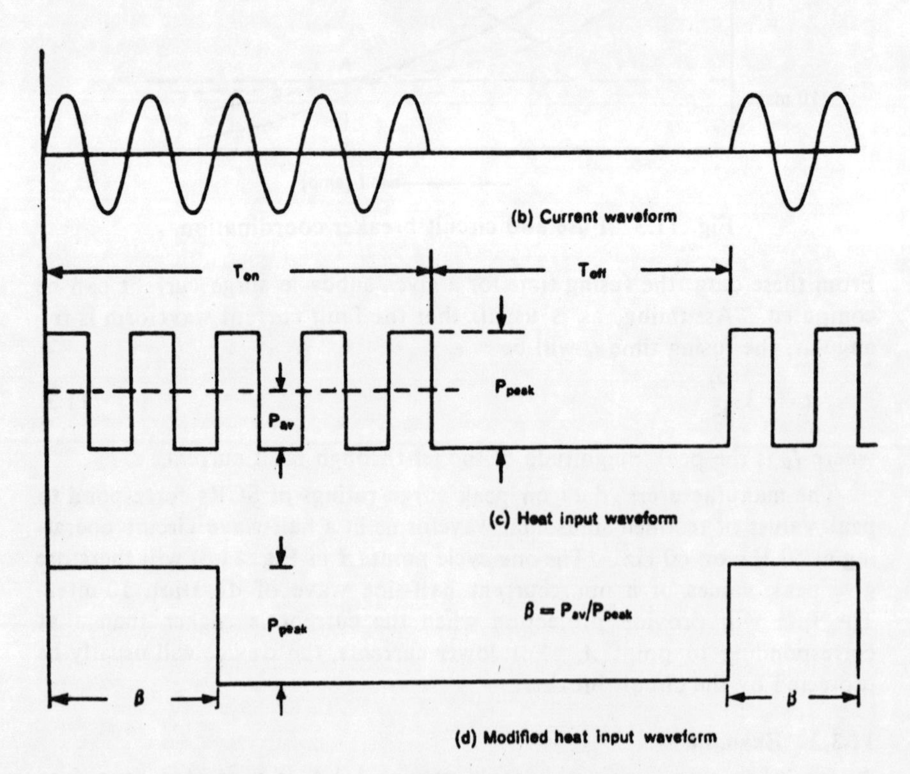

Fig. 11.4 Intermittent current waveform.

current waveform, proper device ratings must be available before adequate protection can be provided. These ratings, which are based on the periodic heat input, are obtained by using the transient thermal impedance of the SCR and the maximum permissible junction temperature. Figure 11.4 shows an SCR controlled full-wave circuit operating in the on-off mode. The schematic diagram is shown in Fig. 11.4a and the line current waveform in Fig. 11.4b. The heat input waveform is given in Fig. 11.4c, where P_{peak} is the peak power dissipation in the device during each half-cycle of conduction, and P_{av} the average power dissipation during on-time. The modified heat input pulse (Fig. 11.4d) to the device is used to evaluate its current rating.

The type of current waveform given in Fig. 11.4b occurs in welding control. It will be noticed that the current rating is a function of the duty cycle $T_{on}/(T_{on} + T_{off})$, and of the number of conducting cycles during each T_{on}. The higher this number, the closer will the junction temperature be to its steady-state value. Therefore, the recurrent rating also will be close to the continuous rating.

11.4 DESIGN OF SNUBBER CIRCUITS

A voltage suppression network, commonly called a ***snubber circuit***, consists of a series-connected resistor and a capacitor placed in shunt with the SCR, as shown in Fig. 11.5. The *RC* network controls the rate of change

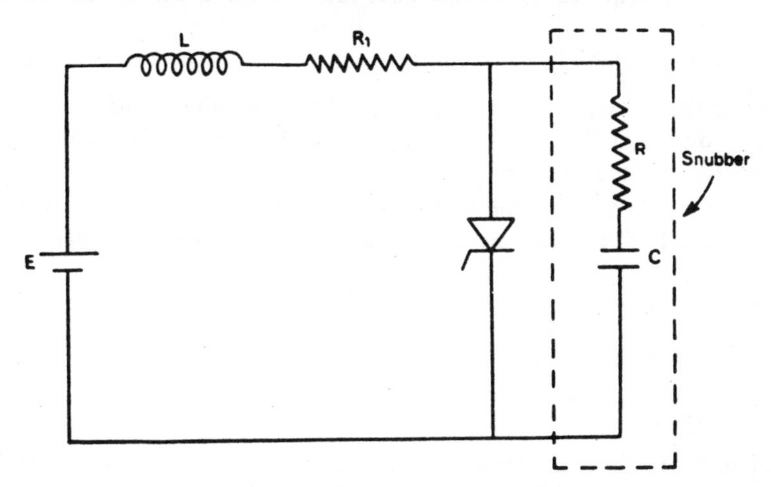

Fig. 11.5 Snubber circuit.

of voltage across the SCR during its blocking state. In the figure, R_l is the load resistance and L the source inductance. When the circuit is energised, assuming that the capacitor is initially uncharged and that the circuit is critically damped, the maximum dv/dt across the SCR will be

$$dv/dt = ER/L, \qquad R + R_l = 2\sqrt{(L/C)}. \tag{11.4}$$

Using the permissible value of dv/dt as the basis for avoiding any maloperation of the SCR, and the given values of L and R_L, we can compute R and C from Eq. (11.4). The stipulation that the maximum discharge current E/R of the capacitor through the SCR, when it is turned on, should be less than the peak repetitive current rating of the device, must also be considered when designing the snubber circuit.

In AC circuits, the maximum input voltage (peak amplitude) can be substituted for E in Eq. (11.4) to calculate the required values of R and C. Another equation that has proved useful in selecting the value of the capacitance required for keeping the voltage transients within the device rating is

$$C = 10\,\frac{\mathrm{VA}}{V_s^2} \times \frac{60}{f}, \tag{11.5}$$

where

C = minimum capacitance required (in microfarads),

(VA) = full load volt-ampere rating of the power circuit,

V_s = voltage applied to the circuit (if a transformer is used, this is the secondary voltage), and

f = operating frequency.

The resistance required to ensure adequate damping can be calculated from

$$R = 2\sigma\sqrt{(L/C)}, \tag{11.6}$$

where σ is the damping factor, normally taken to be about 0.65, and L the effective commutating circuit inductance (transformer leakage inductance for line-commutated circuits).

If the maximum dv/dt for the SCR is specified, the equation used in place of Eq. (11.5) to calculate the required value of the capacitance is

$$C = \frac{1}{2L}\left(\frac{0.564E_m}{dv/dt}\right)^2, \tag{11.7}$$

where E_m is the peak input line-to-line voltage.

11.4.1 Example

Calculate the required parameters for a snubber circuit to provide reliable dv/dt protection to an SCR used in a single-phase fully-controlled bridge. The SCR has a maximum dv/dt capability of 40 V/μsec. The input line-to-line voltage has a peak value 325 V, and the source inductance is 0.1 mH.

Using Eq. (11.7), we have

$$C = \frac{10^4}{2} \times \left(\frac{0.564 \times 325 \times 10^{-6}}{40}\right)^2$$

$$\approx 0.1\ \mu\mathrm{F}.$$

From Eq. (11.6), we have

$$R = 2 \times 0.65 \times \sqrt{10^3}$$
$$= 1.3 \times 31.6 = 41\ \Omega.$$

11.5 SCR MOUNTING

Since the SCR handles substantial amounts of power, it is subjected to high thermal stresses whenever it is conducting. Such stresses are repetitive (because in all control applications each conducting period is followed by an off-period) and result in internal mechanical forces. The SCR has therefore to be properly braced to provide the required strength to withstand the mechanical forces. Further, the mounting of the SCR must be so designed as to be able to carry away the internal heat, and thereby limit the rise in junction temperature. For high-power SCRs, the PNPN pellet is braced by two molybdenum plates (see Fig. 1.2), and the anode is hard-soldered to an aluminium plate with a threaded stud. The SCR is bolted to a heat sink (a large metallic plate, coated black to increase emissivity) through this stud. At normal power ratings, the natural convection and conduction processes provide the required cooling. The heat sink may have fins and is equipped with water cooling or forced-air cooling. (Some typical heat exchanger configurations are shown in the Appendix.) For effective thermal contact between the SCR and the heat sink, spring washers are used with the bolt, and appropriate pressure applied while tightening the nut. A good electrical contact between the SCR and the heat sink may not always be necessary. If electrical isolation is required, thin mica or fibreglass disc-type washers are fixed on the anode stud on either side of the heat sink, as shown in Fig. 11.6. This results in effective thermal but very poor electrical contact between the anode and the heat sink.

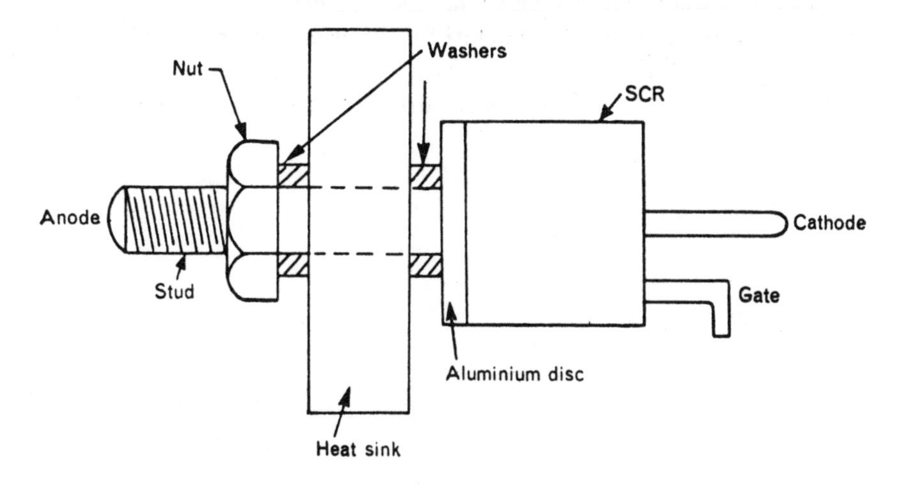

Fig. 11.6 SCR mounting.

11.6 SCR RELIABILITY

With suitable protection circuits and normal operation within the permissible limits of ratings, the SCR can operate with a high degree of reliability. The failure rate claimed for the SCRs manufactured by the General Electric Company is 0.41 per cent for a batch of 950 in a working period of 1000 hours at 90 per cent confidence limits. Derating the device prolongs its life and improves its reliability index.

The majority of "failures" of a device are due to the occurrence of an open-circuit or a short-circuit. An open-circuit, which occurs when ohmic contacts between the terminals and the pellet are broken due to mishandling of the device or internal stresses, causes *mechanical failure*. A short-circuit, which occurs when there is junction breakdown because the device ratings are exceeded or due to flaws in fabrication, results in *electrical failure*. Sometimes, failure is caused by inaccurate design of the heat sink or inadequate cooling arrangements; this is called *thermal failure* Since a device that has failed will have no switching ability, it cannot be used for power control. Therefore, care should be exercised first in the selection of the device (i.e., the one chosen should have the required ratings) and then for its necessary protection.

REFERENCES

Dyer, R. F., The rating and application of SCRs designed for switching at high frequencies, *IEEE Trans.* (IGA), 1966, p. 5.

Jalbert, B. W., Mounting press pack semiconductors, Application Note 200.50, G. E. Company, Auburn, USA.

Newbery, P. G., The correct protection of power thyristors by HRC fuse links, IEE Conference Publication, No. 53, 1969, p. 125.

Rice, J. B., Design of snubber circuits for thyristor converters, IEEE Conference Record (IGA), 1969, p. 485.

Stahl, B., Interaction between SCR drives, *IEEE Trans.* (IGA), 1968, p. 185.

General References

Books

Bedford, B. D., Hoft, R. G., Principles of Inverter Circuits, Wiley, New York, 1964.

Centry, F. E., Gutswiller, F. W., Holonyak, W., Von Zatrov, E. E., Semiconductor Controlled Rectifiers, Prentice-Hall, Englewood Cliffs, New Jersey, 1964.

Chilikan, M., Electric Drive, MIR Publishers, Moscow, 1970.

Davis, R. M., Power Diodes and Thyristor Circuits, Cambridge University Press, London, 1971.

Dewan, S. B., Stranghen, A., Power Semiconductor Circuits, Wiley, New York, 1975.

Kusko, A., Solid-State DC Motor Drives, M.I.T. Press, Cambridge, Mass., 1969.

Mazda, F. F., Thyristor Control, Wiley, New York, 1973.

McMurray, W., The Theory and Design of Cycloconverters, M. I. T. Press, Cambridge, Mass., 1972.

Möltgen, G., Line Commutated Thyristor Converters, Heyden, London, 1972.

Murphy, J. M. D., Thyristor Control of Motors, Pergamon Press, Oxford, 1973.

Pelly, B. R., Thyristor Phase Controlled Converter and Cycloconverter, Wiley, New York, 1971.

Ramshaw, R. S., Power Electronics and Rotating Electric Drives, Chapman and Hall, London, 1973.

Reports and Handbooks

G. E. Company, SCR Manual, Syracuse, USA, 1972.

IEE Conference Publication (No. 53), Power Thyristors and Their Applications, Parts 1 and 2, 1969.

IEEE Press, IEEE Standard Practices and Requirements for Thyristor Converters for Motor Drives, New York, 1974.

IEEE Press, Power Semiconductor Applications, Vols. 1 and 2, New York, 1971.

Motorola Company, Semiconductor Power Circuits, Phoenix, USA, 1968.

RCA, Power Circuits D.C. to Microwave, Princeton, New Jersey, 1969.

Westinghouse, SCR Designers Handbook, Youngwood, USA, 1970.

Problems

1 Calculate the RMS and average values of current $i(t)$ given by the waveforms in Fig. P.1. [*Ans.* I_{RMS}: 70.7 A, 70.7 A, 50.0 A, 50.0 A, 54.8 A; I_{av}: 0.0 A, 63.6 A, 31.8 A, 0.0 A, 0.0 A.]

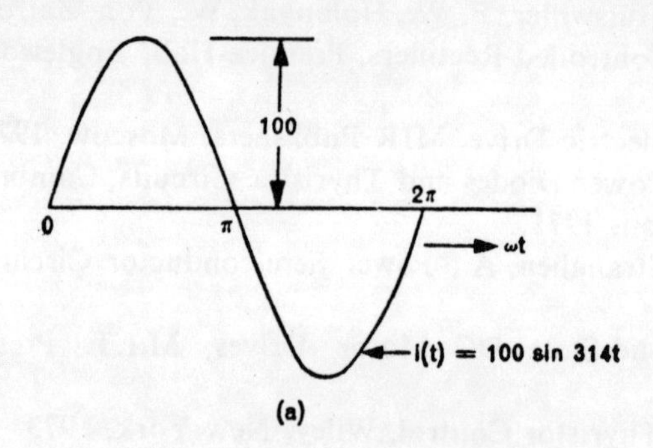

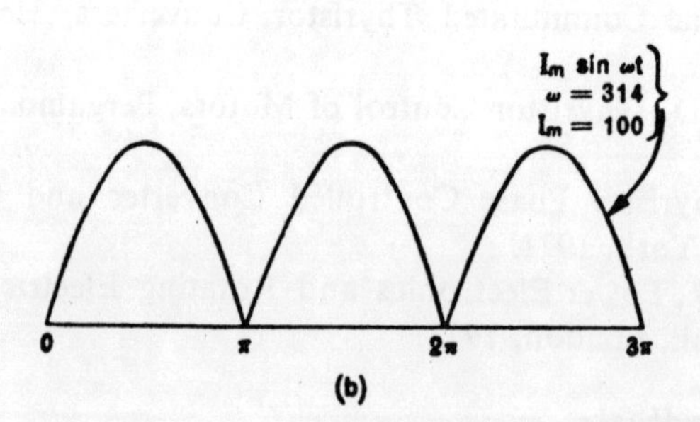

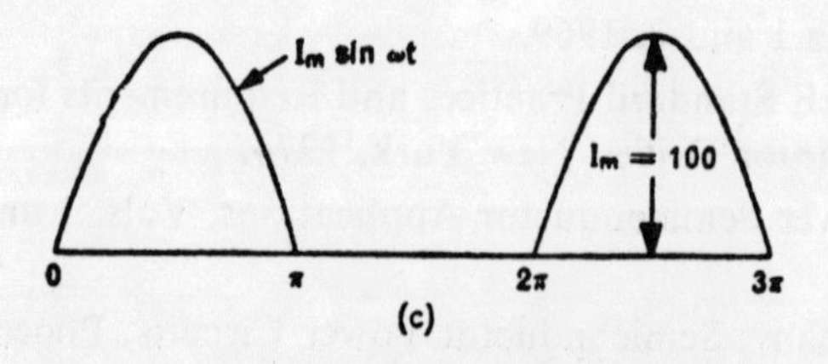

Fig. P.1 Current waveforms (*cont.*).

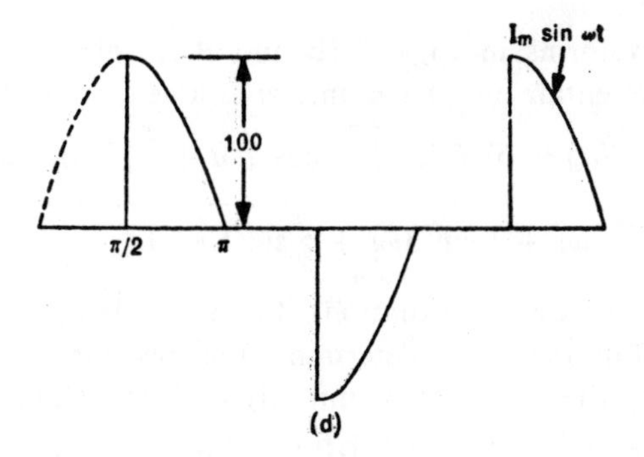

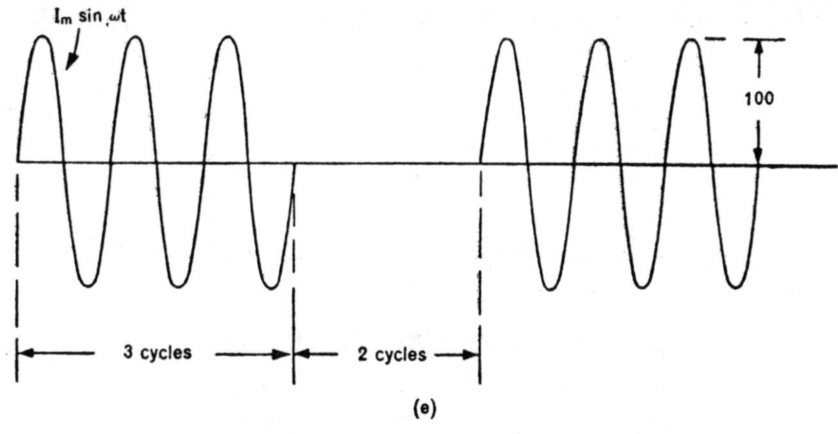

Fig. P.1 Current waveforms.

2 Compute the RMS values of the voltage given by the waveforms in Fig. P.2. [*Ans.* V_{RMS}: (Fig. P.2a) V; (Fig. P.2b) $0.677V$.]

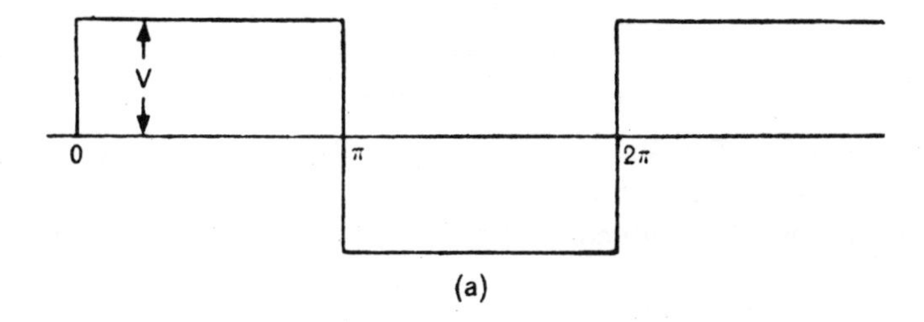

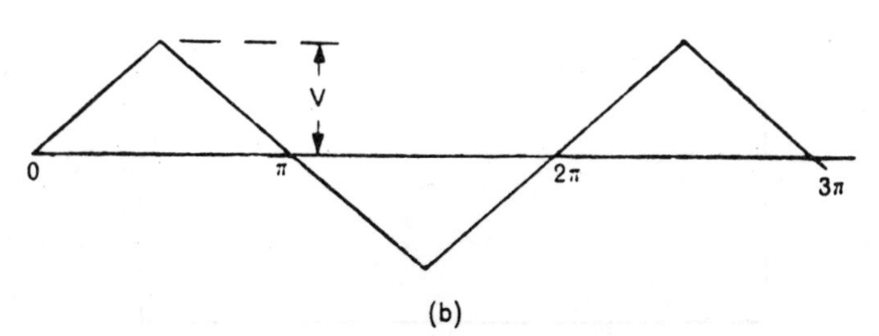

Fig. P.2 Voltage waveforms.

3 For the waveforms in Figs. P.1b and P.2a, obtain the amplitudes of the DC, fundamental, and harmonic components upto the fifth order. [*Ans.* (Fig. P.1b) $F(t) = 63.6 + \frac{400}{3\pi} \cos 2\omega t + \frac{400}{15\pi} \cos 4\omega t$; (Fig. P.2a) $F(t) = 0.0 + \frac{4V}{\pi} (\sin \omega t + \frac{1}{3} \sin 3\omega t + \frac{1}{5} \sin 5\omega t)$.]

4 Obtain the necessary equations for underdamped conditions in the circuits in Fig. P.3 and determine the resonance frequency. In each case, assuming the circuit to be initially relaxed, sketch the waveform of current i when switch S is closed. [*Ans.* (Fig. P.3a) $R < 2\sqrt{L/C}$, $\omega_0 = \sqrt{1/(LC) - R^2/(4L^2)}$; (Fig. P.3b) $R > \frac{1}{2}\sqrt{L/C}$, $\omega_0 = \sqrt{1/(LC) - 1/(4R^2C^2)}$.]

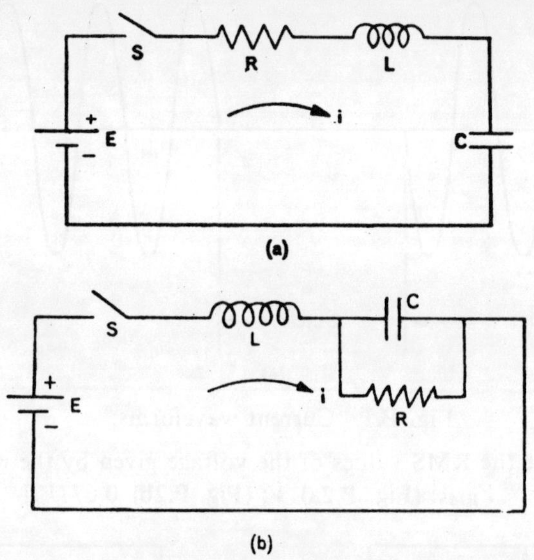

Fig. P.3 *RLC* circuits.

5 If switch S in Fig. P.4 is closed at $t = 0$, obtain an expression for current $i(t)$. Capacitor C is initially charged to 100 V. Sketch the voltage waveform across the inductor. [*Ans.* $i(t) = \{100/(\omega L)\} \sin \omega t$, where $\omega = 1/\sqrt{LC}$.]

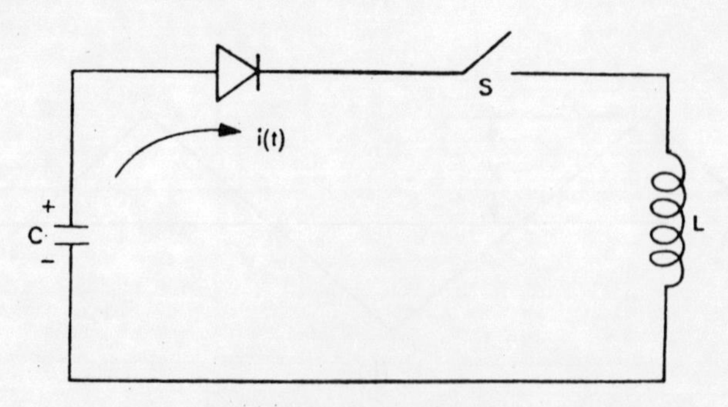

Fig. P.4 *LC* circuit.

6 What will be the average power across the load for the circuit shown in Fig. P.5 when the SCR is fired at an angle $\pi/4$ in every positive half-cycle of the applied voltage? [*Ans.* 248 W.]

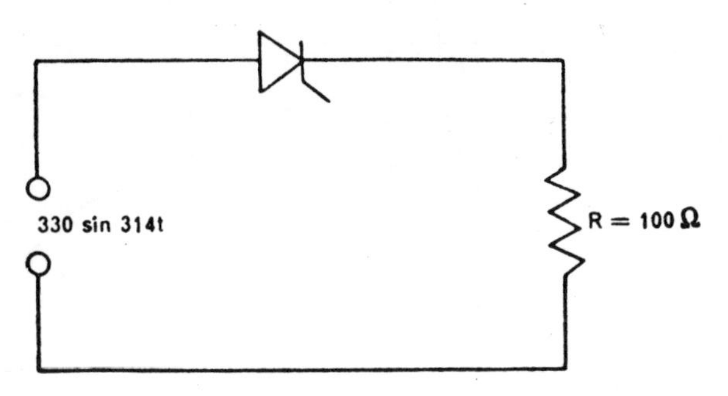

Fig. P.5 Phase-control circuit.

7 If the latching current in the circuit shown in Fig. P.6 is 4 mA, obtain the minimum width of the gating pulse required to properly turn on the SCR. [*Ans.* 4 μsec.]

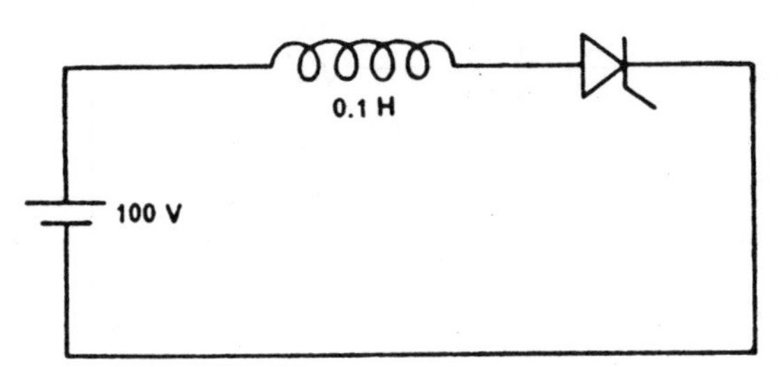

Fig. P.6 Inductive circuit.

8 Compute the minimum value of C (see Fig. P.7) so that the SCR turned off by forced commutation will not turn on due to reapplied dv/dt; the commutating circuit is not shown. The SCR has a junction capacitance 20 pF and a minimum charging current 4 mA to turn it on. [*Ans.* 0.025 μF.]

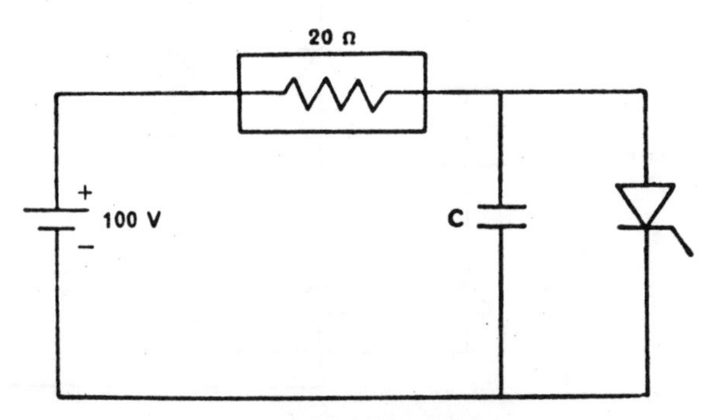

Fig. P.7 Effect of dv/dt.

9 If the SCRs in Fig. P.8 are fired symmetrically in both half-cycles at an angle $\alpha = 75°$, obtain an expression for current $i(t)$. [*Ans.* $i(t) = -0.474e^{-37.7t} + 3.95 \sin (314t + 7°)$.]

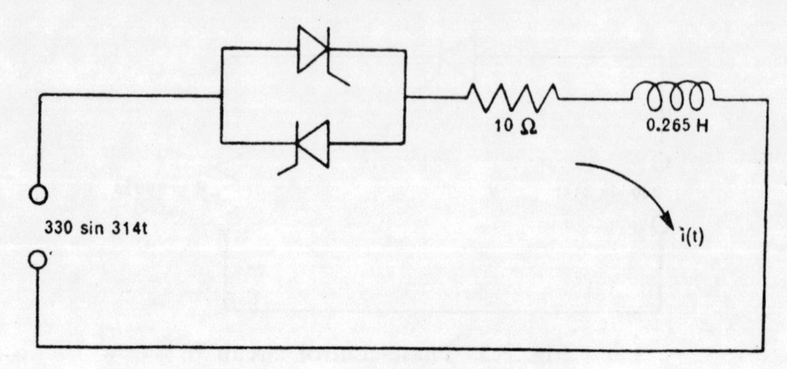

Fig. P.8 AC phase-control circuit.

10 If the SCR in Fig. P.9 is continuously fired by a DC signal, what will be the average value of current $i(t)$? [*Ans.* 9.1 A.]

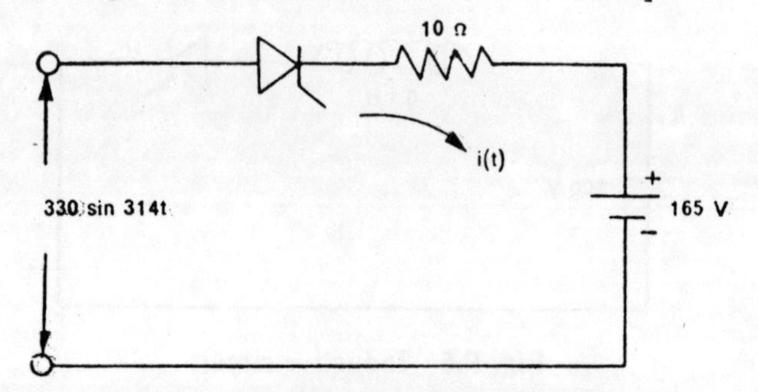

Fig. P.9 Battery-charging circuit.

11 Compute the DC and harmonic voltage amplitudes for the full-

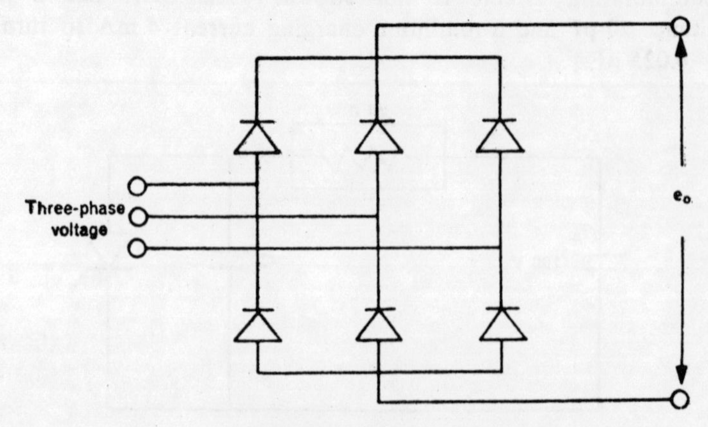

Fig. P.10 Three-phase full-wave bridge circuit.

wave rectified output voltage waveform e_o obtained from the bridge circuit in Fig. P.10.

[*Ans.* $e_o = \frac{3\sqrt{3}}{\pi} E_m\{1 + \frac{2}{\sqrt{3}} \cos n\omega t[\frac{\sin (n-1)(\pi/3)}{(n-1)} + \frac{\sin (n+1)(\pi/3)}{(n+1)}]\}$,
where $n = 6, 12, 18, \ldots$.]

12 For an SCR, the gate-cathode characteristic is given by a straight line with a gradient of 16 volts per ampere passing through the origin; the maximum turn-on time is 4 μsec and the minimum gate current required to obtain this quick turn-on is 500 mA. If the gate source voltage is 15 V, calculate the resistance to be connected in series with the SCR gate. [*Ans.* $R_g = 14\ \Omega$.]

13 With the data in Problem 12, calculate the gate power dissipation. Given that the pulse width is equal to the turn-on time and that the average gate power dissipation is 0.3 W, compute the maximum triggering frequency that will be possible when pulse-firing is used. [*Ans.* $P_g = 4$ W; $f_{max} = 19$ kHz.]

14 An SCR is to be gated by using a relaxation oscillator which has a UJT (see Fig. P.11) with the characteristics: $\eta = 0.7$, $I_p = 0.7$ mA, $V_p = 16.5$ V, normal leakage current with emitter open = 3.7 mA, $V_v = 1.0$ V, $I_v = 6$ mA, and $R_{b1b2} = 5.5$ kΩ. The firing frequency is 1000 Hz. If capacitance $C = 0.1$ μF, calculate the values of R, R_1, and R_2. [*Ans.* $R = 8.3$ kΩ; $R_1 = 600$ Ω; $R_2 = 250$ Ω.]

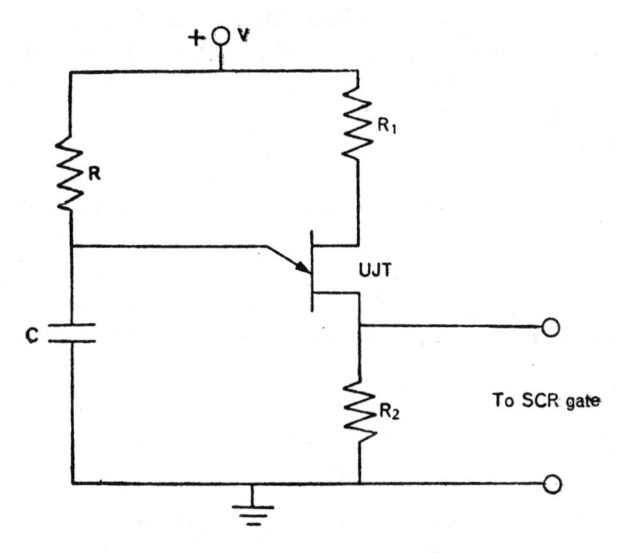

Fig. P.11 Relaxation oscillator.

15 If the firing frequency of the SCR in Problem 14 is changed by varying R, obtain the maximum and minimum values of R and the corresponding frequencies. [*Ans.* $R_{max} = 10$ kΩ, $R_{min} = 3.75$ kΩ; $f_{max} = 2.2$ kHz, $f_{min} = 830$ Hz.]

16 Calculate the values of R and C to be used for commutating the main SCR1 in the circuit shown in Fig. P.12 when it is conducting a full load current of 25 A. The minimum time for which this SCR has to be reverse-biased for proper commutation is 40 μsec. It is given that the auxiliary SCR2 will undergo natural commutation when its forward current falls below the holding current of 2 mA. [*Ans.* $R = 50$ kΩ; $C = 14$ μF.]

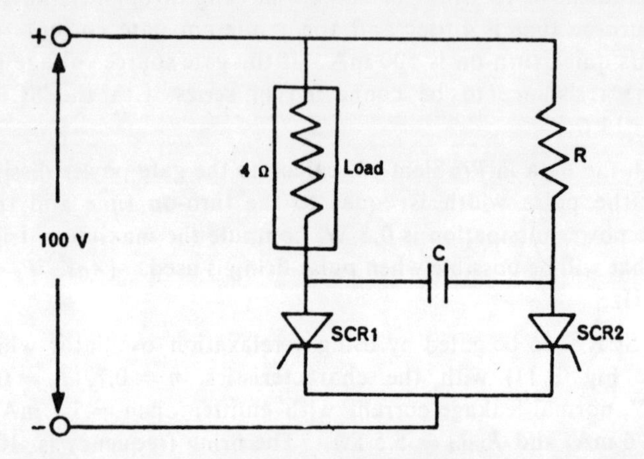

Fig. P.12 DC switch.

17 Obtain the values of leakage inductance L and commutating capacitance C in Fig. P.13. It is assumed that the load current I_L is maintained constant at 10 A by a large inductance in the load circuit and that the peak commutating reactor current is three times the load current. When SCR1 is fired, it will carry the load current and capacitor C_2 will be charged to the full DC supply voltage. When SCR2 is fired, the main SCR1 will be turned off. After SCR1 has been turned off, the load current and the current through SCR2 will be supplied by capacitors C_1 and C_2. It is further assumed that SCR2 will conduct until SCR1 is forward-biased, i.e., when the capacitor voltage becomes 50 V. The time during which SCR1 is reverse-biased is 40 μsec. [*Ans.* $L = 220$ μH; $C = 13.2$ μF.]

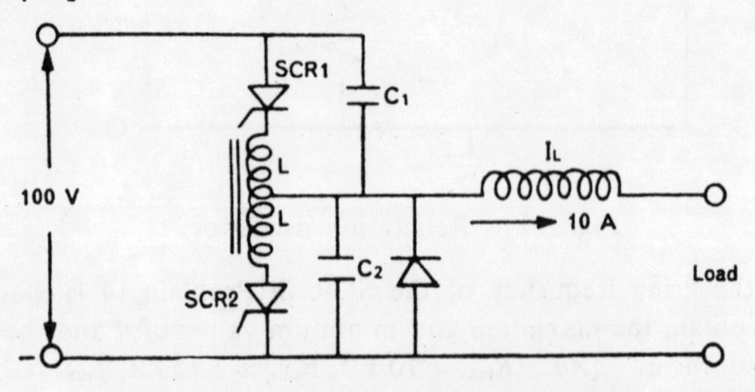

Fig. P.13 Forced-commutated circuit.

18 Obtain the values of L and C in the chopper circuit in Fig. P.14. It is assumed that the load current I_L is constant at 25 A and that the peak capacitor discharge current is twice the load current. The minimum time for which reverse voltage must be applied across the SCR for proper turn-off is 40 μsec. [*Ans.* $L = 53.3$ μH; $C = 10.0$ μF.]

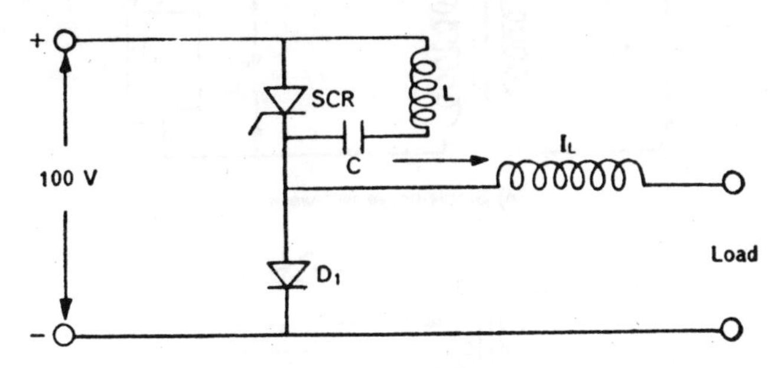

Fig. P.14 DC chopper circuit.

19 The circuit in Fig. P.15 is used for controlling furnace temperature. If the pedestal voltage V_P is zero, calculate the firing angle α and the average load voltage. If the pedestal voltage is changed to 3 V, calculate the change in firing angle. [*Ans.* $\alpha = 8.7°$; 209 V; $\Delta\alpha = 1.3°$.]

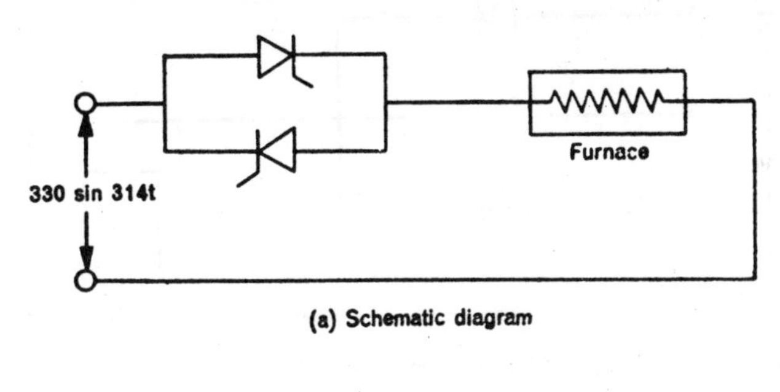

(a) Schematic diagram

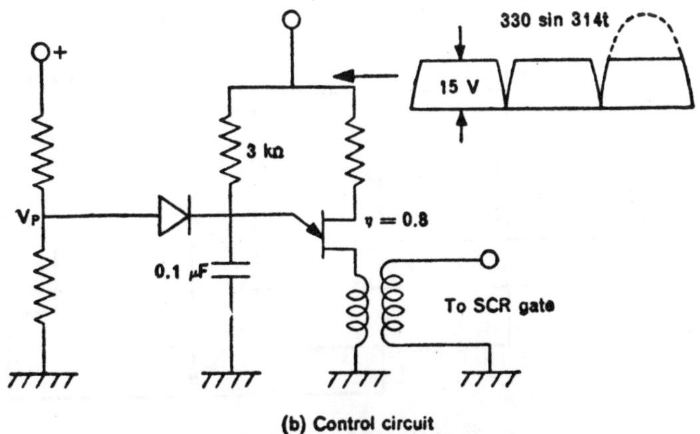

(b) Control circuit

Fig. P.15 Furnace-control circuit.

20 Assuming that the load current in the rectifier circuits shown in Fig. P.16 is continuous, calculate the ripple factor in the output voltage e_o. [*Ans.* Fig. P.16a: 1.22; Fig. P.16b: 0.5; Fig. P.16c: 0.22; Fig. P.16d: 0.076.]

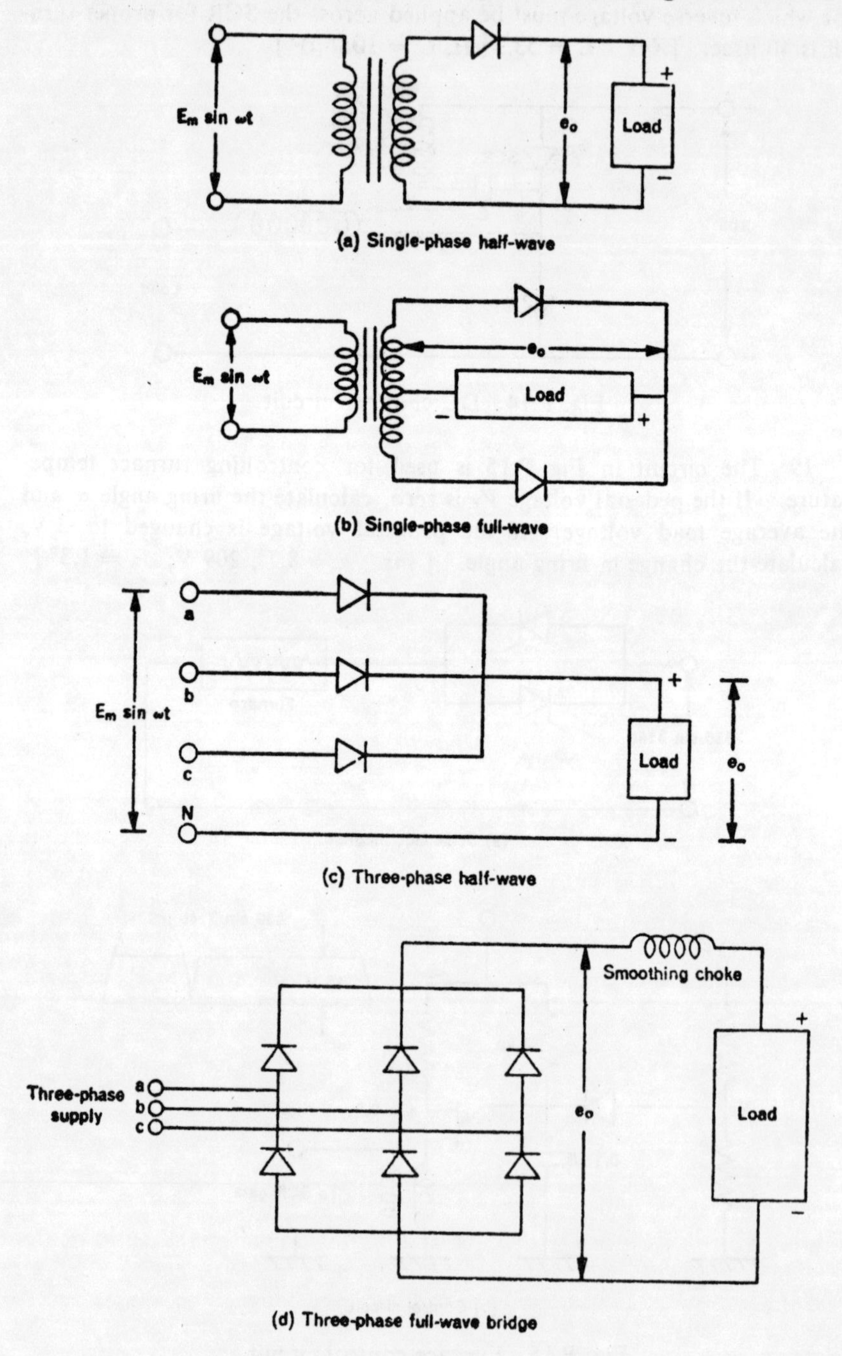

Fig. P.16 Rectifier circuits.

21 If the diodes in the three-phase bridge circuit shown in Fig. P.16d are replaced by SCRs which are fired at an angle $\alpha = \pi/4$, what will be the average DC output voltage when the three-phase supply voltage $e_a = 566 \sin 314t$? [*Ans.* 640 V.]

22 If in Problem 21 the source reactance per phase is 0.8 Ω and the load current is 10 A, calculate the DC output voltage and the overlap angle. [*Ans.* 632.35 V; 1°18′.]

23 Calculate the average current for the circuits in Fig. P.17 when the supply voltage is $330 \sin 314t$. The firing angle α for the SCRs is $\pi/4$. Assume the load current to be constant and continuous and neglect source reactance. [*Ans.* Fig. P.17a: 1.485 A; Fig. P.17b: 1.79 A.]

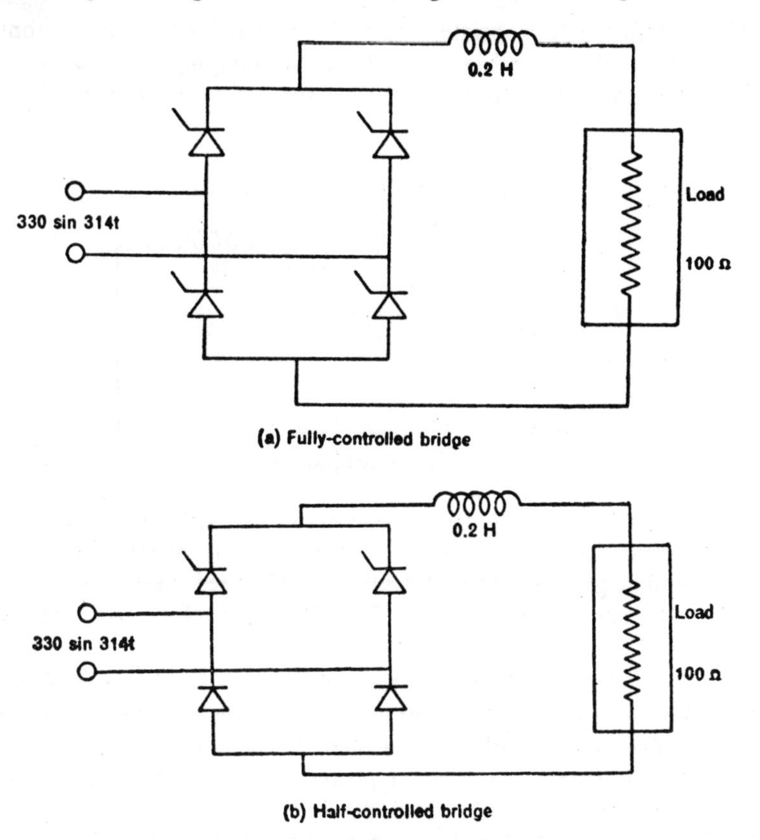

Fig. P.17 Controlled-rectifier circuits.

24 A three-phase full-wave bridge circuit with controlled rectifiers is to be used for rectification and inversion. The leakage inductance of each phase of the input transformer winding is 2.0 mH. The three-phase input voltage is balanced and sinusoidal, and has an RMS magnitude of 230 V per phase and a frequency 50 Hz. The load current on the DC side is 15 A.

(a) Obtain the drop in the DC output voltage caused by the internal reactance drop. [*Ans.* 9 V.]

(b) Calculate the firing angle required for the SCRs (when the bridge is to be used for rectification) when the DC output voltage is 200 V. [*Ans.* 67°.]

(c) Using the condition stated in (b), calculate the overlap angle for the phase currents. [*Ans.* 2°48′.]

(d) Calculate the margin angle required when the bridge is to function as an inverter with a load current 15 A and DC output voltage 200 V. [*Ans.* 67°.]

(e) Obtain the necessary firing angle for the case stated in (d). [*Ans.* 110°12′.]

25 Obtain the appropriate values of R and C for the snubber connected to a triac as shown in Fig. P.18. Take the damping factor δ to be 0.65 and the peak current rating to be 8 A. [*Ans.* $R = 40\ \Omega$; $C = 0.034\ \mu F$.]

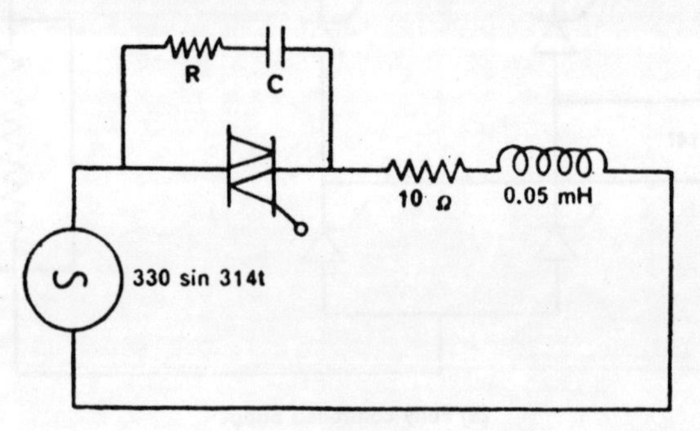

Fig. P.18 Snubber circuit.

26 A particular SCR has a one-cycle nonrepetitive surge current rating of 250 A on a 50-Hz basis. If the circuit is interrupted by a fuse before the SCR gets damaged, calculate the required fusing time when the peak short-circuit current is 500 A. Assume the waveform to be sinusoidal. [*Ans.* $t_c = 2.5$ msec.]

27 An SCR has a full load average current rating of 65 A when the temperature of its heat sink is kept at 65°C and the maximum permissible junction temperature is 120°C. The thermal resistance θ_{JC} of the device is 0.92 °C/W. Calculate the internal power loss, assuming that all of it is due to the dynamic resistance of the device. What will be the percentage increase in the device rating if the temperature of the heat sink is brought down to 55°C? [*Ans.* 60 W; 8.5%.]

28 An SCR carrying a half-wave sinusoidal current has a continuous average current rating of 10 A. Calculate the peak power dissipation when the dynamic resistance is 1 Ω. Assuming that the SCR rating is in direct proportion to the average power dissipation, calculate the intermittent peak current rating for it when the SCR is used in an on-off control scheme for an AC supply at 50 Hz. The on-time is 4 cycles and the off-time 3 cycles. [*Ans.* 940 W; 55 A.]

29 A 100-A SCR is to be used in parallel with a 150-A SCR. The on-state voltage drops of the SCRs are 2.1 V and 1.75 V, respectively. Calculate the series resistance that should be connected with each SCR if the two SCRs have to share the total current 250 A in proportion to their ratings. [*Ans.* 0.007 Ω.]

30 Calculate the number of SCRs, each with a rating 500 V and 75 A, required in each branch of a series-parallel combination for a circuit with a total voltage and current rating 7.5 kV and 1 kA. Assume a derating factor of 14 per cent. [*Ans.* $N_s = 18$; $N_p = 16$.]

31 Calculate the values of R and C that will divide the static and dynamic voltages equally between the series-connected SCRs in Problem 30. These SCRs have the maximum difference in their off-state leakage current $\Delta I_b = 1$ mA and the maximum difference in their reverse recovery charge $\Delta Q = 30$ microcoulombs. [*Ans.* $R = 0.9 \times 10^6$ Ω; $C = 0.33$ μF.]

32 What should be the values of L and C when an RF filter is used for an SCR controlled circuit with a load resistance $R_L = 100$ Ω? Assume the break-point frequency to be 50 kHz. [*Ans.* $L = 320$ μH; $C = 0.03$ μF.]

33 Calculate the maximum possible frequency of the SCR controlled series inverter in Fig. P.19. [*Ans.* $f_{max} = 137$ Hz.]

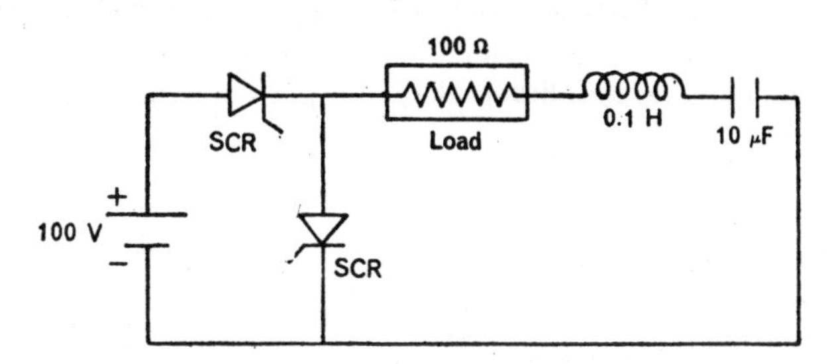

Fig. P.19 Series inverter circuit.

34 Design a series inverter for operation in the frequency range 1–5 kHz. The load resistance may vary from 25 Ω to 100 Ω. The peak load current is limited to 3 A, and the supply voltage is 100 V. [*Ans.* $L = 3$ mH; $C = 0.26$ μF.]

35 Given that a parallel inverter has a DC input voltage 100 V and a maximum commutating current 5 A, obtain the values of the commutating components L and . The minimum turn-off time required for the SCRs is 20 μsec. [*Ans.* $L = 255$ μH; $C = 0.63$ μF.]

36 A DC shunt motor, operating from a single-phase half-controlled bridge at a speed of 1450 rpm, has an input voltage 330 sin 314t and a back-emf 75 V. The SCRs are fired symmetrically at $\alpha = \pi/4$ in every half-cycle and the armature has a resistance of 5 Ω. Neglecting armature inductance, calculate the average armature current and compute the torque. [*Ans.* 20.4 A; 10.0 N-m]

37 A chopper, used for on-off control of a DC shunt motor, has a supply voltage 220 V, an on-time of 10 msec and an off-time of 12 msec. The armature resistance is 3 Ω. Neglecting armature inductance and assuming continuous conduction of the motor current, calculate the average load current when the motor runs at a speed of 1400 rpm and has a voltage constant k_v of 0.495. (Voltage constant k_v is defined as back-emf divided by the motor speed expressed in radians/sec.) [*Ans.* 9.2 A.]

38 A DC series motor has the parameters

$$R_a = 3\ \Omega, \qquad R_f = 3\ \Omega, \qquad M_{af} = 0.15\ \text{H}.$$

The motor speed is varied by a phase-controlled bridge. The firing angle is $\pi/4$, and the average speed of the motor is 1450 rpm. The applied AC voltage to the bridge is 330 sin ωt. Assuming continuous motor current, calculate the steady-state average motor current and torque. [*Ans.* 4.6 A; 3.18 N-m.]

39 A frequency tripler operating from a three-phase input has input phase a voltage $e_a = 330 \sin 314t$. The firing angle is 150°. Assuming a purely resistive load and neglecting source inductance, obtain the magnitude of the fundamental component of the output voltage. [*Ans.* 79 V.]

40 Figure P.20 shows the circuit of a single-phase cycloconverter and the conducting sequence of SCRs. If the supply voltage is 330 sin 314t, obtain the fundamental amplitude of the output voltage. [*Ans.* 182 V.]

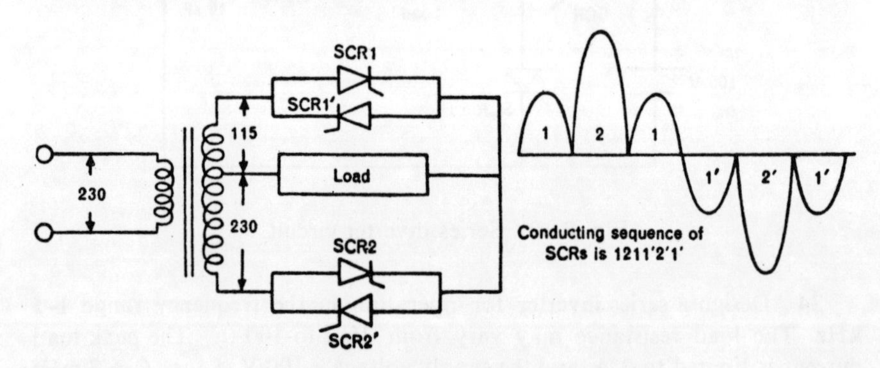

Fig. P.20 Cycloconverter circuit.

41 A phase-controlled circuit with a triac is used for the speed control of a single-phase induction motor. The input voltage is 330 sin 314t. The triac is fired at $\pi/4$ in both half-cycles. Assuming that the motor torque, which is 1.0 pu at normal voltage, varies as the square of the RMS value of the voltage applied to the motor, calculate the torque developed by the motor. Assume further that the conduction angle is 150° and neglect the effect of the induced voltage during the interval the stator is open-circuited. [*Ans.* 0.865 pu.]

42 In Problem 41, assuming a linear torque-slip relation and keeping the load torque constant at 1.0 pu, calculate the slip of the motor. The motor speed at normal voltage is given to be 1400 rpm and the conduction angle is assumed constant at 150°. [*Ans.* 7.7%.]

43 Obtain the results for Problems 41 and 42 if the motor speed is varied by introducing on-off control. The on- and off-time are respectively 4 cycles and 3 cycles and the supply voltage is 330 sin 314*t*. The applied voltage during off-time is assumed to be zero. [*Ans.* 0.57 pu; 11.7%.]

44 A three-phase two-pole slip-ring induction motor has the parameters

$$R_s = 1.3\ \Omega, \qquad X_s = 5.7\ \Omega, \qquad R_r' = 2.8\ \Omega, \qquad X_r' = 5.7\ \Omega.$$

The input supply is three-phase AC at 400 V and 50 Hz. The speed of the motor is varied by rotor on-off control. If the on- and off-time of the chopper are 10 msec and 12 msec, respectively, and the average speed of the motor is 1400 rpm, calculate the average developed torque, neglecting magnetising reactance X_m and mechanical losses. [*Ans.* 5.1 N-m.]

45 A three-phase 400-volt 50-hertz reluctance motor is driven by a variable frequency bridge inverter whose output frequency can be varied from 15 Hz to 65 Hz. Assuming that the output waveform of the bridge inverter is a rectangular pulse with a 120° maximum width at 65 Hz, obtain the magnitude of the DC input voltage required. [*Ans.* 181.5 V.]

46 The speed of the induction motor described in Problem 45 is controlled by a variable frequency inverter. Assuming a constant air gap flux, calculate the torque developed by the motor at different slips for input frequencies of 25 Hz and 50 Hz.

Ans.	Slip		0.05	0.1	0.2	0.3	0.4	0.5
	Torque (in N-m)	For 50 Hz:	8.2	14.15	19.3	19.1	16.35	15.45
		For 25 Hz:	4.27	7.65	11.65	13.2	12.85	12.35

47 Two six-pulse fully-controlled converters are connected in parallel through an interphase reactor as shown in Fig. P.21. The maximum circulating current is to be limited to 20 per cent of the full load current I_L of 20 A. The input to the converter is three-phase AC at 400 V and 50 Hz. Obtain the inductance required for the interphase reactor, neglecting harmonics higher than six in the circulating-current path. [*Ans.* 21.9 mH.]

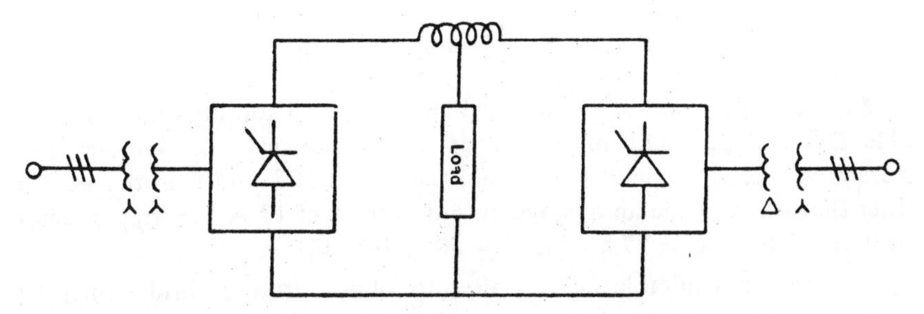

Fig. P.21 Twelve-phase converter.

48 Using the criterion for the maximum circulating current given in Problem 47, calculate the value of the inductance required in the circulating-current path for a dual converter scheme operating in the circulating-current mode and supplied with single-phase AC at 230 V and 50 Hz. [*Ans.* 220 mH.]

49 If the armature of a separately-excited DC motor is connected to a single-phase fully-controlled bridge with a 230-volt 50-hertz supply, calculate the minimum value of the inductance required in the armature circuit to provide continuous armature current of average value 8 A. The firing angle is $\pi/4$ and the counter-emf in the armature is 146.5 V. Neglect source inductance and armature resistance. [*Ans.* 217 mH.]

50 If the single-phase bridge circuit in Problem 49 is half-controlled, what will now be the value of the inductance? [*Ans.* 177 mH.]

51 The voltage waveform shown in Fig. P.22 is obtained from an AC chopper. Calculate the value of inductance L required for providing continuous load current i_L. In the free-wheeling periods (PQ, RS), the minimum value of i_L necessary for maintaining conduction is 0.5 A. [*Ans.* 1.83 H.]

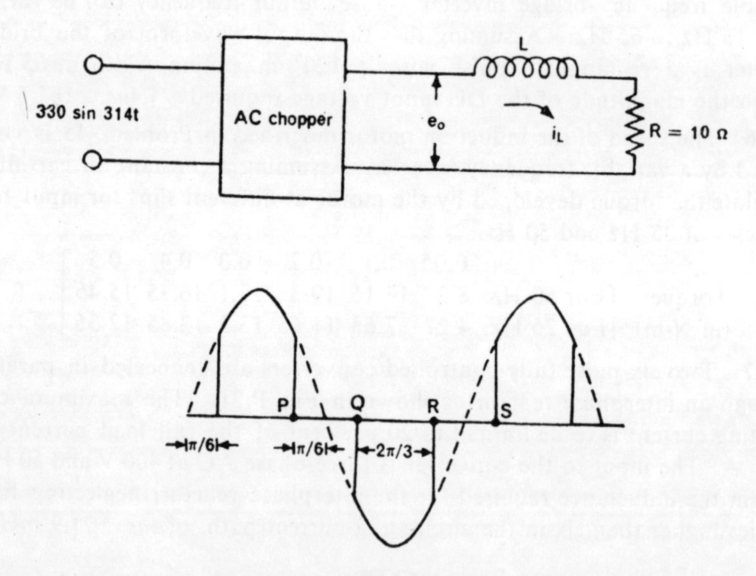

Fig. P.22 AC chopper circuit.

52 A single-phase half-controlled circuit is used for obtaining a variable DC voltage. The input voltage to the converter is 330 sin 314*t*. Assuming a firing angle of $\pi/4$, obtain the values of L and C in the output filter that will provide an average output current of 10 A at a ripple factor of 0.01. [*Ans.* $L = 27.8$ mH; $C = 50 \times 10^4$ μF.]

53 Design a suitable LC Ott filter to obtain from a bridge inverter 50 Hz frequency output which is to be fed to a load of resistance 50 Ω. [*Ans.* $L_1 = 0.36$ H, $L_2 = 0.08$ H; $C_1 = 21.3$ μF, $C_2 = 42.6$ μF.]

APPENDIX

Selection of SCRs

A.1 Sample Specification Chart for a Thyristor

Device	SCR
Type Number	GE C38
Rating	25–800 V
Maximum RMS on-state current	35 A
Maximum average on-state current at 180° conduction and case temperature 70°C	22.5 A
Maximum peak nonrepetitive one-cycle surge current	225 A
Maximum I^2t for fusing for 5–8.3 msec	100 A^2sec
Maximum rate of rise of on-state current	150 A/μsec
Junction operation temperature range	−65–125°C
Minimum critical rate of rise of off-state voltage	20 V/μsec
Specification sheet number (G.E. Company, USA)	160–30

For other members of the thyristor family, details similar to those listed here can be obtained from data sheets of the manufacturer.

A.2 Voltage and Current Ratings of Typical SCRs Available from Indian Manufacturers

A.2.1 SCRs manufactured by Semiconductors Ltd., Pune

Identification Number	*Current/Voltage rating*
SN 101–105	1 A 100–500 V
SN 501–505	5 A 100–500 V

A.2.2 SCRs manufactured by Hind Rectifiers, Bombay

Identification Number	*Current/Voltage rating*
26 TB2–TB12	16 A 200–1200 V
36 TB2–TB12	45 A 200–1200 V
42 TB2–TB12	85 A 200–1200 V
80 TB2–TB12	325 A 200–1200 V

A.2.3 SCRs manufactured by Solid State Electronics, Bombay

Identification Number	*Current/Voltage rating*
SS 1596–1603	1.6 A 100–800 V
SS–C220A–N	10 A 100–800 V
SS–C685–696	25 A 200–1200 V
SS–C51P–PB	140 A 1000–1200 V
SS–C180BX–PBX	235 A 200–1200 V

A.3 Terminal Configurations of SCRs

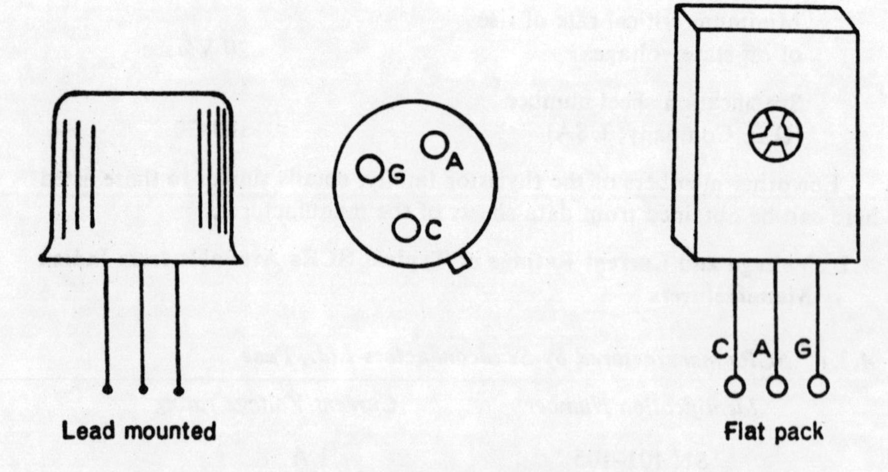

(a) Low-current SCRs

Fig. A.1 Terminal configurations of typical SCRs (*cont.*).

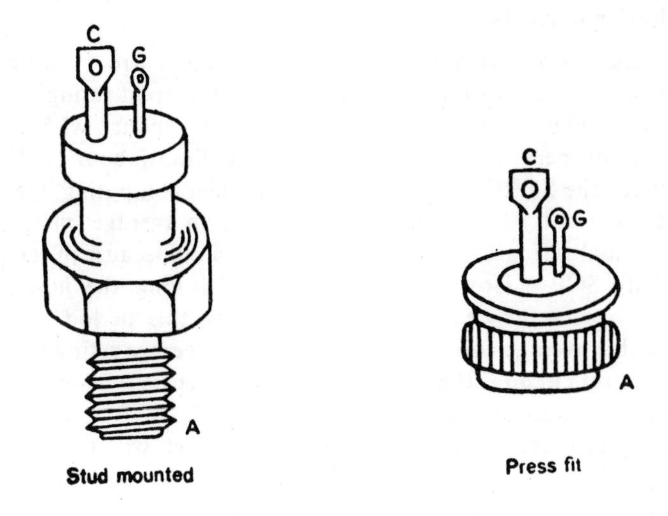

(b) Medium-current SCRs

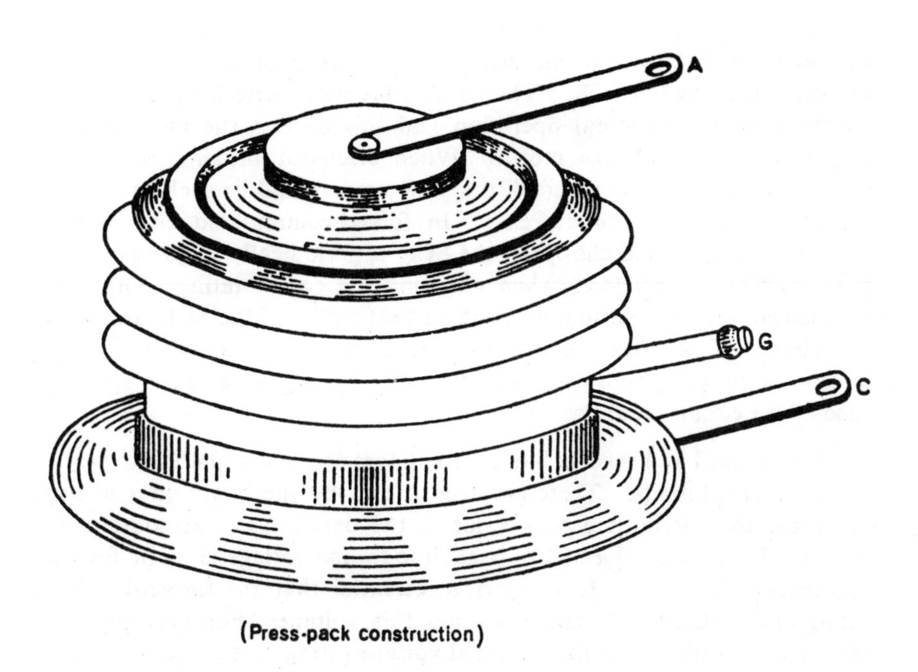

(c) High-current SCRs

Fig. A.1 Terminal configurations of typical SCRs.

A.4 Selection of SCRs

SCRs are selected on the basis of the voltage and current ratings of the circuits to which they have to be connected. The current rating of an SCR is the average value of the forward current it can conduct without raising the junction temperature above the permissible limit (about 125°C). The duty cycle for the SCR has to be considered when computing the average value of the current. In on-off control circuits, an average current higher than the normal rating of the SCR can be tolerated because of the periodic cooling of the SCR. However, if the on-period is long, the junction temperature may reach the steady-state value during this period and restrict the permissible average current to the continuous average rating of the SCR. In phase-control circuits, the form factor of the current waveform determines the average current-carrying capability of the SCR. Thus, the permissible average current for small conduction angles will be considerably lower than that for half-cycle (180°) conduction. In high-frequency operations, the switching (turn-on and turn-off) losses increase substantially and the SCR needs derating. Reduction factors for the current rating of an SCR (expressed in terms of the duty cycle, the conduction angle, and the frequency) are specified by the manufacturer and should be used while selecting the appropriate SCR for a given application.

For motor applications, the intermittent peak current that results when starting the motor or from momentary overloads must be within the safe limits imposed by the SCR. The permissible peak current depends on the duration of the abnormal operation and this decides the method to be employed for starting the motor. When electrical braking is used for rapid stopping or speed reversal, the SCR used should be such that it can carry the resulting high currents. In forced-commutated circuits, e.g., those for inverters and choppers, the SCR is periodically subjected to high peak currents. Therefore, when choosing the commutating components for such circuits the intermittent peak current rating of the SCR should be considered. The multicycle and subcycle surge-current ratings of the SCR have to be properly coordinated with the I^2t fuse rating and the operating time of the circuit breaker.

The forward and reverse blocking voltages applied to the SCR must be within its rated limits. These are defined also by the continuous average value and the intermittent peak value of the device. In line-commutated circuits, the SCR is subjected to peak line-to-line voltage in both forward and reverse directions. It is therefore essential that the forward voltage rating of the SCR be at least 1.5 times this voltage. For example, in a 400-volt three-phase circuit, the usual voltage rating of the SCR is about 1000 V. For large voltage and current ratings, the series and parallel operation of SCRs can be employed, keeping in view all the requirements of each SCR as discussed in Chapter 3. Suitable equalising circuits for voltage and current sharing should be provided. The commutating circuits for the SCR have to be so designed that it experiences a reverse voltage for a duration greater than its maximum required turn-off time. The reapplied and static *dv/dt* for the circuit to which the SCR is connected can be calculated, and the necessary protection using a snubber circuit provided (see Chapter 11) to prevent maloperation of the SCR. Further, the use of suitable RF filters will reduce the probability of unnecessary firing of the SCR due to spurious signals.

The choice of appropriate heat sinks and method of cooling (to improve heat dissipation) depends on the nature of load and location of the SCR. The characteristics of various types of heat sinks are specified by the manufacturer. The size of a heat sink which would be compatible with the SCR in use can be determined from these characteristics.

A.5 Heat Exchangers for SCRs

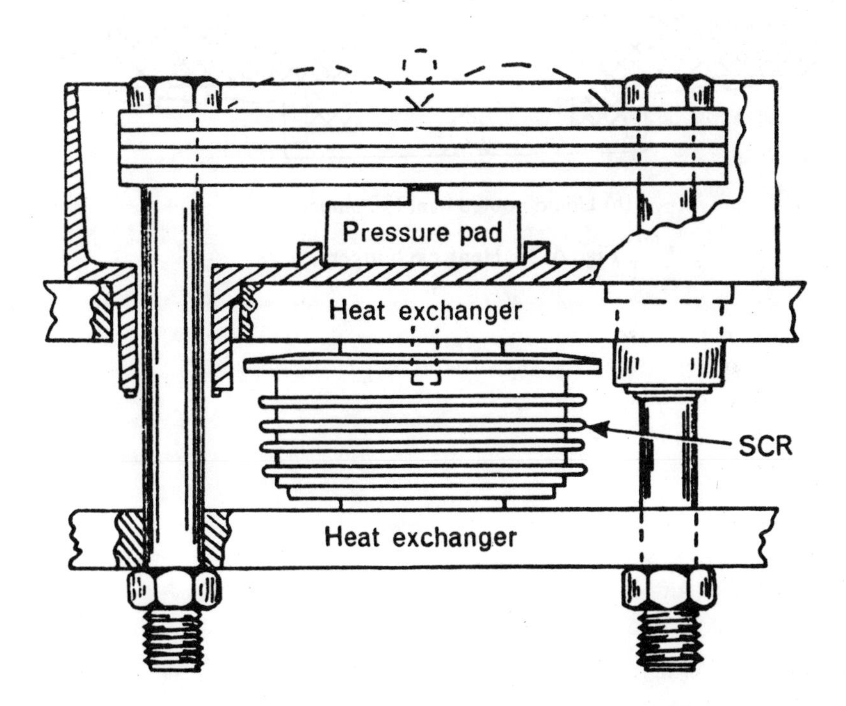

(a) Air-cooled heat exchanger

Fig. A.2 Heat exchangers (*cont.*).

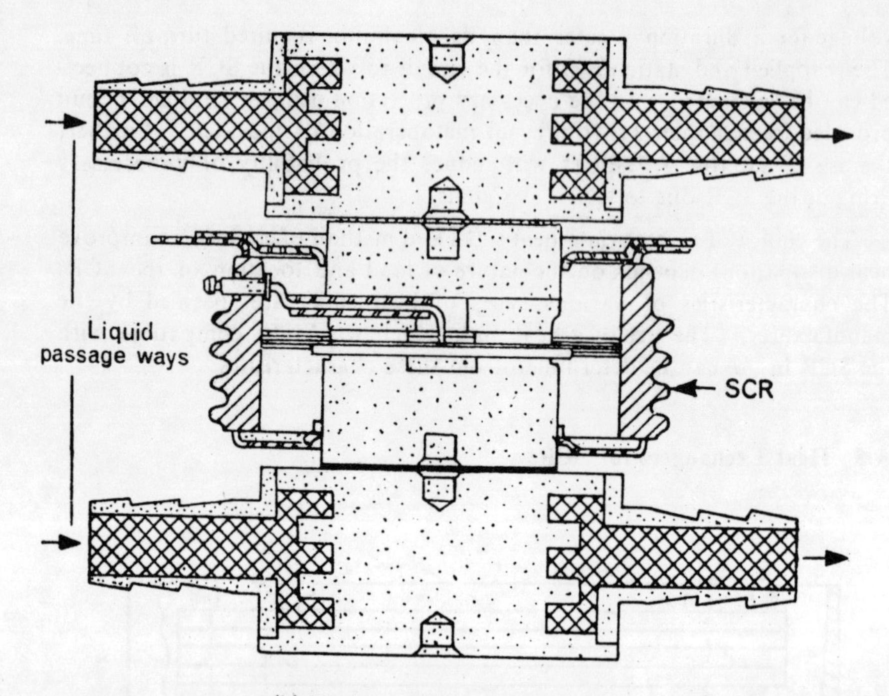

(b) Liquid-cooled heat exchanger

Fig. A.2 Heat exchangers.

Index